Advanced Biological Processes for Wastewater Treatment

Márcia Dezotti • Geraldo Lippel
João Paulo Bassin

Advanced Biological Processes for Wastewater Treatment

Emerging, Consolidated Technologies and Introduction to Molecular Techniques

Springer

Márcia Dezotti
Chemical Engineering Program, COPPE
Federal University of Rio de Janeiro
Rio de Janeiro, Rio de Janeiro, Brazil

Geraldo Lippel
Chemical Engineering Program, COPPE
Federal University of Rio de Janeiro
Rio de Janeiro, Rio de Janeiro, Brazil

João Paulo Bassin
Chemical Engineering Program, COPPE
Federal University of Rio de Janeiro
Rio de Janeiro, Rio de Janeiro, Brazil

Original Portuguese edition published by Editora Interciência, Rio de Janeiro, 2011
ISBN 978-3-319-58834-6 ISBN 978-3-319-58835-3 (eBook)
DOI 10.1007/978-3-319-58835-3

Library of Congress Control Number: 2017946857

Figures/Illustration: Some figures and illustrations were edited by Erimar Pontes Santiago.
Email: erimarpontes@hotmail.com

Printed on acid-free paper

This Springer imprint is published by Springer Nature
The registered company is Springer International Publishing AG
The registered company address is: Gewerbestrasse 11, 6330 Cham, Switzerland

Contents

Chapter 1
Introduction

Geraldo Lippel and Márcia Dezotti

Considering the large number of advances in the area of biological processes for wastewater treatment, to approach them all in a single book would be a rather daunting task. Therefore, we decided it would be best to focus on processes and techniques whose application appears to be spreading and which could alter, in the near future, the scenario of wastewater treatment.

The subject of biological wastewater treatment never attained the status of consolidated knowledge. In fact, new approaches frequently appear which often render established concepts obsolete and allow the development of new technologies for the treatment of different types of wastewater. This is exemplified by the advancement of knowledge related to anaerobic treatment, starting in the late 1970s, as well as more recent findings which verified the versatility of nitrifying bacteria and new forms of nitrogen transformation and also a deepening of our understanding of microbiology and biochemical processes for the removal of phosphorus and sulfur.

These findings and the advancement of knowledge have led to important technological developments. In the case of anaerobic treatment, there has been a remarkable increase in the implementation of reactors which promote microbial retention (UASB, IC, EGSB), and in the case of nitrogen removal, different processes have been proposed including Sharon-Anammox, which has found an interesting niche for its application. Insights regarding the removal of phosphorous and sulfur have provided a new perspective in relation to wastewater treatment, which aims not only at the removal but also the recovery of these elements for reuse, since they are important chemical consumables.

As previously mentioned, in this book we approach some selected themes which we consider to be extremely relevant to the current context of biological treatment processes, these being membrane biofilm reactors (MBR), moving bed biofilm

G. Lippel • M. Dezotti (✉)
Chemical Engineering Program, COPPE, Federal University of Rio de Janeiro,
Rio de Janeiro, Rio de Janeiro, Brazil
e-mail: lippel@peq.coppe.ufrj.br; mdezotti@peq.coppe.ufrj.br

M. Dezotti et al., *Advanced Biological Processes for Wastewater Treatment*,
DOI 10.1007/978-3-319-58835-3_1

reactors (MBBR), new biological nitrogen removal processes, aerobic granulation technology, and molecular biology techniques applied to the study of microbial diversity.

We consider that by approaching these themes we will provide our readers with an insight into the current trends in relation to treatment processes. The themes discussed should guide the conception of new treatment plants, which must fulfill the growing demands for treated effluent quality and confront the current challenges to not only remove macropollutants but also the so-called emerging pollutants present in wastewaters in concentrations of micrograms or nanograms per liter.

In this introduction, we would like to highlight the evolution of effluent treatment processes, which has occurred as a result of targets to be reached. Over many years, the treatment processes were simply aimed at removing particulate material in suspension (suspended solids—SS) or biodegradable organic matter (biochemical oxygen demand—BOD). Later, however, the oxygen demand associated with the oxidation of ammonium nitrogen gained attention as an important factor, and, consequently, nitrification was considered in the conception of treatment plants.

Over time, the need to remove not only ammonium nitrogen but also inorganic oxidized forms of this element (nitrite and nitrate), along with phosphorous, was verified, since these nutrients have adverse effects on the receiving water bodies, one of which is the intensification of the eutrophication of aquatic systems. This led to the development of different processes which combine anoxic and oxic environments in different tanks (Bardenpho process and others) or in different periods in the same tank (sequencing batch reactor).

More recently, treatment processes have had to be adjusted to meet the demands for the removal of emerging pollutants and the production of water for reuse. At the same time, as mentioned above, there has been an interest in the recovery of certain pollutants in the form of industrial consumables, as in the case of phosphorus and sulfur, or even in the production of other consumables such as organic acids and hydrogen.

In the current scenario, particularly in developed countries, there are strict demands regarding effluent treatment processes, and in this context it is evident that the combination of biological and physicochemical processes is imposed as a requirement in the conception of modern wastewater treatment plants.

Figure 1.1 shows the trends in the evolution of wastewater treatment processes. The physicochemical processes are also included in this figure. It should be considered that certain processes can still be improved; however, in the case of those which are well established in the market, such as activated sludge, the improvements must be carried out gradually over time. As new processes appear, such as membrane reactors (MBRs) or Anammox, a wave of innovation follows; however, the consistency and applicability of a new approach will only be determined over time. The horizontal bars representing each process in Fig. 1.1 are closed at their right extremity. This means that the process reached a high degree of maturity in a certain period of time, although improvement could and should still occur.

In order to face the challenges associated with removing emerging pollutants or obtaining water for reuse in applications which require high quality, the combining

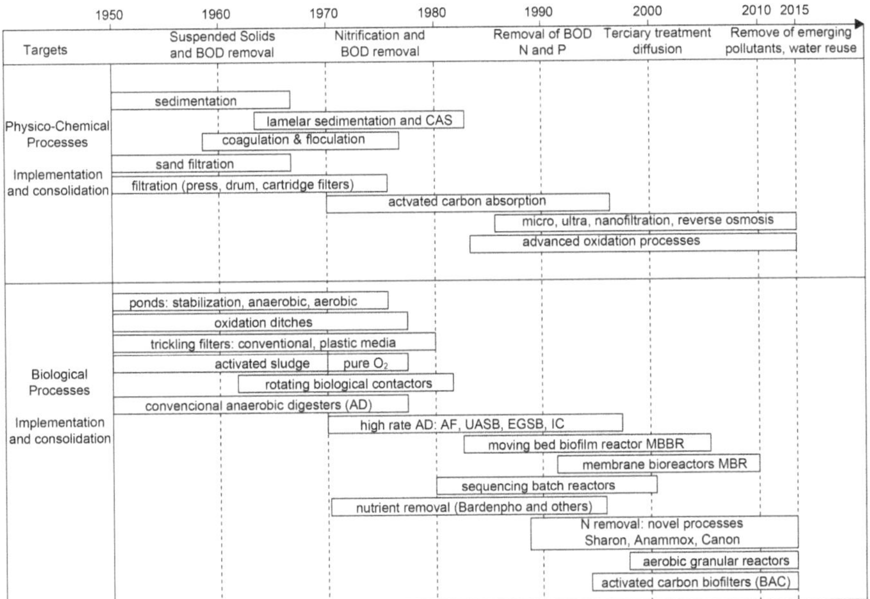

AD anaerobic digester; AF anaerobic filter; BAC biological activated carbon; CAS Chemically assisted sedimentation; EGSB expanded granular sludge bed; IC internal circulation; MBR membrane bioreactor; MBBR moving bed biofilm reactor; UASB upflow anaerobic sludge blanket reactor.

Fig. 1.1 Trends in the evolution of wastewater treatment processes

of processes is imperative. In this regard, the combination of biological processes, membrane filtration, advanced oxidation processes (AOPs), and adsorption onto activated carbon, in different sequential configurations, has been adopted in several countries by companies and local governments.

The appropriate conception of a process is a fundamental, but not the only, step to ensure that the treatment aims are achieved. The monitoring and control of the process are essential for operational success. A treatment process needs to be continuously monitored; however, unfortunately this aspect does not, in general, receive due attention. Although several types of sensors and measuring instruments are available in the market, the degree of instrumentation and automation of many wastewater treatment plants is still deficient.

Although there are sensors and other measuring equipment available for the online or offline monitoring of several variables of interest (pH, temperature, dissolved oxygen, redox potential, flow rate, methane production, ammonia, nitrite, nitrate, phosphorus, and others), challenges remain regarding the characteristics of the main agent of biological processes: the microbial community.

A useful, albeit extremely limited, parameter used to determine the quantity or concentration of microbial agents present in a given process is the volatile suspended solid (VSS) content. The number of design and operational parameters of biological treatment systems which are associated with this highly inconsistent indicator is impressive. Specific removal rates for nitrogen, phosphorus, and organic matter are commonly expressed in specific terms, and the variable used to quantify the biomass of interest is generally volatile solids.

It can be argued that for a given biological sludge sample, it is very difficult to quantitatively determine the presence of each microbial group. For instance, in a sludge sample taken from a reactor which promotes the removal of BOD and nutrients, it is difficult to evaluate the percentages of some microbial groups of interest, such as heterotrophic, ammonia-oxidizing, nitrite-oxidizing, and phosphate-accumulating bacteria. The same applies to a sample taken from an anaerobic reactor regarding methanogenic, homoacetogenic, acetogenic, and sulfate-reducing organisms.

An additional factor which renders the challenge even more complex is that besides the quantification it is important to obtain information regarding the physiological state of the microorganisms of these groups. In relation to obtaining more detailed information on the microbial communities typically present in wastewater treatment processes, the complexity of the problem is enormous and the challenges immense.

Nevertheless, science and technology have contributed to broadening our knowledge regarding these complex microbial communities. This has been achieved due to the development of different molecular techniques and advances on microscopy. In this book, a chapter specifically on molecular techniques has been included. As an introductory and illustrative aid, some of these techniques are detailed in Fig. 1.2. The evolution of most of these techniques was based on fundamental knowledge regarding DNA and RNA molecules obtained during the second half of the twentieth century. In particular, the development of the polymerase chain reaction (PCR) and later the use of thermostable DNA polymerase enabled researchers to gain an insight into microbial diversity. The area known as metagenomics relates to the knowledge which permits the functional analysis of the nucleotide sequences of the collective microbial genomes present in a certain environmental sample. Thus, it allows studies to be carried out on the microbial ecology and an exploration of the diversity of complex communities such as those associated with biological treatment processes.

Although the PCR technique allows the amplification of the number of DNA fragments of a given sample, other operations are still required to further elucidate the diversity. It is necessary to separate the fragments, identify them, and compare the sequences obtained with databanks available on the Internet. The technique of denaturing gradient gel electrophoresis (DGGE) developed by Fischer and Lerman (1983) is the most widely used in the separation stage and in the obtainment of molecular fingerprints. Muyzer et al. (1993) were pioneers in the use of this technique applied to environmental samples.

As shown in Fig. 1.2, several techniques have been developed and refined to study the microbial ecology and diversity, such as respiratory quinone profile (RQP), fluorescent in situ hybridization (FISH), and microautoradiography (MAR). It should be noted that the FISH technique is currently widely employed in academic studies on the microbial groups present in biological treatment systems. This method has become popular due to the development and commercialization of specific

Period	Development
1970–1980	Establishment of the basic principle of PCR (Polymerase Chain Reaction) technique. Kleppe *et alii* (1971)
1980–1990	Utilization of DGGE (Denaturing Gel Gradient Electrophoresis) to separate DNA fragments. Fisher and Lerman (1983)
	Amplification, detection and/or cloning nucleic acid sequences - US patent. Mullis *et alii* (1987)
	Detection of polymorphims of human DNA by gel electrophoresis as single-strand conformation polymorphisms (SSCP). Orita *et alii* (1989)
1990–2000	Development of ARDRA (Amplified Ribosomal DNA Restriction Analysis). Veneechoutte *et alii* (1992)
	Development of FISH (Flouorescent *in-situ* Hybridization). Lengauer *et alii* (1992)
	DGGE utilization for profiling complex microbial populations. Muyzer *et alii* (1993)
	Genome analysis with DNA microarrays. Schena *et alii* (1995)
	Utilization of RQP (Respiratory Quinone Profile) for microbiota characterization. Hu *et alii* (1997)
	Pyrosequencing. Ronaghi *et alii* (1996, 1998)
	Development of TRFLP (Terminal Restriction Fragment Length Polymorphism) technique. Liu *et alii* (1997)
	Analysis of microbial diversity by using RISA (Ribosomal Intergenic Spacer Analysis) method. Borneman and Triplett (1997)
	Combination of FISH and MAR (Micro Autoradiography) for microbial ecology analysis. Lee *et alii* (1999)
2000–2015	Development of high-throughput next-generation sequencers. Shendure *et alii* (2005)
	Development and commercialization of high-throughput sequencers by different companies (2005-2015)
	Nanopore DNA sequencing. Stoddart *et alii* (2009)
	Utilization of Gene FISH technique for linking gene presence and cell identity in environmental microorganisms. Kawakami *et alii* (2010)
	DNA sequencing using electrical tunnelling currents. Di Ventra (2013)

Fig. 1.2 Evolution of different molecular techniques and microscopy applied in the study of microbial diversity

probes for different groups of interest. Its combination with the MAR technique provides a powerful tool for the study of microbial communities.

The use of the abovementioned techniques is still restricted to research of an academic nature or industrial R&D. However, their use as a monitoring and control tool for industrial-scale processes will certainly become a reality in the near future.

New processes, the improvement of established processes, and new tools for the monitoring and control of the variables and parameters of interest are appearing in the area of effluent treatment. These will be of great value in terms of obtaining treated effluent with the best possible quality.

This book is committed to this vision of the future.

References

BORNEMAN, J., TRIPLETT, E. Molecular microbial diversity in soils from eastern Amazonia: evidence for unusual microorganisms and microbial population shifts associated with deforestation, *Appl. Env. Microb.*, v. 63, p. 2647-2653, 1997.

DI VENTRA, M. Fast DNA sequencing by electrical means inches closer. *Nanotechnology*, v. 24, n. 34, 2013.

FISHER, S. G., LERMAN, L. S. DNA fragments differing by single base-pair substitution are separated in denaturing gradient gels: correspondence with melting theory. *Proc. Natl. Acad. Sci., USA*, v. 80, n. 6, p. 1579-1583, 1983.

HU, H-Y., FUGIE, K., TANAKA, H., MAKABE, T., URANO, K. Respiratory quinone profile as a tool for refractory chemical biodegradation study. *Water Sci. Technol.*, v. 35, n; 8, p. 103-110, 1997.

KAWAKAMI, S., KUBOTA, K., IMACHI, H., YAMAGUCHI, T., HARADA, H., and OHASHI, A. Detection of single copy genes by two-pass tyramide signal amplification fluorescence in situ hybridization (Two-Pass TSA-FISH) with single oligonucleotide probes. *Microbes Environ*, v. 25, p. 15-21, 2010.

KLEPPE K., OHTSUKA, E., KLEPPE, R., MOLINEUX, I., KHORANA, H. G., Studies on polynucleotides. XCVI. Repair replications of short synthetic DNAs as catalyzed by DNA polymerases. *J. Mol. Biol.*, v. 56, p. 341-361, 1971.

LENGAUER, C., GREEN, E. D., CREMER, T. Fluorescence in situ hybridization of YAC clones after Alu-PCR amplifications. *Genomics*, v. 13, p. 826-828, 1992.

LEE, N., NIELSEN, P. H., ANDREASEN, K. H., JURETSCHKO, S., NIELSEN, J. L., SCHLEIFER, K. H., WAGNER, M. Combination of fluorescent in situ hybridization and microautoradiography – a new tool for structure – function analyses in microbial ecology. *Appl. Env. Microb.*, v. 65, p. 1289-1297, 1999.

LIU, W.T., MARSH, T. L., CHENG, H., FORNEY, L. J. Characterization of microbial diversity by determining terminal restriction fragment length polymorphisms of genes encoding 16S rRNA. *Appl. Env. Microb.*, v. 63, p. 4516-4522, 1997.

MULLIS, K. US Patent 4,683,202; 1987.

MUYZER, G., DE WALL, E. C., UITTERLINDEN, A. G. Profile of complex microbial populations by denaturing gradient gel electrophoresis analysis of polymerase chain reaction-amplified genes coding for 16S rRNA. *Appl. Env. Microb.*, v. 59, p. 695-700, 1993.

ORITA, M., IWAHANA, H., KANAZAWAT, H., HAYASHI, K., Sekiya, TAKAO Detection of Polymorphisms of Human DNA by Gel Electrophoresis as SSCPs., *Proc. Natl. Acad. Sci.*, USA, v. 86, p. 2766-2770, 1989.

RONAGHI, M., KARAMOHAMED, S., PETTERSSON, B., UHLÉN, M., NYRÉN, P. Real-time DNA sequencing using detection of pyrophosphate release. *Anal. Biochem.*, v.242, n. 1, p. 84-89, 1996.

RONAGHI, M., UHLÉN, M., NYRÉN, P. A sequencing method based on real-time pyrophosphate. *Science*, v. 281, n. 5375, p. 363-365, 1998.

SCHENA, M., SHALON, D., DAVIS, R. W., BROWN, P. O. Quantitative monitoring of gene expression patterns with a complementary DNA microarray. *Science*, v. 270, p. 467-470, 1995.

SCHENDURE, J., PORRECA, G. J., REPPAS, N. B., LIN, X., McCUTCHEON, J. P., ROSENBAU, A. M. Accurate multiplex polony sequencing of an evolved bacterial genome. *Science*, v. 309, p. 1728-1732, 2005.

STODDDART, D., HERON, A. J., MIKHAILOVA, E., MAGLIA, G., BAYLEY, H. Single-nucleotide discrimination in immobilized DNA oligonucleotides with a biological nanopore. *Proc. Natl. Acad. Sci.*, USA, v. 106, p. 7702-7707, 2009.
VANEECHOUTTE, M., ROSSAU, R., DE VOS, P., GILLIS, M., JANSSENS, D., PAEPE, N., DE ROUCK, A., FIERS, T., CLAEYS, G., KERSTERS, K. Rapid identification of *Comamonadaceae* with amplified ribosomal DNA-restriction analysis (ARDRA). *FEMS Microbiological Letters,* v. 93, p. 227-234, 1992.

Chapter 2
Membrane Bioreactors (MBRs)

Geraldo Lippel and Ana Claudia Cerqueira

2.1 Introduction

Membrane bioreactors (MBRs) resulted from the development of membrane processes which began to gain recognition in the 1960s in different forms known as microfiltration, ultrafiltration, nanofiltration, and reverse osmosis. Dialysis, also an operation involving membranes, preceded the development of the abovementioned types of processes and remains an important separation and purification technique, with diverse applications including hemodialysis.

The fundamental principles of membrane processes are well established and have been the subject of specific publications (RAUTENBACH and ALBRECHT, 1989, NOBLE and STERN, 1995). The MBRs applied in environmental technology are, in general, microfiltration or ultrafiltration systems, installed in or associated with tanks in which reactions are carried out by microorganisms and substances are permeated through membranes. Most MBRs operate with microorganisms agglomerated in the form of flocs, maintained in suspension by mechanical mixing or by the influx of air. The membrane acts as a selective barrier to these flocs and, depending on the characteristics of the membrane, to substances of high molecular weight or with characteristics which impede their transport through this barrier.

Despite the importance of the permeation which occurs through membranes, it should be noted that the microorganisms are the agents responsible for the degradation of the pollutants and, consequently, for the significant levels of organic matter removal which occur in the MBRs. Therefore, knowledge of the principles which control the biological treatment of wastewaters is of great importance in relation to the design and operation of these bioreactors. In addition, as has been widely reported, the substances produced by the microorganisms, in the form of soluble

G. Lippel • A.C. Cerqueira
Chemical Engineering Program, COPPE, Federal University of Rio de Janeiro,
Rio de Janeiro, Rio de Janeiro, Brazil
e-mail: lippel@peq.coppe.ufrj.br

M. Dezotti et al., *Advanced Biological Processes for Wastewater Treatment*,
DOI 10.1007/978-3-319-58835-3_2

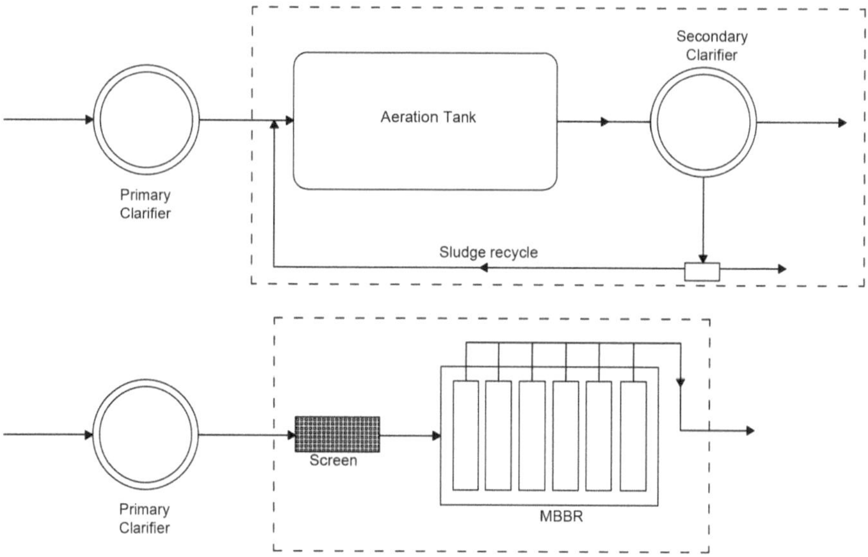

Fig. 2.1 Schematic configurations of the conventional activated sludge process and the MBR system—the dashed line delimits the treatment referred to as secondary

microbial products (SMP) and extracellular polymeric substances (EPS), play an important role in the permeation performance, since they are directly involved in the fouling phenomenon, described later in this chapter, which leads to a drop in the permeation flux during the operational period.

Membrane bioreactors allow the separation of microbial flocs (sludge) from the aqueous phase with relative ease. This is a very valuable feature since it is known that one of the critical points of the biological process most commonly used to treat wastewaters (activated sludge) is the separation of the sludge by sedimentation. Besides ensuring the obtainment of a clear supernatant, sedimentation should promote the production of a denser sludge, which is partially recirculated to the reactor or aeration tank. However, sludge sedimentation is a relatively delicate operation, and the settling tank occupies considerable space, due to its high diameter/height ratio.

Figure 2.1 shows the changes in the layout of the wastewater treatment plant where the conventional activated sludge process is replaced with an MBR system. It can be observed that the area required for the installation of equipment is smaller when an MBR is used, since secondary settling tanks are not needed in this configuration. On the other hand, a sieve or other device for the retention of fine solids needs to be installed upstream of the MBR.

In addition to obtaining a reduction in the bioreactor volume, since the MBR can operate with higher sludge concentrations, the greatest gain achieved with the MBR configuration is the quality of the treated effluent, as will be described later in this chapter.

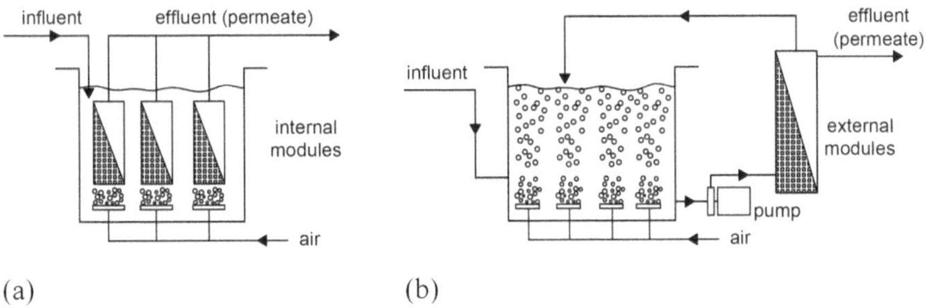

Fig. 2.2 Common configurations of MBRs used for the treatment of wastewaters: (**a**) internal (submerged) modules and (**b**) external (sidestream) modules

One example of a reduction in the area required for wastewater treatment is the sewage treatment plant (STP) in Kaarst, Germany. This plant was designed to treat 48,000 m^3/d, corresponding to a population of 80,000 inhabitants (JUDD, 2006). The area occupied by the plant, which includes an MBR, is approximately half of that which would be required by a conventional activated sludge plant.

The better quality of the treated effluent and lower space requirement for the installation are very attractive characteristics of the MBR system. However, it was the development of more robust membrane modules, with appropriate permeation fluxes, which were, most importantly, commercialized at accessible cost, which enabled the widespread application of MBR systems.

There are reports in the literature that the first MBR was developed by the company Dorr-Oliver in 1966 (YANG et al., 2006). This system was comprised of a bioreactor with suspended biomass, and the content was continuously removed and fed to a rotary sieve and then to an ultrafiltration membrane module. The studies carried out in later decades contributed to improving MBRs and reducing their costs, and at the end of the 1980s and through the 1990s, there was widespread application of these reactors.

2.2 Types of Membrane Bioreactors

In this chapter only, the bioreactors designed for the treatment of wastewaters will be discussed, although a variety of configurations have been proposed for enzymatic reactors, extraction systems, and the cultivation of animal cells.

The first MBRs had a configuration with an external module, as shown in Fig. 2.2, employing ceramic membranes in some cases. In this type of MBR, the circulation in the module needed to be carried out at high speeds in order to reduce the tendency toward fouling. Consequently, the energy consumption was high (2.5–6 kWh/m^3). The external "airlift" configuration of the company Norit reduces significantly the energy consumption, on operating with low recirculation speeds and the injection of air into the module to avoid sludge deposition.

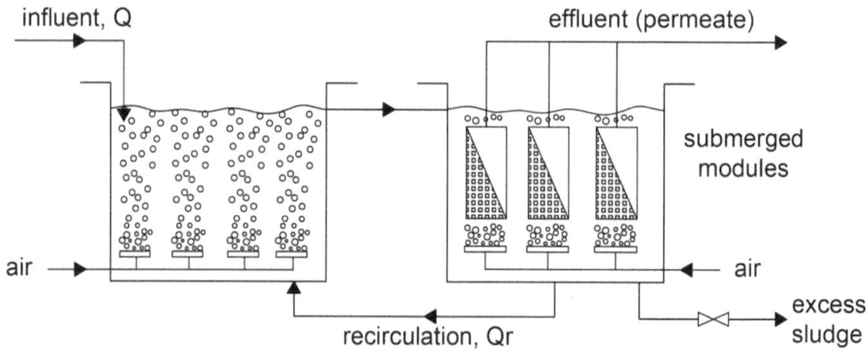

Fig. 2.3 Configuration of MBR with two tanks

During the 1990s, the configuration which employs submerged membranes, attributed to YAMAMOTO et al. (1989), became very attractive and was implemented in large scale in several countries, mainly for the treatment of domestic sewage. However, the configuration with external modules with a high recirculation speed has been employed to treat industrial wastewaters with high temperature, high organic matter content, pH value which is not close to neutral, and high toxicity (YANG et al., 2006).

More recently, the installation of membrane modules in tanks which are separate from the bioreactor has been proposed. In this regard, some configurations in use operate according to the scheme shown in Fig. 2.3, with the bioreactor in series with the tank in which the membrane modules are installed. In large-capacity plants, the modules are distributed in tanks which operate in parallel, which increases the operational flexibility during the cleaning and maintenance procedures. The transfer of sludge from the bioreactor to the membrane tank is normally carried out by pumping and the return by gravity or vice versa. The recycle ratio is defined based on a balance between the suspended solids and the desired concentration in the membrane tanks.

This modification facilitated the cleaning and maintenance of the system with the possibility of allowing the membrane module to operate in a more protected environment than that which receives the wastewater to be treated and which has higher solids and organic matter contents. This can reduce the degree of fouling.

This configuration has gained acceptance and may be appropriate for wastewaters which contain solids, colloids, and macromolecules, which would tend to intensify the fouling if they were fed close to the membrane modules. In the first tank, the partial or total transformation of these pollutants can occur, protecting the membrane modules installed in the second tank, since in this environment the abovementioned pollutants would be present in lower concentrations.

The systems with multiple tanks used for the removal of organic matter and nitrogen, which combine anoxic and oxic environments, allow the installation of membrane modules in the last stage of the process, ensuring the obtainment of a treated effluent which has a higher standard of quality. Figure 2.4 illustrates the

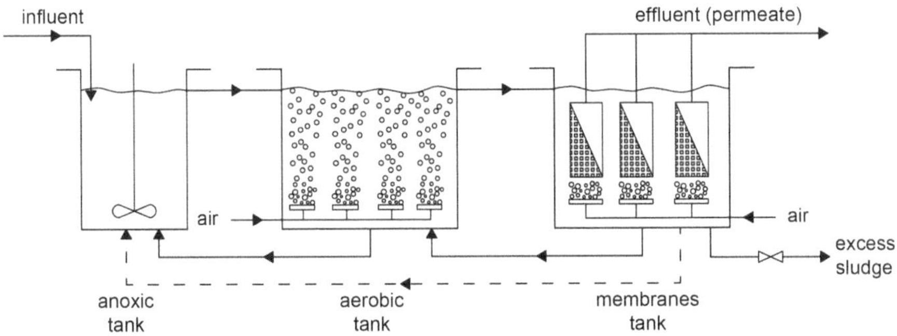

Fig. 2.4 MBR coupled with a system designed for the removal of BOD and nitrogen

incorporation of membranes in the final stage of a system designed for the removal of ammoniacal nitrogen in the aerobic tank (nitrification) and nitrate in the anoxic tank (denitrification), in addition to the biochemical oxygen demand (BOD).

Besides recycling to maintain and control the sludge concentration in the tanks, the sludge needs to be recycled from the aerobic to the anoxic tank to remove nitrate. In Fig. 2.4, two possibilities for the recycling are presented. Recycling directly from the membrane tank to the anoxic tank offers greater simplicity from the operational point of view, reducing the amount of equipment installed and the energy consumption. Recycling from the membrane tank to the aerobic tank and then to the anoxic tank allows greater operational flexibility with the possibility to adjust the recycle flow.

Another distinct MBR configuration is that known as the membrane-aerated biofilm reactor (MABR). In this type of reactor, biofilms form over permeable membranes, and oxygen is transferred to the biofilm through the membrane without forming bubbles in the liquid phase. The oxygen transfer is very efficient in this system, and MABRs are appropriate for effluents whose treatment is associated with high oxygen consumption rates. Figure 2.5 shows a schematic diagram of this type of reactor.

The application of hollow fibers microporous membranes in MABRs has been investigated employing pure oxygen, which is transferred through the hydrophobic membrane pores (BRINDLE et al., 1997; PANKHANIA et al. 1999; SEMMENS et al., 2003). However, dense membranes, which can be operated with oxygen-enriched air, are not associated with the operational problems encountered with the use of porous membranes, which include a low bubble point and interruption of the oxygen transfer due to saturation of the pores with water (AHMED et al., 2004). Despite these advantages, dense homogenous membranes can have a high resistance to the transfer of oxygen. Highly permeable dense composite membranes in the form of hollow fibers can operate with oxygen-enriched air and represent an interesting alternative for MABRs, as verified in a study by CERQUEIRA (2005). The biofilm deposited on the hollow fiber can be seen in Figs. 2.6 and 2.7. Despite their advantages, MABRs have not yet gained a place in the market.

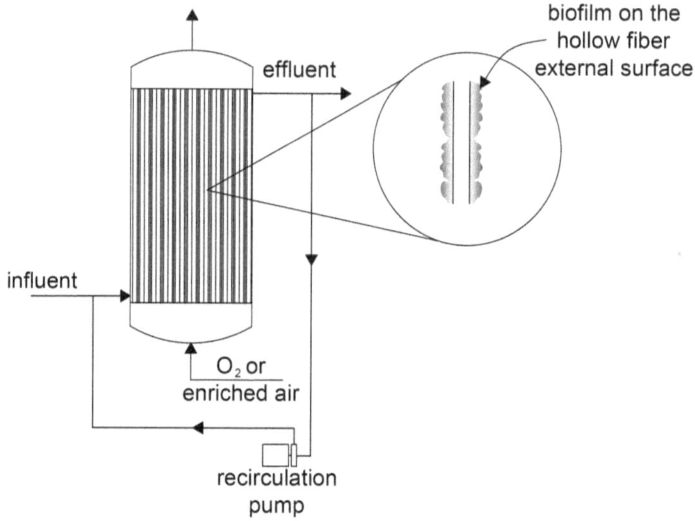

Fig. 2.5 Schematic diagram of an MABR

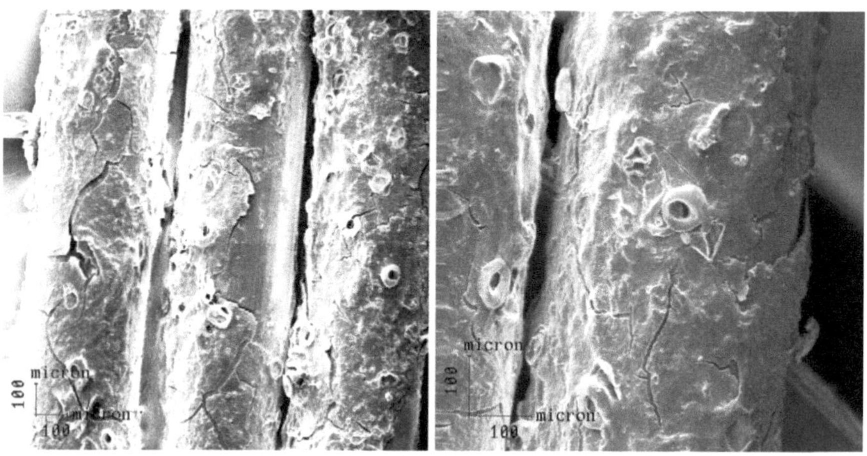

Fig. 2.6 Scanning electron microscopy photomicrographs of fiber samples with biofilm (CERQUEIRA 2005)

Fig. 2.7 Photomicrograph of fiber sample surrounded by biofilm—magnification 100X (CERQUEIRA 2005)

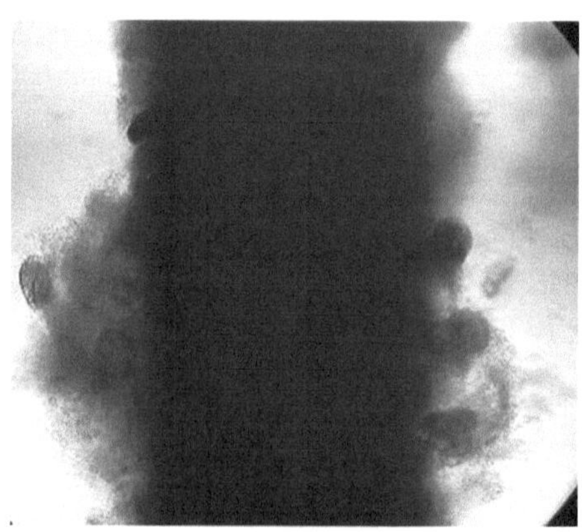

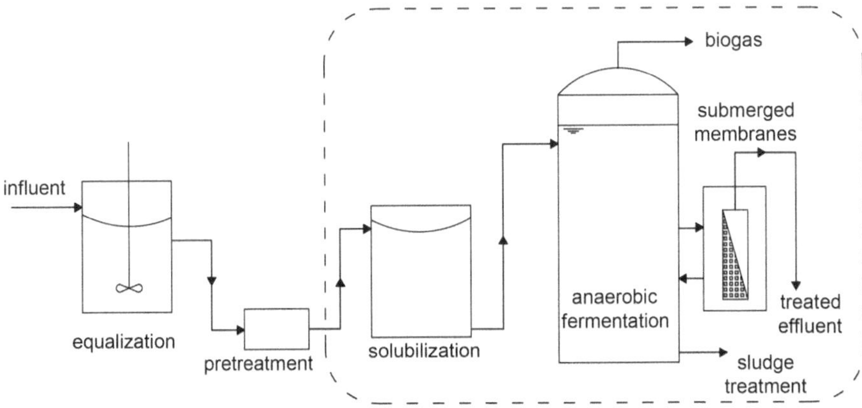

Fig. 2.8 Anaerobic MBR system developed by the company Kubota

Another type of bioreactor with membranes has been investigated for anaerobic treatment processes. The development of these reactors did not occur as rapidly as that of the aerobic MBRs. In the latter, the aeration has been shown to reduce the degree of fouling. In anaerobic processes, the use of biogas to perform the function carried out by air in aerobic MBRs is more complex and requires special attention. In addition, information is still lacking regarding the characteristics and mechanisms of fouling in anaerobic media. However, some commercial success has been obtained with an anaerobic membrane bioreactor (AnMBR) developed by the company Kubota. This system, shown in Fig. 2.8, consists of a tank used to solubilize the wastewater components and, possibly, permit some degree of hydrolysis of the organic substances of high molecular weight, followed by an anaerobic tank operated under thermophilic conditions. This tank, in which methanogenesis occurs, has a sub-compartment where the membranes are installed, in the form of submerged modules. The holders of the patent for this process claim that the membranes are very effective for the removal, by permeation, of ammoniacal nitrogen, the presence of which adversely affects the process, and for maintaining the concentration of volatile solids in the bioreactor at around three to five times higher than that found in commonly employed anaerobic digesters.

Additional details regarding this process can be found in the literature (KANAI et al., 2010), where it is reported that in 2008 there were 15 industrial units in operation, 14 in Japan and 1 in North America. This technology is mainly used by the food industry and in drink distilleries (treatment of stillage).

Fig. 2.9 shows a schematic diagram of a model of the AnMBR proposed by Wu and WANG (2004) (*apud* WANG et al., 2008), in which the biogas is partially recirculated in the reactor via a gas lift inside the digester, where the membrane modules are located. According to the authors, the gas bubbles help to control the fouling and promote mixing of the suspended solids.

The types of MBRs described above are those which have featured prominently on the wastewater treatment market or have the potential to do so. However, other models and other configurations have been studied and proposed for a variety of

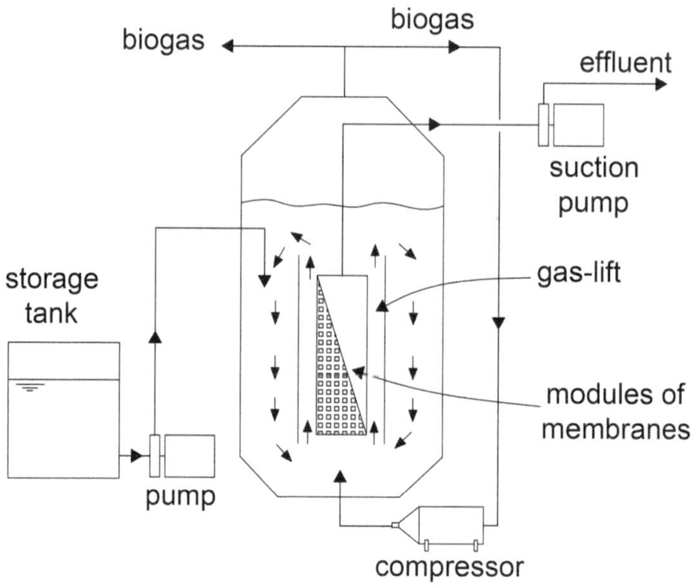

Fig. 2.9 Model of AnMBR with gas lift and submerged membrane module (adapted from WANG et al., 2008)

applications. It is worth mentioning MBR models which have been investigated for the removal of nitrate from potable water, which involve extraction, ion exchange, and gas transfer. They were described in a review by McADAM and JUDD (2006).

2.3 MBR Technology: Current Scenario

Several authors have investigated the application of MBR technology around the world or in specific regions (JUDD, 2006; YANG et al., 2006; WANG et al., 2008). Reports published internationally reveal a predominance of investigations on MBRs which employ submerged modules. A study published in 2006 reported the installation of 258 MBRs in real scale in North America and approximately 2200 around the world (YANG et al., 2006). In most cases, these installations correspond to systems with submerged modules applied to the treatment of domestic sewage.

The use of MBRs in China has increased considerably with the participation of international and national companies. The critical review published by WANG et al. (2008) reported 254 MBRs installed, of which 117 were for the treatment of industrial wastewaters and landfill leachates.

According to the abovementioned authors, the installed MBR plants have a wide range of capacities, varying between 1 and 1000 m^3/d for industrial plants in the USA and between 5 and 25,000 m^3/d in China, the Chinese plant with the largest capacity being installed at a petrochemical site. The municipal plants can reach

Table 2.1 Characteristics of membrane modules commercialized by some international companies (sources: JUDD, 2006; STOWA, 2010; WANG et al. 2008)

	Kubota	Puron	Zenon	Toray
Type of module	Flat sheet	Hollow fiber	Hollow fiber[b]	Flat sheet
Configuration	Submerged	Submerged	Submerged	Submerged
Membrane material	PE[a]	–	PVDF[c]	PVDF[c]
Pore size (μm)	0.4	0.05	0.04	0.08
Internal diameter (mm)	–	1.2	0.8	–
External diameter (mm)	–	2.6	1.9	–
Fiber length (mm)	–	1800	2000	–
Sheet width (mm)	490	–	–	515
Sheet height (mm)	1000	–	–	1608
Area (m^2)	0.8/sheet	30/element	31.6/module	1.4/sheet
Design flux ($L/m^2.h$)	25–40	–	–	8.3–62

[a]Polyethylene chloride
[b]ZeeWeed model 500d
[c]Polyvinylidene fluoride

capacities of 80,000 m^3/d, as in the case of that in Beijing (up to 2010 the largest in the world). These numbers confirm the global spread of MBR technology.

Several companies commercialize MBRs with submerged modules, including the international companies Kubota, Zenon, Siemens, Mitsubishi Rayon, Toray, and Koch. Table 2.1 shows the characteristics of the membrane modules commercialized by some of these companies. However, the data in this table should be considered with caution, since companies have altered the characteristics of their products seeking greater efficiency and lower costs.

Figure 2.10 shows some commercial systems of submerged modules, with configurations of membranes in the form of flat sheets or hollow fibers. In one of these modules, the fibers are sealed but free at one end, and according to the patent holders of this technology (Koch/Puron), this avoids the accumulation of solids and sludge in the spaces between the fibers.

The volumes of air used are significant, as revealed by data from pilot and full-scale installations which indicate consumptions in the ranges of 10–90 and 10–65 m^3 of air/m^3 of permeate, respectively (JUDD, 2006). The aeration costs influence considerably the operational costs of aerobic MBRs. One of the main modifications made to sheet modules was the stacking in two or three superimposed stages. In the fiber modules, the adoption of intermittent aeration reduces the air consumption.

In general, the MBR configuration with an external module makes use of multitubular membranes. Some configurations operate with high-speed recirculation through the module, while others operate with air injection to avoid fouling. Figure 2.11 shows the Pentair X-Flow tubular module and the Pentair Crossflow and Pentair Airlift configurations. Some characteristics of the modules of multitubular membranes are provided in Table 2.2.

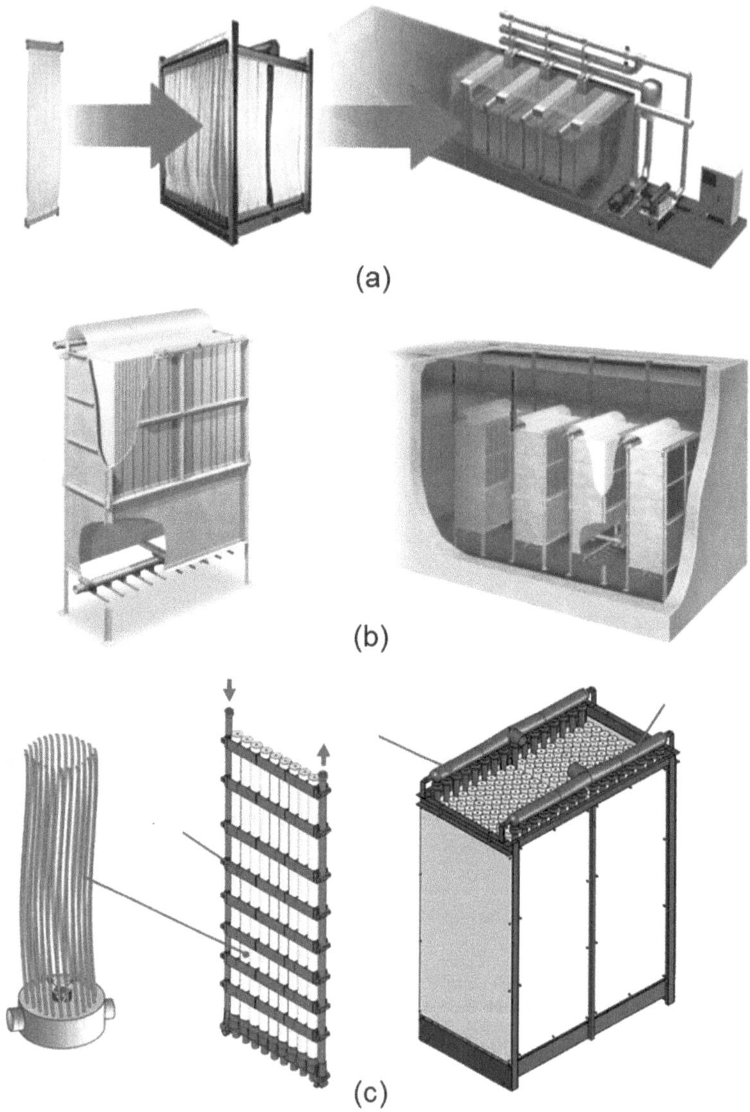

Fig. 2.10 Membrane modules employed in MBRs: (**a**) hollow fibers—GE/Zenon, (**b**) flat sheets—Kubota, and (**c**) hollow fibers with free and sealed ends—Koch

Despite the widespread use of MBRs, the technology still present challenges, and many operational problems encountered with these reactors are associated with fouling. Other problems reported in the literature are the low efficiency of oxygen transfer in the bioreactor, the energy costs associated with aeration, the formation of foam, and the complexity of some membrane cleaning procedures.

(a) (b) (c)

Fig. 2.11 (**a**) Pentair X-Flow tubular membrane module employed in MBRs, (**b**) crossflow configuration, (**c**) airlift configuration

Table 2.2 Characteristics of some external modules for MBRs (source: JUDD, 2006)

	Berghof	Pentair X-Flow	Novasep
Type of module	tubular	tubular	flat sheet
Membrane material	PES, PVDF	PVDF	PAN[a]
Pore diameter (μm)	–	0.03	–
Cutoff (Daltons, Da)	250	–	40
Module dimensions			
Length (mm)	–	3000	2610
Diameter or width (mm)	225	200	438
Tube diameter (mm)	11.5	5.2–8	–
Area/module (m^2)	11.8	33–27	70[b]

[a]Polyacrylonitrile
[b]For a module with a depth of 1710 mm

Although MBRs are increasingly being used for the treatment of industrial wastewaters, their application is predominantly in the food, pharmaceutical, and cellulose/paper sectors and in the processing of landfill leachates. The treatment of complex industrial wastewaters, such as those generated by petroleum refineries, organic solvent production plants, and chemical synthesis plants, can present significant challenges. The presence of suspended solids, oily material, salts, or solvents, for instance, can contribute to increasing the fouling or even damage the components of the membrane modules. Thus, the use of MBRs has not been fully established for the treatment of wastewaters in some industrial sectors. However, it is expected that over time technical solutions will become available to solve the problems and increase the range of use of these reactors.

In the domestic sewage treatment market, MBRs became rapidly and widely established, and the data on the performance of these reactors drove this increase in their application. One notable example is the MBR installed in Porlock (United Kingdom), which in 2008 completed 10 years of operation. With a flow rate of

1970 m^3/d, the MBR of this sewage treatment plant contains 3600 membrane panels, with an area of 2880 m^2. The average operational results for this period reveal a BOD of the treated effluent of less than 5 mg/L and average COD of 22 mg/L. The coliform removal was greater than 5.8 log. During this period, chemical cleaning was carried out every 8 months, on average, and only 6% of the panels were replaced (KUBOTA, 2008).

The characteristics of the 37 largest MBR plants in Europe were summarized by LESJEAN et al. (2009). An installed capacity of 5000 m^3/d was the criteria adopted for the classification of the plants in this category. Of these plants, 32 treated domestic sewage and 5 processed industrial wastewaters. The implantation (capital) costs varied between 200 and 400 euros/inhabitant equivalent, and the energy consumption was between 0.6 and 1 kWh/m^3 treated effluent. Of the 32 plants which treated domestic sewage, 20 were supplied by the company Zenon (GE/Zenon) and 12 by the company Kubota. In agreement with the tendency mentioned above regarding the confinement of the membrane modules in a tank specifically for this purpose, as illustrated in Fig. 2.3, 18 of the Zenon and 8 of the Kubota installations adopted this configuration.

The authors mention the planned construction of large-capacity plants in Oman (78,000 m^3/d) and the United Arab Emirates (269,000 m^3/d), by the companies Kubota and Zenon, respectively. The plant designed for the United Arab Emirates has a much larger capacity than the one in Beijing which was considered to be the largest in the world at the beginning of 2010. It can be observed that other companies have implemented large-capacity MBR plants, such as Pentair X-Flow (17,000 m^3/d, Dubai) and Koch/Puron (15,000 m^3/d, France; 30,000 m^3/d, Australia).

From these reports, it can be observed that MBR technology has advanced considerably in recent years, and it should also be mentioned that this technology is of great importance in the sequential treatment of wastewaters aimed at water reuse. The downstream processes of an MBR, which will be necessary to obtain water with appropriate quality for reuse, can be simplified considering that the quality of the MBR effluent is superior to that of conventional biological treatment processes.

2.4 Fouling in MBRs

2.4.1 Operational Factors

The gradual increase in the permeation resistance is a well-known phenomenon, and it tends to be enhanced in plants operating with complex fluids, as in the case of wastewaters. MBR membranes are inserted in media which contain suspended solids such as biological flocs, microbial cells, cellular debris, colloids, macromolecules, and a varied range of soluble organic substances. In order to ensure prolonged operation, without the need for frequent chemical cleaning, MBRs operate with moderate permeate fluxes.

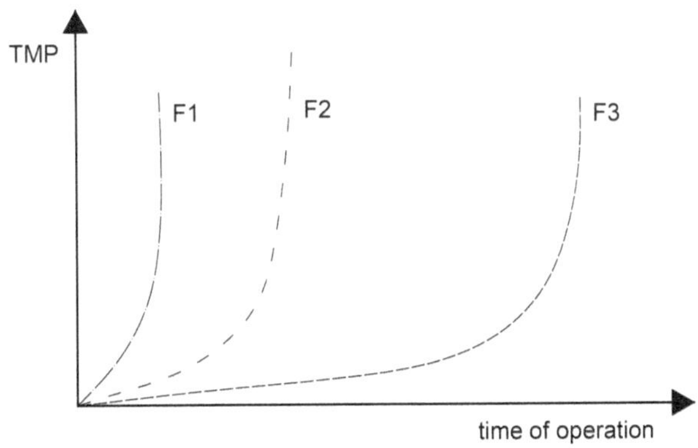

Fig. 2.12 Variation in TMP with operation time at different fluxes (F1 > F2 > F3)

Some authors employ the notion of a critical flux, or even a sustainable flux, to designate operational conditions which ensure long-term operation without the need for interventions involving invasive cleaning.

When the operation is carried out with a constant flux, a gradual increase in the transmembrane pressure (TMP) is observed, due to fouling. With low fluxes, the operation can be prolonged, without a premature and sharp increase in the TMP, as shown in Fig. 2.12.

According to LE-CLECH et al. (2006a), the critical flux corresponds to that at which any additional increase in the flux causes a rapid increase in fouling (sharp increase in the TMP). The sustainable flux is the result of a compromise between fouling and productivity (daily volume of wastewater effectively treated). Since the operation with higher fluxes enhances the fouling, most MBRs operate with low fluxes. In this context, the sustainable flux is that at which the TMP increases at an acceptable rate, and thus chemical cleaning is not required.

Many studies have been carried out with respect to fouling in MBRs; however, despite these investigative efforts, there is still no consensus regarding the causes and mechanisms of this phenomenon. Furthermore, a complex interaction between the factors considered to lead to fouling has been noted.

One parameter which merits particular attention in this regard is the concentration of suspended solids in the reactor. One of the notable advantages of MBRs was initially that they could operate with high concentrations of biological sludge, leading to a reduction in the reaction volume. However, a tendency toward a reduction in the recommended levels of biosolids content, from up to 30 g/L in the first MBRs proposed to levels in the range of 8–12 g/L, has been observed. This reduction is due to some operational problems which have been observed, such as an increase in the viscosity of the reaction medium, an increase in the air supply required to maintain the solids in suspension, and enhanced fouling.

Despite the debate regarding the effect of the suspended solids (SS) concentration on the fouling of the membranes, an increase in SS is generally considered to adversely affect the permeate flux (lower flux or greater TMP). SOMBATSOMPOP et al. (2006) observed a significant increase in the fouling rate when the volatile suspended solid (VSS) content increased from 6 to 15 g/L. A decrease in the permeate flux with an increase in the VSS content was observed by SCHARZ et al. (2006) for levels of up to 5 g/L. For values commonly found in MBRs (>5 g/L), the authors note that the control of the flux so as to avoid the intense formation of a cake layer on the membranes was found to be more important than the variation in the VSS concentration. On the other hand, operation with greater VSS concentrations adversely affected the oxygen transfer.

A parameter of great importance in the biological process is the sludge age or the sludge retention time (SRT). One of the advantages of MBRs is that they operate with high values for this parameter, which leads to a lower production of biomass in the process and also ensures a longer time for the action of microbial strains which have a slow growth rate but are effective for the degradation of certain classes of pollutants. However, operation of the system with high values for the SRT leads to higher VSS contents in the reactor, with the consequences mentioned above.

There is also some uncertainty regarding the effect of sludge age on fouling in MBRs. LE-CLECH et al. (2006a) reported a tenfold increase in the fouling rate when the sludge age was decreased from 10 to 2 days. HAN et al. (2005) observed that there was an increase in the fouling when the sludge age was increased from 30 to 100 days. Thus, it appears that there are negative effects on the process when the sludge age assumes extreme values. The idea associated with MBRs at the beginning of their development, that is, that the sludge age could have values tending toward infinity, has not proved to be viable in practice. It is necessary to remove sludge from the system since, besides the accumulation of biological solids in the reactor, inert and nonbiodegradable materials will also accumulate which can damage the membranes and the modules.

A variable of great relevance in relation to the control of fouling is the flow of air injected at the base of the membrane modules. The movement of air bubbles close to the membrane surface causes shear stress and turbulence which increase the back transport of the materials and substances deposited to the liquid medium. In the case of modules with hollow fibers, the air bubbles also cause movement of the fibers with the possibility of dislodging adhered material. However, above a limit value, the air supply no longer contributes to suppressing the fouling and starts to increase the energy costs for the treatment process. As mentioned above, the companies which commercialize MBRs have invested in superimposed modules, to improve the use of the injected air, as well as in more efficient air distribution systems.

In the case of MBRs with external modules, the tangential velocity of the fluid is an important parameter in relation to reducing fouling. In this case, the positive effect of the velocity can also be restricted to a certain range, since the action of shearing which occurs in the pumping operation and which intensifies with high flows leads to the breaking of flocs and the generation of materials and substances which can act as agents which cause fouling. One alternative which has been adopted is the injection of air into the discharge line of the pump, so that the flow close to the membranes is biphasic (air and liquid), and in this case the system can operate with lower liquid flow rates.

Another operational parameter which affects fouling is the level of dissolved oxygen (DO) in the reactor. In fact, the DO content influences several factors which govern the process and, in particular, the structure of the biofilms, the distribution of the floc sizes, and the secretion of exopolymeric substances (EPS) and soluble microbial products (SMP). In general, high levels of DO allow the system to operate with good filterability and low fouling rates, due to the formation of cake layers with lower resistance, because flocs have a larger size and a more porous structure (LE-CLECH et al., 2006a).

The relation between fouling and the membrane properties is very complex, and there are discrepancies in the results reported in the literature. On altering the characteristics of the membranes in a certain experiment, other variables need to be modified, which makes it difficult to identify and differentiate the effects associated solely with the membranes.

In general, the concept that hydrophilic membranes have a lower tendency toward fouling than hydrophobic membranes prevails. In experiments in bench scale with different types of ultrafiltration membranes (hydrophilic and hydrophobic), MAXIMOUS et al. (2009) concluded that the hydrophilicity of the membrane did not appear to be an advantageous characteristic in relation to the tendency toward fouling. However, it is advantageous with regard to the reversibility of the cake resistance. The nature of the membrane can have an effect during the initial phase of fouling, but this parameter may have little influence over longer periods of operation. As will be discussed later in this chapter, substances and particulates accumulate on the surface of the membrane, which alters its original characteristics over time.

The effects of the pore size and distribution in the membrane are also controversial, mainly considering the variety in the characteristics of wastewaters. The formation of dynamic layers on the membranes as the operation proceeds also hinders an evaluation of the tendency toward fouling of the membranes with different pore size distributions. However, fouling, which is caused by the deposition of organic and inorganic material at the entrance to and inside the pores, is irreversible, and this contributes significantly to the low performance of membranes with larger pores (LE-CLECH et al., 2006a).

2.4.2 *Occurrence of Fouling in MBRs*

As mentioned above, the complex nature of the liquid media in MBRs hinders an understanding of the occurrence of fouling. Nonetheless, a consensus has been established among researchers that there is a strong interaction between the main parameters involved in the occurrence of fouling. Thus, this phenomenon results from complex interactions and is not the product of the action of a single parameter or operational variable.

Figure 2.13 illustrates the gradual clogging of the membrane pores which, during operation with a constant flux, leads to an increase in the TMP. According to ZHANG et al. (2006), fouling occurs in three stages: (1) conditioning of the clean

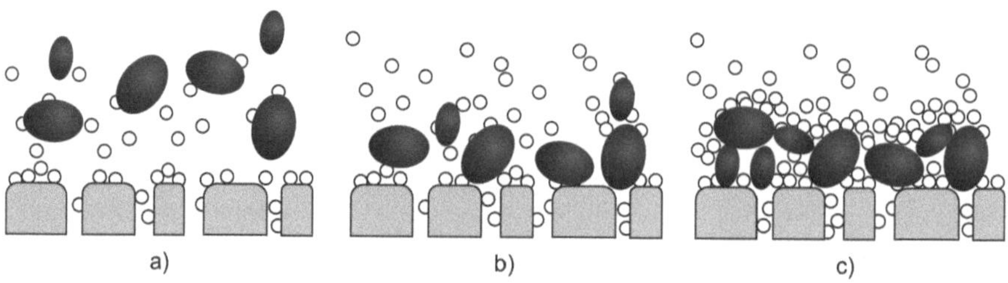

Fig. 2.13 Illustration of the gradual clogging of pores through the action of macromolecules (*small open circles*) and microbial flocs (*dark ellipses*). (**a**), (**b**) and (**c**) sequential stages of gradual membrane fouling. Adapted from LE-CLECH et al. (2006a)

membrane with a concomitant increase in the TMP for a short period, (2) slow fouling during which the TMP increases linearly or shows a weak exponential increase, and (3) rapid fouling with a sharp increase in the TMP.

Several events contribute to the occurrence of these stages. Microbial flocs, EPS, and SMP are essential agents for the fouling phenomenon to occur.

During the first stage, conditioning of the membrane surface by macromolecules occurs, which is an attractive factor for microorganisms, particularly those which are not in flocs (planktonic cells). The flocs can also reach the surface and adhere reversibly to it, leaving traces of exopolymeric material on detaching, which contributes to the conditioning of the membrane surface. Thus, the pores begin to become clogged.

In the second stage, several events take place: restriction and clogging of the pores, deposition of flocs in regions where there is a low level of shearing, and the formation of thin biofilms and cake layers in some regions of the membrane. Fouling occurs even when operating with fluxes below the critical flux, and in this phase the contribution of EPS is significant.

In the third stage, the phenomenon of fouling becomes self-accelerating. Some of the pores become completely clogged, and the flow through the pores which are still open becomes very high and greater than that corresponding to the critical flux, which leads to the blocking of these channels. Fouling does not occur in a uniform manner on the membrane surface, and accentuated growth of the cake layer can occur in some places, which can cause the cake layer to collapse leading to the dispersion of debris on the membranes. The result of these events is a rapid and almost exponential increase in the TMP, as shown previously in Fig. 2.12. Some models proposed in the literature to describe the events which lead to an abrupt increase in the TMP have been summarized by LE-CLECH et al. (2006a).

A model commonly adopted to describe the drop in the permeate flux as the operation proceeds is that of resistance in series. The total resistance (R_t) is the sum of the following: membrane resistance (R_m), measured with the clean membrane and deionized water; reversible resistance of the cake layer (R_c), caused by the deposition of cake layer on the membrane surface; and resistance to irreversible fouling (R_f), caused by the adsorption of materials and substances in the pores and on the membrane surface.

On investigating the effects of placing a mechanical mixer in the tank containing the membrane module, KHAN and VISVANATHAN (2008) verified, under stable operational conditions, that R_c corresponded to approximately 98% of the total resistance. An increase in the mixing velocity, at the levels investigated, did not promote a significant drop in the total permeation resistance.

SOMBATSOMPOP et al. (2006), in experiments on MBRs operating with VSS concentrations of 6, 10, and 15 g/L, obtained R_c/R_t ratios of 0.95, 0.97, and 0.97, respectively. In addition, the authors concluded that the above ratios were not notably affected by the concentration of solids in the reactor. CHANG and KIM (2005) observed a reduction in the R_c value with a decrease in the solid concentration.

The fact that the cake resistance (R_c) is a highly significant factor does not reduce the importance of the role of EPS and SMP in the occurrence of fouling, as will be observed below.

2.4.3 Polymeric Extracellular Substances and Soluble Microbial Products

Firstly, it is important to describe the two classes of substances known as polymeric extracellular substances (EPS) and soluble microbial products (SMP). EPS are a relatively broad class of substances of high molecular weight, which are found in the constituent matrices of biofilms or microbial flocs. With regard to their chemical nature, these substances are predominantly polysaccharides, proteins, phospholipids, and nucleic acids. Some of them, in particular the polysaccharides and proteins, can be firmly anchored on the surface of cells, weakly bound to them or retained in the interstitial spaces of the matrices. They are determined using extraction or separation techniques.

The same techniques can be used to determine SMP, and their differentiation from EPS is dependent on the methods employed in their characterization. SMP are comprised of various soluble substances in the liquid phase, such as polysaccharides and proteins secreted by microbial cells, decomposition products of floc matrices, and cells and substances present in the wastewater (modified or not).

A scheme for the method which can be employed for the extraction and determination of EPS and SMP was presented by LE-CLECH et al. (2006a) and is shown in Fig. 2.14.

As illustrated in the figure, this approach can be used to determine extracellular polymeric substances which are susceptible to extraction by the method (eEPS) and the substances which are free in aqueous medium, considered in this case to be soluble microbial products (SMP).

Once the eEPS and SMP fractions have been obtained, their contents can be expressed as protein (eEPSp, SMPp) or carbohydrate (eEPSc, SMPc), employing the classical analytical protocols proposed by Lowry (proteins) and Dubois (carbohydrates).

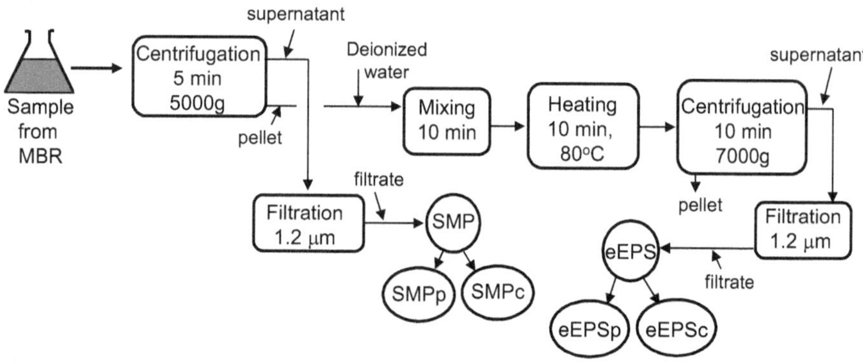

Fig. 2.14 Scheme for the extraction method to determine EPS and SMP (source: LE-CLECH et al., 2006a)

On investigating the exopolymers present in activated sludge with the aid of size-exclusion chromatographic techniques and infrared microspectroscopy, GÖMER et al. (2003) verified that EPS have several peaks. The molecular weights of proteins lie between 45 and 670 kDa, while those of polysaccharides (with three peaks) were lower, of the order of 0.5–1 kDa. In addition, the authors observed a strong association between proteins and polysaccharides.

The role of substances with high molecular weight in the occurrence of fouling should always be considered. The barrier imposed by the membrane allows these substances to be retained in the reaction medium. Since the biodegradation of these molecules is relatively slow, there is sufficient time for them to reach the pores and the surface of the membranes.

A study carried out by ZHANG et al. (2006) allowed substances with molecular weights above 5 kDa to be identified in the medium (referred to as the supernatant), corresponding to a total organic carbon (TOC) value of 20.8 mg/L. In the permeate, the TOC reached only 0.3 mg/L. In the case of substances with a molecular weight below 5 kDa, the TOC values of the supernatant and the permeate were 1.6 and 1.2 mg/L, respectively. Thus, the membrane had a high capacity to retain macromolecules during operation, even though it was a flat sheet microfiltration membrane with a nominal pore size of 0.2 μm. The authors used liquid chromatography with organic carbon detection (LC-OCD) to characterize the soluble compounds in terms of molecular weight and the respective carbon contents mentioned above.

The layers of fouling (second stage) removed from the membranes had higher contents of EPS of a polysaccharide nature than of a proteinaceous nature. In addition, the EPS/VSS ratio was greater for the fouling layer than for the biomass in suspension in the reactor. ZHANG et al. (2006) concluded that polysaccharides appear to be the main fouling agents.

VIERO et al. (2007) verified greater concentrations of polysaccharides than proteins in the supernatant of an MBR and concluded that polysaccharides play an important role in fouling via the formation of a gelatinous layer over the hollow

Table 2.3 Protein and polysaccharide contents in the supernatants of MBRs

eEPSp	eEPSc	SMPp	SMPc	Reference
11–116 (mg/gVSS)	7–40 (mg/gVSS)	0.5–34 (mg/L)	3–33 (mg/L)	Le-Clech et al., 2006a
–	–	47.1 (mg/L)	21.5 (mg/L)	Maximous et al., 2009
22.1–36.4 (mg/L)	6.2–7.5 (mg/L)	2.1–3.1 (mg/L)	8.4–9.5 (mg/L)	Khan et al., 2008
59.5–78.7 (mg/gVSS)	19.7–31.9 (mg/gVSS)	2.2–7.7 (mg/gVSS)	7.4–12.9 (mg/gVSS)	Sombatsompop et al., 2006
–	–	0–20 (mg/L)	0–200 (mg/L)	Viero et al., 2008
–	–	0–18 (mg/L)	10–80 (mg/L)	Viero et al., 2007
4.5–9.4 (mg/gVSS)	3.7–7.3 (mg/gVSS)	20–45 (mg/L)	39–65 (mg/L)	Arabi et al., 2009

membrane fibers. Later, VIERO et al. (2008) confirmed the higher polysaccharide contents in the supernatant of an MBR applied for the treatment of a petroleum refinery effluent and highlighted the strong contribution of these substances to the fouling of membranes.

OKAMURA et al. (2009) investigated the occurrence of fouling on ultrafiltration membranes by way of tests on samples collected from different reactors and filtered through filter paper (1 μm). These researchers studied the occurrence of fouling by SMP in tests of short duration (70 min). The results revealed a positive correlation between the contents of saccharides and uronic acids in the aqueous phase and the filtration resistance. In the case of the proteinaceous material, no correlation was observed. The authors reported the formation of a gelatinous layer on the membranes and that the concentrations of saccharides and uronic acids in this layer were around 50 times higher than the respective concentrations in the liquid phase. They concluded that the cause of the membrane fouling was the polysaccharides containing uronic acids in their composition.

In the literature, however, there is some discrepancy in the results related to the role of proteins and polysaccharides in membrane fouling. There are even inconsistencies regarding the contents of proteins and polysaccharides found in the supernatants of MBRs. Table 2.3 shows some results selected from different publications, which demonstrate the variation in the concentrations reported by researchers.

ARABI and NAKHLA (2009) investigated the effect of monovalent and bivalent cations on the production of EPS and SMP, the floc size, and the fouling rate in bench MBRs equipped with modules with hollow fiber membranes with an average pore size of 0.047 μm. An increase in the solid concentration from 140 to 645 mg/L, for a monovalent to bivalent cation ratio of 3, led to a reduction in the floc size and an increase in the fouling rate. Under these conditions, there was a decrease in the eEPS and an increase in the SMP contents.

2.4.4 Operation Modes and Fouling Control

Operation with constant flux is more common than operation with constant TMP. The occurrence of fouling with constant flux has been investigated and was discussed in Section 2.4.2. In the first stage of fouling, there is a gradual, although small, variation in the pressure. A survey carried out by LE-CLECH et al. (2006a) indicated, for this stage, durations of between 48 and 1200 h for operational fluxes of 4–30 $L/m^2.h$. The ranges are wide because the data originate from distinct studies conducted under different conditions.

Depending on the characteristics associated with the occurrence of fouling in a certain MBR, several measures can be taken. Physical cleaning of the membranes is carried out to restore the permeate flux and remove the fouling layer. In principle, this is defined by the MBR manufacturer, since the devices used to perform this task, which are mostly automatic, should be available in the unit. Chemical cleaning is carried out with less frequency, but has been shown to be essential when physical cleaning no longer restores an adequate flux for the operation of the MBR.

One form of physical cleaning is known as "relaxation" and consists of stopping the permeate flow for short periods (1–2 min). With the continuation of the aeration, substances which cause the fouling are returned back to the liquid phase from the membrane surface. This procedure is frequently repeated, generally at intervals of 8–15 min. Relaxation can be combined with the so-called backwashing.

As the term suggests, backwashing involves the injection of treated effluent in the opposite direction to that of the permeate flow. The injected fluxes are around one to three times the permeate flux and occur intermittently. Each backwashing episode can last some seconds, and the procedure is repeated at intervals of a few minutes. Typical durations for filtration/backwashing periods of 10 min/45 s, 3 min/15 s, and 8–16 min/25–45 s have been reported by LE-CLECH et al., 2006b.

It is important to note that when backwashing is carried out with the permeate, it affects the productivity of the process and the energy consumption. Permeate volumes of 5 to 30% of the treated effluent volume can be required when backwashing is frequently carried out for relatively long periods.

One way to reduce the permeate consumption is to perform backwashing with air (VISVANATHAN et al., 1997, VIERO et al., 2007). Operation of an MBR employing filtration periods of 15 min and backwashing with air for 30 s has been shown to be suitable for controlling fouling, as verified by VIERO et al. (2007).

The efficacy of physical cleaning tends to decrease over time during MBR operation, and irreversible fouling becomes significant leading to the need for chemical cleaning. This can be performed employing low concentrations of chemical agents in the form of daily backwashing. It can also be carried out weekly with a higher concentration of chemical agents as a preventative measure or even in an intense way once or twice a year. Preventative cleaning can last 30 min employing around 0.01% NaClO, and intensive cleaning can be carried out with NaClO in a concentration of 0.2 to 0.3% and citric acid (0.5 to 1%) or oxalic acid (0.5 to 1%) (LE-CLECH et al., 2006a).

The cleaning frequency during the operation of MBRs is defined according to multiple variables, such as the characteristics of the wastewater, the sludge, and the membrane system used. The transmembrane pressure is the main parameter for defining the protocol and the cleaning frequency which in practice can vary from every 3 months to once a year.

2.5 Use of Activated Carbon in MBRs

Many studies have been carried out with the use of powdered (PAC) or granular (GAC) activated carbon in activated sludges, and, in general, the results obtained have been positive in terms of the process performance. The advantages associated with the use of activated carbon in activated sludge include: (1) biofilms grow on the particles and microorganisms are protected against inhibitory and toxic substances present in the medium; (2) microorganisms which are slow growing but active in the degradation of more recalcitrant substances remain in the reactor; (3) the adsorption of pollutants is intensified when activated carbon and biofilms are present; and (4) the sedimentability and dewatering characteristics of the sludges generated are improved.

However, the cost of using activated carbon, associated with the high doses employed in many studies, has not received as much attention from researchers as the advantages cited above. As observed by MUNZ et al. (2007), the cost of using activated carbon has limited its applications in real-scale activated sludge installations, although good results have been obtained in bench and pilot scales.

In the case of MBRs, the viability of using activated carbon appears to be greater, since its contribution to the filterability, verified in several studies, leads to a decrease in the requirement for physical and chemical cleaning, with consequent savings in energy and chemical consumables.

PIRBAZARI et al. (1996) reported that the use of 1% of PAC reduced the drop in the flux, attributed to fouling, in an MBR with an external ultrafiltration module used for the treatment of landfill leachate.

KIM et al. (1998) observed greater permeability in an ultrafiltration module when PAC was incorporated into the sludge. They verified a reduction in the size of the biological flocs and lower EPS content in their interior when PAC was incorporated into the system. The addition of PAC led to a decrease in the compressibility of the flocs and an increase in the porosity of the cake layer formed with a consequent increase in the permeate flux.

The beneficial effects of the use of PAC in an anaerobic reactor coupled to an ultrafiltration module were reported by PARK et al. (1999). The resistance of the cake layer and the fouling decreased with an increase in the PAC dosage (up to 5 g/L). According to the authors, the use of PAC contributed to generating an incompressible cake layer, and the carbon particles exerted abrasive action and friction on the membrane surface which aided the removal of deposits accumulated on it. In addition, the activated carbon acted as an adsorbent and coagulant of organic substances and colloidal material. The authors also observed greater operational stability in response to shock loads when PAC was added.

LI et al. (2005) compared MBRs with submerged modules operating under similar conditions in experiments with and without the addition of PAC (1.2 g/L). The use of powdered activated carbon contributed significantly to increasing the permeate flux (32% greater in the system with PAC) and decreased the rate of TMP increase, which allowed a 1.8-fold increase in the period of operation without the need for cleaning.

YING and PING (2006) verified that the use of PAC in a concentration of 0.75 g/L had positive effects on the MBR operation applied to the treatment of domestic sewage including a reduction in the resistance of the cake layer and irreversible fouling. A decrease in the fouling rate was also observed by MUNZ et al. (2007) in a reactor applied in the treatment of industrial effluent (tannery) when PAC was used in concentrations of 1.5 and 3.0 g/L.

LESAGE et al. (2008) used 1 g/L of PAC in an MBR and verified an increase in the filterability of the supernatant. The authors postulated that the activated carbon altered the properties of the biological flocs and, most importantly, reduced the concentrations of carbohydrates and proteins in the liquid phase from 81 and 71 mg/L (without the addition of PAC) to 20 and 40 mg/L (with the addition of PAC). In addition, the authors observed that in the system with PAC, the biomass had good resistance to a toxic shock load represented by the addition of 2,4-dimethylphenol to the reactor.

In the studies published in the literature, powdered activated carbon (PAC) has been primarily applied rather than granular activated carbon (GAC). This is because one of the main reasons for using activated carbon is that it affects the rate of fouling which occurs in the narrow regions close to the membranes. In particular, powdered carbon alters the compressibility and permeability characteristics of the cake layer which would not easily be achieved with the granular form. Moreover, there is concern that the membrane could be damaged by larger carbon particles colliding with it. Abrasion reduces the effective life of the membranes, and thus a compromise has to be sought between the benefits of the use of activated carbon and the associated damage and costs.

As noted above, the use of PAC has been shown to provide benefits in the operation of MBRs, but the gains achieved need to be weighed against the cost.

2.6 Combined MBBR-MBR

The evolution of MBBR (moving bed biofilm reactor) and MBR (membrane bioreactor) technologies has led to the proposal of their combination, known as MBBR-MBR. The combination of the two types of reactors or processes was suggested in 2001, as can be observed in a later publication (LEIKNES and ODEGAARD, 2007). The proposal is based on the concept that when the biomass is predominantly fixed on supports, as occurs in the MBBR, the tendency toward fouling of the membranes can be reduced. Furthermore, with the use of MBBRs the installations could be more compact since they operate with high organic loads.

In the above-cited publication, the authors submitted the MBBR (equipped with AnoxKaldnes carriers) to different organic loads through varying the hydraulic retention time (HRT) and maintained the HRT of the MBR (in its own tank or compartment) at a low value, since this would predominantly promote biomass separation. The results from this study showed that the fouling rate was greater when the MBBR was operated with a high organic load and this finding was attributed to a variation in the size distribution of the particles present in the MBR compartment. There was a relative decrease in the fraction of particles smaller than 1 μm when higher organic loads were applied. According to the authors, if these particles have a significant role in the occurrence of fouling, the intensification of this phenomenon would be expected when the process is operated with higher loads.

LEE et al. (2006) carried out experiments with a module of hollow fibers immersed in a vessel containing commercial polyurethane carriers (13 mm cubes) impregnated with activated carbon. They investigated the effects resulting from variations in the air flow (5–9 L/min) and the amount of carriers, expressed as a volumetric percentage (5, 10, and 20%), on the floc size and the eEPS concentration. The results revealed that the floc size decreased when the aeration or the volumetric fraction of the carriers was increased. The increase in the TMP during the operation was greater when lower air flows and volumetric fractions were applied. In relation to the eEPSp and eEPSc, none of the variables investigated had a significant influence on the contents of these constituents. An interesting result was observed when a steel mesh was installed around the membrane module in order to avoid collision between the carriers and the hollow fibers. The rate of variation in the TMP during the operation was much greater when the steel mesh was placed around the membrane module. A TMP of 30 kPa (operational limit) was reached in 33 min when the mesh was present and in 155 min when it was absent. The authors concluded that the collision of the carriers with the membranes creates friction forces which hinder the cake layer formation, and, consequently, the permeability of the membranes is maintained for a longer period.

SOMBATSOMPOP et al. (2006) studied an MBR with two interconnected compartments. In the second compartment, the module with hollow fibers was installed, and in the first compartment, two conditions were investigated: the use of biomass in suspension and the use of biomass fixed on supports (polypropylene rings). In this study, the carriers did not have access to the compartment containing the module, to avoid their collision with the hollow fibers. In relation to the fouling, it was observed that the resistance of the cake layer decreased significantly when the first compartment was operated as an MBBR. The viscosity of the sludge also decreased in this mode of operation.

The promising results obtained suggest that the MBBR-MBR combination could become more widely used in the future. The variants investigated to date involve the use of two interconnecting tanks and the use of a single tank, as shown in Fig. 2.15.

Despite the good results reported for a single tank, illustrated in Fig. 2.15b, it should be noted that the carriers employed in the study by LEE et al. (2006) were

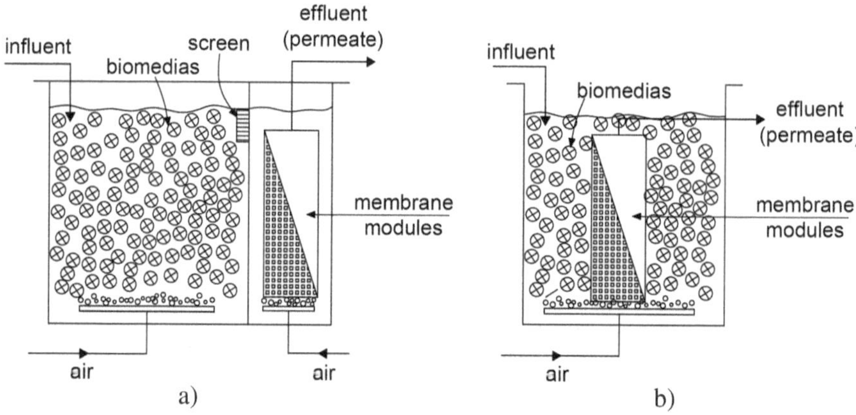

Fig. 2.15 MBBR-MBR models: (**a**) two tanks and confined carriers and (**b**) a single tank

made of nonrigid material, in contrast with the commercial carriers used in MBBRs. Further investigation is required, in particular experiments of long duration, in order to evaluate the possible damage caused to the membranes exposed to the constant collisions of the carriers. On the other hand, the variant shown in Fig. 2.15a is more suited to the wastewater treatment market.

2.7 Future Advances in MBR Technology

The progress achieved in recent years, in particular in relation to reactors which make use of submerged modules, reveals impressive advances in the evolution of MBR technology. Nonetheless, further improvements are certainly required, especially with regard to reducing the operational costs (energy and chemical consumables). In this regard, attention has been focused on reducing the fouling, which is considered to be the most critical aspect of MBR technology and directly related to the operational costs.

The challenges associated with MBRs, as reported by YANG et al. (2006), include improving our understanding of the fouling phenomenon and developing easily applied methods to control and reduce it, developing pretreatment methods which ensure safe operation, increasing the effective life of the membranes (greater chemical and mechanical stability), reducing the costs (membranes, labor, and energy), and designing and implementing large-capacity plants.

Due to the close relation observed between the flux and fouling, the most commonly used MBRs (submerged membranes) operate with modest fluxes. As reported by LESJEAN et al. (2009), although the maximum flux for operation with domestic sewage considered in the design is within the range of 14 to 50 $L/m^2.h$, with averages of 29 to 32 $L/m^2.h$, in practice the fluxes are within the range of 8–25 $L/m^2.h$. Although these fluxes are viable for the installation and operation of large-capacity

MBRs (>5000 m^3/d), the development of new membranes which ensure greater flux, without increasing the degree of fouling, is a theme to be pursued in order to make this technology more attractive.

Advances have been achieved in relation to the aeration, with the use of superimposed modules, through the designing of more efficient aerators, or with the use of cyclic or intermittent aeration.

The addition of chemical products to the reactor to modify the biomass characteristics and increase the filterability is also a theme which requires further investigation. In this regard, aluminum sulfate, ferric chloride, and different polymers have been tested. It is widely known that these substances act as coagulants/flocculants and can be used to remove liquid-phase colloids, particles in suspension, and some types of SMP. Certainly, new products will be launched in the future for this purpose, since the fouling caused by colloids, EPS, and SMP is critical to the functioning of MBRs.

The addition of adsorbents, such as activated carbon, to MBRs has been the subject of several studies, but the beneficial effect of their use during prolonged operation needs to be verified and a detailed evaluation of the associated costs should be carried out.

Finally, the success of the MBBR-MBR combination, which until now has been the subject of relatively few studies, in pilot or real scale, should be noted. By combining two technologies whose application has spread and become well established in recent years, the coupling of biofilm reactors with MBRs shows promise in terms of its future commercial success.

References

AHMED, T., SEMMENS, M.J., VOSS, M.A. Oxygen transfer characteristics of hollow-fiber composite membranes. *Advances in Environmental Research*, vol. 8, p. 637-646, 2004.

ARABI, S., NAKHLA, G. Impact of cation concentrations on fouling in membrane bioreactors. *Journal of Membrane Science*, v. 343, p. 110-118, 2009.

BRINDLE, K., STEPHENSON, T., SEMMENS, M.J. Enhanced biological treatment of oxygen demanding wastewaters by a membrane bioreactor capable of bubbleless oxygen mass transfer. *Proceedings of Water Environment Federation 70th Annual Conference and Exposition*, Chicago, Illinois, USA, p. 63-72, 1997.

CERQUEIRA, Ana Claudia Ferreira Pereira de. *Avaliação de membrana composta para transferência de oxigênio em biorreatores*, Doctoral thesis, COPPE/UFRJ, 2005.

CHANG, I.S., KIM, S-N., Wastewater treatment using membrane filtration – effect of biosolids concentration on cake resistance. *Process Biochemistry*, v. 40, p. 1307-1314, 2005.

GÖMER, T., de DONATO, P., AMEIL, M-H., MONTARGES-PELLETIER, E., LARTIGES, B.S. Activated sludge exopolymers: separation and identification using size exclusion chromatography and infrared micro-spectroscopy. *Water Research*, v. 37, p. 2388-2393, 2003.

HAN, S.S., BAE, T.H., JANG, G.G., TAK, T.M. Influence of sludge retention time on membrane fouling and bioactivities in membrane bioreactor system. *Process Biochemistry*, v. 40, p. 2393-2400, 2005.

JUDD, S. *The MBR book: principles and applications of membrane bioreactors in water and wastewater treatment*. Elsevier Ltd., 2006

KANAI, M., FERRE, V., WAKAHARA, S., YAMAMOTO, T., MORO, M. A novel combination of methane fermentation and MBR – Kubota submerged anaerobic membrane bioreactor process. *Desalination*, v. 250, p. 964-967, 2010.

KHAN, S.J., VISVANATHAN, C. Influence of mechanical mixing intensity on a biofilm structure and permeability in a membrane bioreactor. *Desalination*, v. 231, p. 253-267, 2008.

KIM, J-S., LEE, C-H., CHUN, H-D. Comparison of ultrafiltration characteristics between activated sludge and BAC sludge. *Water Research*, v. 32, p.3443-3451, 1998.

KUBOTA, Kubota Membrane Europe. Europe's oldest municipal MBR celebrates ten years, *Water 21*, April, p. 36, 2008.

LE-CLECH, P., CHEN, V., FANE, T.A.G. Fouling in membrane bioreactors used in wastewater treatment. *Journal of Membrane Science*, v. 284, p. 17-53, 2006a.

LE-CLECH, P., CHEN, V., FANE, T.A.G. Fundamentals, In: Judd, S., *The MBR book: principles and applications of membrane bioreactors in water and wastewater treatment*. Elsevier Ltd., Chapter 2, 2.33 to 2.3.9, 2006b.

LEE, W-N., KANG, I-J., LEE, C-H. Factors affecting filtration characteristics in membrane-coupled moving bed biofilm reactor. *Water Research*, v. 40, p. 1827-1835, 2006.

LEIKNES, T., ODEGAARD, H. The development of a biofilm membrane bioreactor. *Desalination*, v. 202, p. 135-143, 2007.

LESAGE, N., SPERANDIO, M., CABASSUD, C. Study of a hybrid process: adsorption on activated carbon/membrane bioreactor for the treatment of an industrial wastewater. *Chemical Engineering and Processing*, v. 47, p. 303-307, 2008.

LESJEAN, B., FERRE, V., VONGHIA, E., MOESLANG, H. Market and design considerations of the 37 larger MBR plants in Europe. *Desalination and Water Treatment*, v. 6, p. 227-233, 2009.

LI, Y-Z., HE, Y-L., LIU, Y-H., YANG, S-C., ZHANG, G-J., Comparison of the filtration characteristics between biological powdered activated carbon sludge and activated sludge in submerged membrane bioreactors. *Desalination*, v. 174, 305-314, 2005.

MAXIMOUS, N., NAKHLA, G., WAN, W. Comparative assessment of hydrophobic and hydrophilic membrane fouling in wastewater applications, *Journal of Membrane Science*, v. 339, p. 93-99, 2009.

McADAM, E.J., JUDD, S.J., A review of membrane bioreactor potential for nitrate removal from drinking water. *Desalination*, v. 196, p. 135-148, 2006.

MUNZ, G., GORI, R., MORI, G., LUBELLO, C. Powdered activated carbon membrane bioreactors (MBR-PAC) for tannery wastewater treatment: long term effect on biological and filtration performances. *Desalination*, v. 207, p. 349-360, 2007.

NOBLE, R. D., STERN, S. A., *Membrane separations technology*, Membrane Science and Technology Series, Vol. 2, 1st Edition, Elsevier, 1995.

OKAMURA, D., MORI, Y., HASHIMOTO, T., HORI, K. Identification of biofoulants of membrane bioreactors in soluble microbial products. *Water Research*, v. 43, p. 4356–4362, 2009.

PANKHANIA, M., BRINDLE, K., STEPHENSON, T. Membrane aeration bioreactors for wastewater treatment: completely mixed and plug-flow operation. *Chemical Engineering Journal*, v. 73, p. 131-136, 1999.

PARK, H., CHOO, K-H., LEE, C-H. Flux enhancement with powdered activated carbon addition in the membrane anaerobic bioreactor. *Separation Science and Technology*, v. 34, p. 2781-2792, 1999.

PIRBAZARI, M., RAVINDRAN, V., BADRIYHA, B.N., KIM, S-H. Hybrid membrane filtration process for leachate treatment. *Water Research*, v. 30, p. 2691-2706, 1996.

RAUTENBACH, R., ALBRECHT, R., *Membrane processes*, John Wiley and Sons, 1989.

SCHARZ, A.O., RITTMANN, B.E., CRAWFORD, G.V., KLEIN, A.M., DAIGGER, G.T. Critical review on the effects of mixed liquor suspended solids on membrane reactor operation. *Separation Science and Technology*, v. 41, p. 1489-1511, 2006.

SEMMENS, M.J., DAHM, K., SHANAHAN, J., CRISTIANSON, A. COD and nitrogen removal by biofilms growing on gas permeable membranes. *Water Research*, v. 37, p. 4343-4350, 2003.

SOMBATSOMPOP, K., VISVANATHAN, C., BEN AIM, R. Evaluation of biofouling phenomenon in suspended and attached growth membrane bioreactor systems. *Desalination*, v. 201, p. 138-149, 2006.

STOWA., http://www.stowa-selectedtechnologies.nl, em 15/03/2010.

VIERO, A.F., NOBREGA, R., SANT'ANNA JR., G.L., The use of polyetherimide hollow fibres in a submerged membrane bioreactor operating with air backwashing, *Journal of Membrane Science*, v. 302, p. 127-135, 2007.

VIERO, A.F., de MELO, T.M., TORRES, A.P.R., FERREIRA, N.R., SANT'ANNA JR., G.L., BORGES, C.P., SANTIAGO, V.M.J. The effects of long-term feeding of high organic loading in a submerged membrane bioreactor treating oil refinery wastewater. *Journal of Membrane Science*, v. 319, p. 223-230, 2008.

VISVANATHAN, C., YANG, B.S., MUTTAMARA, S., MAYTHANUKHRAW, R. Application of air backflushing technique in membrane bioreactor. *Water Science and Technology*, v. 36, p. 259-266, 1997.

WANG, Z., ZHICHAO, W., MAI, S., YANG, C., WANG, X., AN, Y., ZHOU, Z. Research and applications of membrane bioreactors in China: progress and prospect. *Separation and Purification Technology*, v. 62, p. 249-263, 2008.

WU, Z.C., WANG, Z.W. Study of integrated flat-sheet style membrane coupled anaerobic bioreactor for treatment of high strength alcohol distillery wastewater. *Proceedings of the 8th Seminar of JSPS-MOE Core University Program on Urban Environment*, Shanghai, China, p. 105-112, 2004.

YAMAMOTO, K., HIASA, M., MAHMOOD, T., MATSUO, T., Direct solid-liquid separation using hollow fiber membrane in an activated-sludge aeration tank. *Water Science and Technology*, v. 21, p. 43-54, 1989.

YANG, W., CICEK, N., ILG, J. State-of-the-art of membrane bioreactors: worldwide research and commercial applications in North America. *Journal of Membrane Science*, v. 270, p. 201-211, 2006.

YING, Z., PING, G. Effect of powdered activated carbon dosage on retarding membrane fouling in MBR. *Separation and Purification Technology*, v. 52, p. 154-160, 2006.

ZHANG, J., CHUA, H.C., ZHOU, J., FANE, A.G. Factors affecting membrane fouling performance in submerged membrane bioreactors. *Journal of Membrane Science*, v. 284, p. 54-66, 2006.

Chapter 3
Moving Bed Biofilm Reactor (MBBR)

João Paulo Bassin and Márcia Dezotti

3.1 Contextualization of and Introduction to MBBR Process

The exponential and sudden growth of large urban centers always generates the question: will there be space for the installation of treatment plants for the wastewater generated by the population and by the developing industrialization? Although this is currently a difficult question to answer, it is likely that in the near future emphasis will be given to the construction of treatment plants which enable compact installations, occupying the least possible physical space and which manage to maintain stable operation with reduced environmental impact. In this context, it is crucial that decision-making in relation to the design is preceded by discussions involving diverse sectors of society, considering not only the technical aspects but also the administrative factors and financial support associated with them, aiming to reach the targeted objectives.

In recent years, an increase in the interest in biofilm processes for the treatment of municipal and industrial wastewaters has been observed, due to the fact that these processes fulfill the future expectations mentioned above. There are several reasons why biofilm reactors have been preferred over other conventional processes with suspended biomass. One of the main reasons is the possibility to work with high biomass concentrations, which allows the reactor to operate with a higher load, good removal efficiency of organic compounds, greater stability in relation to variations in the influent composition and abrupt changes in load, temperature and toxicity, and easier separation of solids downstream of the reactor.

The attractiveness of biofilm processes is evident from the more compact form of these systems, occupying less space, which often represents a critical

J.P. Bassin (✉) • M. Dezotti
Chemical Engineering Program, COPPE, Federal University of Rio de Janeiro,
Rio de Janeiro, Rio de Janeiro, Brazil
e-mail: jbassin@peq.coppe.ufrj.br; mdezotti@peq.coppe.ufrj.br

M. Dezotti et al., *Advanced Biological Processes for Wastewater Treatment*,
DOI 10.1007/978-3-319-58835-3_3

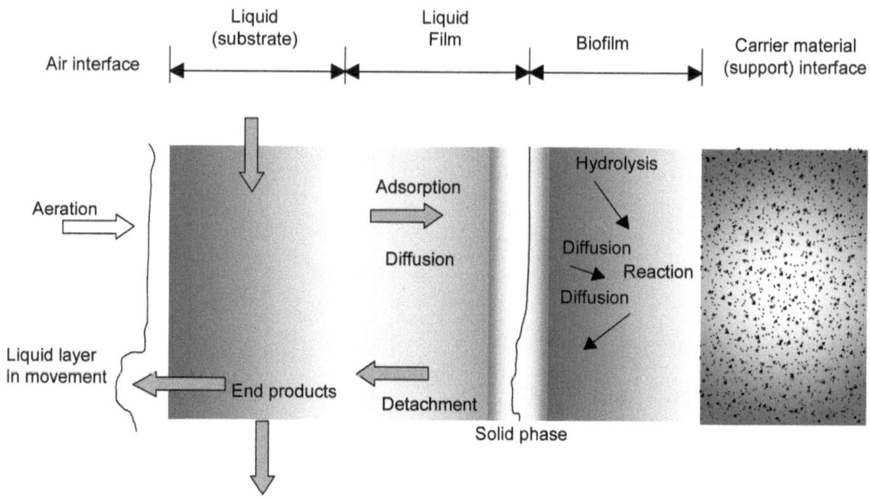

Fig. 3.1 Illustration of the transport and fate of the main components in biofilms (Adapted from GONÇALVES et al. 2001)

factor for the wastewater treatment plants. ØDEGAARD et al. (1994) stated that the systems which employ biomass adhered to a support medium, as in the case of the moving-bed biofilm reactor (MBBR), besides not requiring the traditional sludge recirculation of conventional systems, allow that the biomass always remains inside the reactor, a factor which makes these systems more specialized to the function for which they are destined.

A biofilm can be understood as a complex structure of cells and cellular products present in an immobilized form in a matrix of extracellular polymeric substances which are able to spontaneously form dense agglomerates growing adhered to static solid surfaces or moving carriers. Both the formation and accumulation of biofilms in aqueous media result from physical, chemical, and biological processes (NICOLELLA et al. 2000).

The microbial development in a biofilm relies on the transport of vital components (organic matter, oxygen, and nutrients). These essential components are firstly adsorbed onto the surface of the biofilm and then transported by diffusion processes initially through the liquid film, then the liquid/biofilm interface, and finally the biofilm, where they are metabolized by the microbial community. The final biodegradation products have a reverse flow, being directed toward the exterior of the biofilm. Figure 3.1 shows a schematic diagram of the transport and fate of the main components in biofilms.

Considering the nature of the processes involved in the development of a biofilm, it should be noted that diffusional limitations to mass transfer may occur, which could reduce the global reaction rate in these heterogeneous systems. Therefore, it is crucial to minimize these diffusional effects in order to improve the performance of bioreactors with immobilized biomass.

For a bacterial cell, there are many advantages of growing in the form of biofilms on a support material. First because this may be a protective environment against harmful agents, such as the recalcitrant and inhibitory compounds frequently present in industrial wastewaters. In addition, the microbial community may show resistance to dehydration due to the high level of hydration of the exopolymeric matrix secreted naturally by the microorganisms and also to predators such as protozoans (BASSIN and DEZOTTI 2008). Another point to be considered is that biofilm processes, in general, can have a higher potential for the removal of wastewater components mainly due to the wide variety of the microbial functional groups present in these environments.

The biofilm reactors most commonly used for the removal of organic matter and nutrients (nitrogen and phosphorus) include biological trickling filters, aerated submerged fixed-bed biofilm reactors, fluidized-bed reactors, and rotating biological contactors (RBC). All of these present advantages and disadvantages. The trickling filter does not have an effective volume, and the RBC is susceptible to mechanical failure. The fluidized-bed reactors frequently show hydraulic instability, and difficulties associated with reaching a uniform distribution of biofilm on the support surface are among the drawbacks of aerated submerged fixed-bed biofilm reactors (RUSTEN et al. 1995, 2006).

In order to overcome these operational problems, the compact and innovative moving bed biofilm reactor (MBBR) was designed. This technology was developed in Norway in the late 1980s and early 1990s (European Patent no. 0,575,314; US Patent no. 5,458,779), a period in which the authorities responsible for controlling pollution in Norway opted for the development of small sewage treatment plants but with large capacity, based on biological and chemical processes. The motivation for these new developments was based on the possibility of upgrading the majority of existing treatment plants (around 70% of the total), since these were of small size, serving populations of 50–2000 people.

In this context, the use of biofilms adhered to different carrier elements inside the reactor was initiated. The biological treatment was combined with pretreatment in large septic tanks and with posttreatment units, making use of chemicals for the coagulation/flocculation process. It was during this period that the Norwegian company Kaldnes Miljoteknologi developed the MBBR technology. This was carried out with the collaboration of a university research group represented by NTNU/SINTEF (Norwegian University of Science and Technology (NTNU); Foundation for Scientific and Industrial Research (SINTEF), University of Norway). In addition, the development of the MBBR process received financial support from the Norwegian Pollution Control Agency (SFT) and Norwegian Research Council (NRF) (RUSTEN et al. 1998).

In July 2002, the Norwegian company Kaldnes Miljoteknologi, holder of the MBBR technology, merged with the Swedish company Anox AB, in order to consolidate its position on the world market for wastewater treatment systems. The new company adopted the name of AnoxKaldnes in June 2004. AnoxKaldnes was bought by Veolia in July 2007, although the nomenclature of the MBBR media remained unchanged.

The first experience with this new technology dates back to October 1992, in two existing treatment installations (ØDEGAARD and RUSTEN 1993). The results obtained provided an indication of the capacity of this new system to improve the performance of existing treatment plants. Regarding the removal of nutrients, particularly nitrogen and phosphorus, it was noted that it was necessary to increase the quantity of carriers or insert physicochemical processes to accomplish their removal (ØDEGAARD and RUSTEN 1993).

The MBBR technology has been successfully established in the market, given that great number of large-scale wastewater treatment plants along with many other smaller treatment installations around the world (RUSTEN et al. 2006). It should be noted that, despite the large quantity of plants employing this technology, the number of studies on the removal of organic matter and nutrients (nitrogen and phosphorous) in these systems is small when compared with, for example, the respective studies carried out on activated sludge and other conventional reactors.

3.2 Principle of MBBR Operation

The development of the MBBR process was based on the central idea of gathering, in a single system, the best characteristics of the activated sludge and biofilm processes and eliminating the undesirable characteristics of each process (RUSTEN et al. 2006).

In contrast to most biofilm reactors, the MBBR system uses all of the effective volume of the reactor for the microbial growth, offering some advantages over its competitors. The head loss is considerably reduced, which represents a significant advantage in relation to fixed-bed systems, which exhibits a relatively high head loss. Furthermore, the filter medium of the latter can become blocked or clogged. Unlike activated sludge systems, MBBR does not require recirculation of the sludge from the secondary clarifier, since the biomass growth occurs on carriers which move freely inside the reactor tank. With the biomass fixed on a support media, enhanced solid retention in the biological reactor can be attained in comparison with conventional suspended biomass systems, in which cell loss eventually occurs due to poor sludge settling ability. Consequently, biofilm-based processes may show higher volumetric treatment capacity, i.e., can treat the same amount of wastewater in a lower volume. Furthermore, the size of the secondary tank is largely minimized, and there is no need for periodic cleaning of the media, as required in fixed-bed reactors. Also, existing reactors can be equipped and adapted to the MBBR configuration with relatively small modifications (SALVETTI et al. 2006).

MBBR technology can be applied to aerobic and anaerobic/anoxic systems. Figure 3.2 illustrates the possible configurations. In aerobic systems (Fig. 3.2a), the aeration is responsible for the movement of the carriers. Thus, the aerators perform a dual function, that is, they are responsible for the oxygenation of the microorganisms and for the maintenance of the carriers in movement in the reaction medium. Consequently, a greater input of air is required, which contributes to increasing the

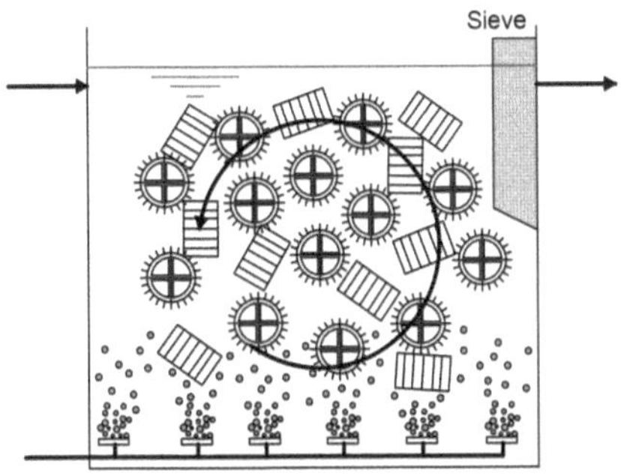

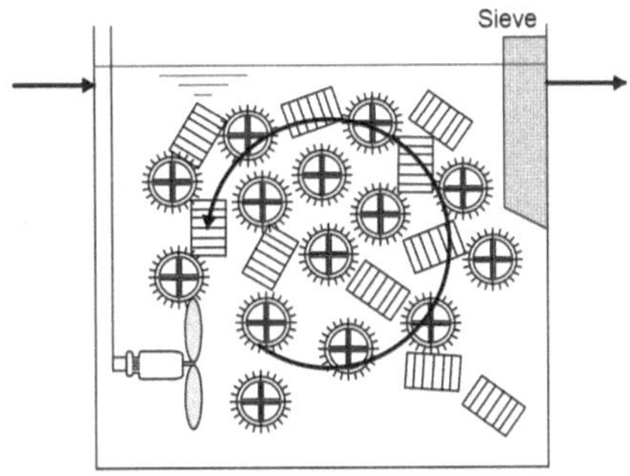

Fig. 3.2 Functioning of the variants of the MBBR process (Adapted from RUSTEN et al. 2006). (**a**) Aerobic (aerated) reactor. (**b**) Anaerobic-anoxic reactor

operational costs, particularly those associated with energy. Furthermore, the need for devices which provide adequate aeration and movement of the moving supports enhanced the cost of the process. In anoxic/anaerobic systems (Fig. 3.2b), a mechanical mixing device is required. In the case of aerobic systems, the appropriate design of the aerators is of crucial importance to improve the performance of the MBBR process (RUSTEN et al. 2006; ØDEGAARD et al. 1994). A device, commonly referred to as a sieve, is installed at the reactor outlet to retain the media within the tank.

A schematic representation of one of the possible configurations of MBBR systems, used in an industrial scale, can be seen in Fig. 3.3. It can be observed that the aeration is provided by perforated stainless steel tubes placed at the base of the reactor. This type of aeration system is commonly used in full-scale reactors as it provides adequate movement of the carriers inside the reactor and allows good conditions for oxygen transfer to the liquid phase.

Figure 3.4 illustrates several MBBR systems used in industrial-scale treatment. In Fig. 3.4a, c, a single large tank destined for the removal of organic matter can be observed. The occurrence of nitrification in this reactor, in particular, will be dependent on several factors, one of which is the concentration of organic matter in the influent and the hydraulic residence time (HRT). When the concentration of organic matter is relatively high, the development of nitrifying bacteria may be adversely affected, while fast-growing heterotrophic bacteria will proliferate. On the other hand, nitrification is favored in systems characterized by extended aeration and operated at high HRT.

In order to overcome the problems associated with the configuration of a single tank, multistage processes can be used, dividing the reaction tank into sections in order to create reactors in series, as shown in Fig. 3.4b. In this configuration, the removal of organic matter is carried out in the first tanks, where the chemical oxygen demand (COD) is greater. In the final tanks of the series, nitrification can occur since most of the organic matter has already been oxidized in the previous reactors. The aeration system, normally consisting of perforated stainless steel tubes located at the base of the reactor, is shown in Fig. 3.5.

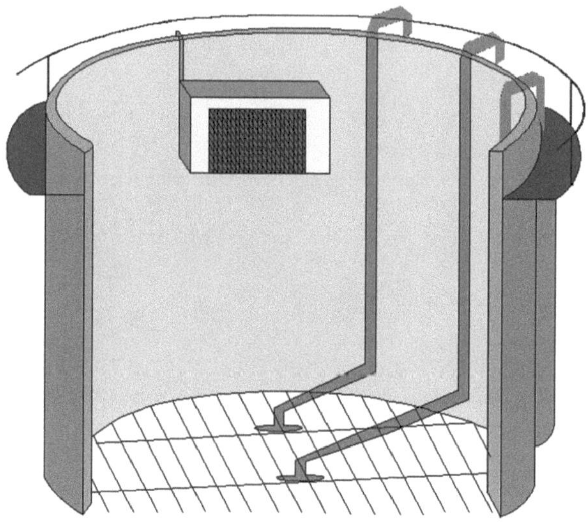

Fig. 3.3 Schematic representation of one of the possible configurations for an MBBR system (Source: adapted from http://www.cleanwatertech.com)

(a) (b) (c) (d)

Fig. 3.4 Photographs illustrating industrial-scale moving bed biofilm reactors (provided by Veolia Water Technologies—AnoxKaldnes). (**a**) Single tank configuration. (**b**) Tanks in series configuration. (**c**) Single tank showing carriers in detail. (**d**) Single-stage process

Fig. 3.5 Photographs of aeration grids used in MBBR processes (provided by Veolia Water Technologies—AnoxKaldnes)

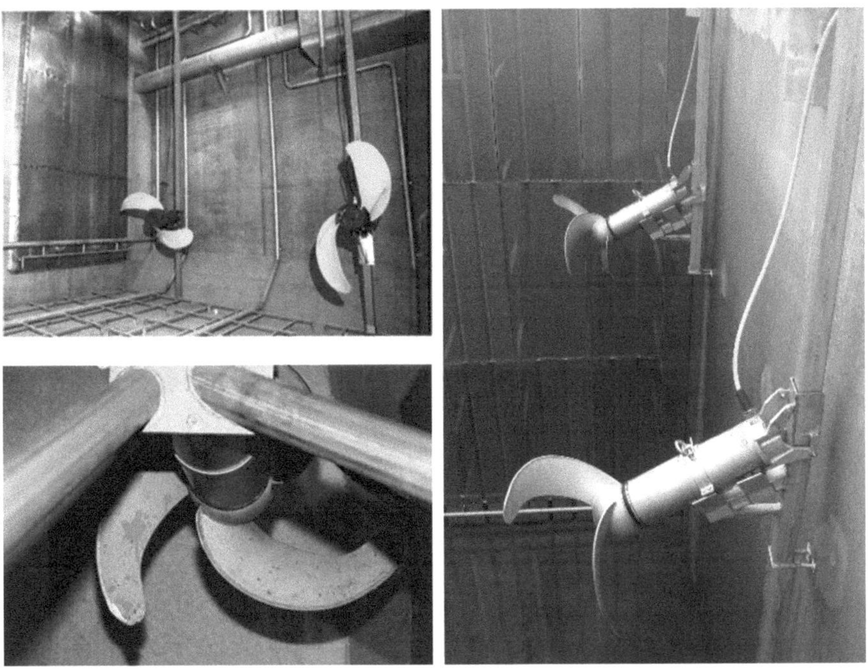

Fig. 3.6 Examples of mixers used in MBBR systems on an industrial scale (Provided by Veolia Water Technologies—AnoxKaldnes)

In systems which require anoxic conditions, the aerators are replaced with mechanical mixers responsible for the movement of the carriers in the reactor. Figure 3.6 shows some types of mixers used in such systems on an industrial scale.

Sieves are installed at the reactor outlet to prevent the carriers from leaving the tank (RUSTEN et al. 2006). They must be appropriately designed not only to retain

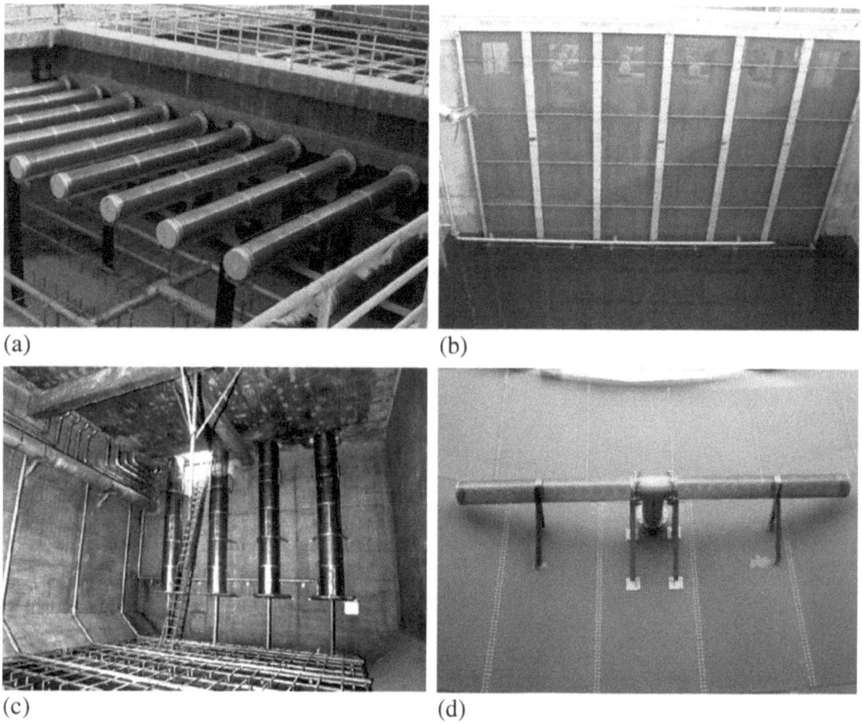

Fig. 3.7 Distinct sieve configurations used in industrial-scale MBBR systems (Provided by Veolia Water Technologies—AnoxKaldnes). (**a**) Cylindrical sieve mounted horizontally. (**b**) Sieve in the form of a mesh. (**c**) Cylindrical sieve mounted vertically. (**d**) Cylindrical sieve fixed on the wall in the center of the reactor

the carriers but also to avoid hydrodynamic problems. Figure 3.7 shows some sieve configurations used in the industrial sector.

Table 3.1 summarizes some of the main advantages and disadvantages associated with the MBBR technology.

3.3 Biofilm Carriers Used in MBBR Systems

The first biofilm carriers used in MBBR systems were developed by the patent holder AnoxKaldnes®. They are made of polyethylene with a density of 0.95 g/cm^3, have different dimensions (diameter and height), and generally have a cross in the center which divides them into circular sections with longitudinal ridges (or "fins") on the external surface (SALVETTI et al. 2006; RUSTEN et al. 1998). There are many types of carriers, and the characteristics of some examples are given in Table 3.2. It should be noted that only the effective surface area for the biofilm adhesion of each type of carrier is described, since the biomass grows mainly on the

Table 3.1 Main advantages and disadvantages associated with moving bed biofilm reactors (MBBRs)

Advantages	Disadvantages
They can be applied in existing treatment installations, often being used to improve their performance	High energy costs associated with aeration, which is responsible not only for supplying the oxygen for the microorganisms but also for the movement of the carriers inside the reactor
In contrast with conventional activated sludge processes, the sludge does not need to be recirculated through the system since the biomass grows adhered to the carriers	If the system is not well designed, problems related to the hydrodynamics may occur, such as the formation of stagnant regions
They allow a reduction in the installation cost, since they dispense with the need for some of the stages required in conventional processes	The initial investment needed to construct the reactor and acquire the patented carriers (biomedias) may hamper the implementation of such systems
In contrast to fixed-bed reactors, there is no clogging of the sludge bed, and therefore periodic cleaning is not needed	
The system footprint can be reduced, and the treatment plant can be built much more compact	
The biofilm adhered to the carriers is more resistant to variations in the influent concentration and abrupt changes in organic and hydraulic loads, pH, temperature, and toxicity	

protected surface on the inside of the carriers. The total surface area is much greater than the protected surface area of the biofilm (RUSTEN et al. 2006; ØDEGAARD et al. 1994). Figure 3.8 shows some of the carrier elements described in Table 3.2.

A comparison of the surface areas (or grid heights) of the Kaldnes® media described in Table 3.2 is given in Fig. 3.9.

Recently, given the widespread application of the MBBR process, a large variety of carriers has been developed by several companies. The design is usually made to provide the best treatment performance for each specific case. The carriers can be adapted to different reactor configurations and types of wastewaters, since they are produced in different forms and sizes. Therefore, the acquisition of the carriers represents an important step of the MBBR technology implementation.

The conventional approach used in carrier development aimed at obtaining larger protected surface area. However, under some circumstances, this may lead to uncontrolled biofilm thickness, bringing uncertainties to the process design based on the carrier surface area. To overcome this drawback, a new type of carrier (Z series, displayed in Fig. 3.8) was developed by AnoxKaldnes. The media design is covered with a grid of defined height, so that the biofilm is allowed to grow outside of the carrier in a protected environment. Due to the collision between different media within the reactor, the biofilm cannot grow higher than the grid height. Therefore, the thickness can be controlled (Table 3.2 presents two types of Z carriers with a

Table 3.2 Characteristics of some Kaldnes® carriers (Adapted from RUSTEN et al. 2006 and http://www.anoxkaldnes.com)

	Type of carrier (Kaldnes®)									
	K1	K2	K3	K5	Natrix C2	Natrix F3	Biofilm Chip M	Biofilm Chip P	Z-200	Z-400
Nominal diameter (mm)	9.1	15	25	25	36	46	48	45	30[a]	30[a]
Nominal height (mm)	7.2	15	10	3.5	30	37	2.2	3	–	–
Density (kg/dm^3)	0.95	0.95	0.95				0.96	0.96	0.95	0.95
Protected surface area (m^2/m^3)[b]	500	350	500	800	220	200	1200	900	–	–
Grid height (μm) for maximum biofilm thickness	–	–	–	–	–	–	–	–	200	400
Protected surface area at 60% filling (m^2/m^3)[c]	300	210	300	480	132	120	720	540	–	–

[a]Indicates the protected diameter of the saddle-shaped media
[b]Total area theoretically available for microbial adhesion/reactor volume, completely filled with support (fixed bed)
[c]Total area theoretically available for microbial adhesion (60% bed)/reactor volume

grid height of 200 and 400 μm). Such characteristic avoids carrier clogging-related problems and is very convenient for MBBR applications where thick biofilms are expected but not desired, such as nitrification at high organic loading rates.

3.4 Operational Aspects

3.4.1 *Filling Ratio (V_S/V_R) or Filling Fraction (%)*

The amount of carriers added to the reactor is commonly referred to as the media filling ratio (ratio of volume occupied by the carriers (fixed bed) to total reactor volume, V_S/V_R) or simply the filling fraction (%).

One advantage of the MBBR system is that this filling fraction can be altered as desired, although values lower than 70% are recommended in order to provide

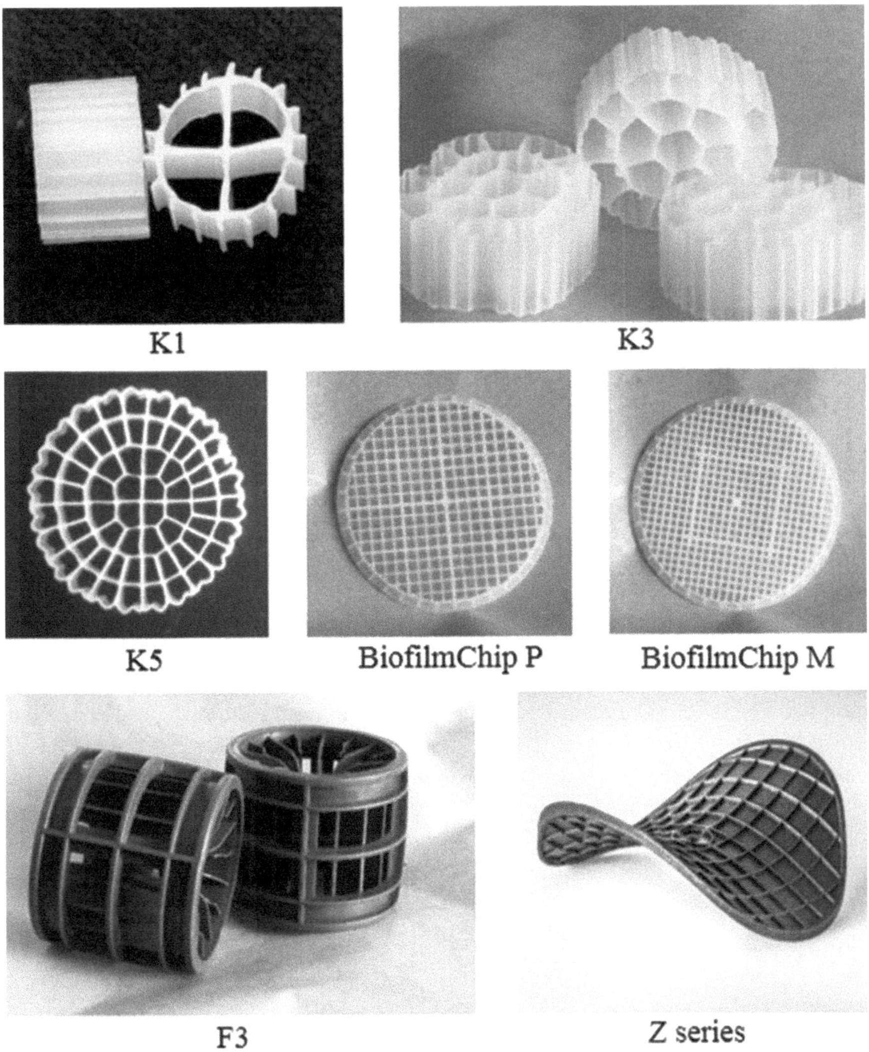

Fig. 3.8 Commercially available Kaldnes® carriers for MBBR systems (Provided by Veolia Water Technologies—AnoxKaldnes)

adequate mixing and allow good movement of the carriers, avoiding hydrodynamics problems (e.g., presence of stagnant areas) (RUSTEN et al. 2006; SALVETTI et al. 2006). According to SOKÓL (2003), the recommended V_S/V_R ratio is 0.55, although some researchers have shown that MBBRs can operate at higher filling ratios (0.6–0.7). However, as noted above, high filling fractions can adversely affect the hydrodynamics of the reactor, which, in turn, has a significant effect on the biofilm thickness and, consequently, on the performance of the process.

In order to determine the appropriate amount of carriers to be introduced into the aeration tank, the specific surface area available for microbial growth needs to be

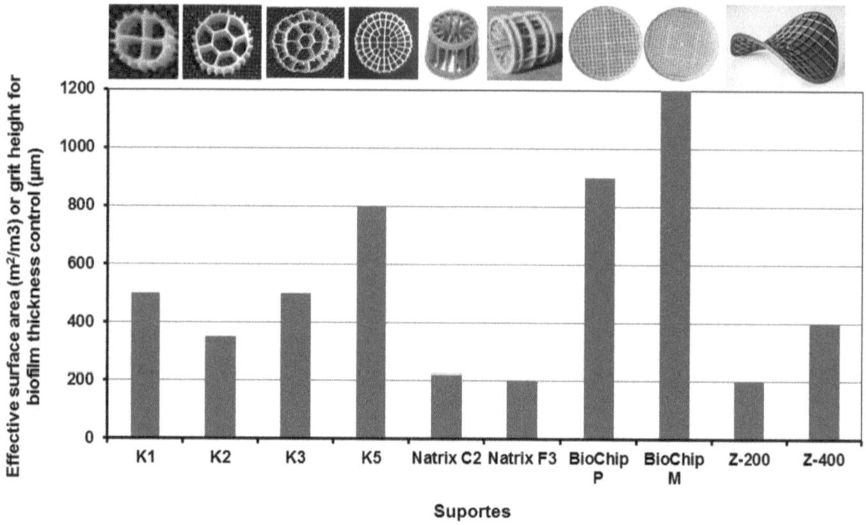

Fig. 3.9 Effective surface area (m^2/m^3) or grit height (μ) to control biofilm thickness of Kaldnes® carriers

known, which is dependent on the size and design of the carrier. Both the filling ratio (V_S/V_R), in terms of aeration tank volume, and the specific surface area of each carrier determine the area available for biofilm adhesion. Therefore, if the treatment plant requires a greater capacity due to an increase in the load, more carriers can be added to the reactor, thus increasing the surface area available for microbial adhesion (RUSTEN et al. 1995). For instance, if the specific area of the support is 500 m^2/m^3 and the filling fraction is 50%, the surface area available for the biofilm growth will be 250 m^2/m^3 of reactor (SALVETTI et al. 2006).

3.4.2 *Hydrodynamics of MBBR*

As in the case of any biofilm process, the diffusion of compounds into and out of the biofilm plays an important role in MBBR systems. In fact, the mass transfer involved in such reactors is directly related to diffusion effects (RUSTEN et al. 2006). Therefore, most characteristics attributed to microbial growth in biofilms can be explained through transfer phenomena (STEWART 2003).

In systems with biomass in suspension, the transport of solutes from the liquid medium to the cell is a relatively rapid process and is usually not the limiting step of the bioprocess which occurs in the cell. On the other hand, microbial aggregates present in biofilms are densely packed environments where the flow of liquid is limited (STEWART 2003). In relatively thick microbial aggregates, the diffusional distances are sufficiently large that the solute transport toward the interior of the microbial cells becomes slow compared with the biodegradation kinetics. In this situation, solute concentration gradients may be established in the biofilm and need to be considered (XAVIER et al. 2005).

Thus, the effective thickness of the biofilm, which corresponds to the depth to which the biofilm is penetrated by the substrate, is of great importance. Since this depth should be less than 100 μm for the complete substrate penetration, the ideal biofilm needs to be thin and uniformly distributed on the carrier surface. To obtain these characteristics, the turbulence in the reactor, along with the action of the shear forces originating from it, is of crucial importance both for the transport of substrates to the biofilm and for the maintenance of a thin biofilm (RUSTEN et al. 2006).

The chaotic movement of the carriers in the bioreactor caused by the turbulence originating from the air flow allows the natural sloughing of the biofilm, which leads to the renovation of the biomass. Therefore, dead bacteria can be removed, leaving space for the carrier to be colonized by new bacteria. This scenario highlights the importance of ensuring good hydrodynamic conditions, which not only involves the model for the liquid phase mixture (perfect mixing or not) but is also related to the segregation of the carriers and the appearance of stagnant zones.

With regard to the model for the liquid phase mixing, preliminary hydrodynamic tests are generally carried out to determine whether or not the bioreactor has a perfect mixing behavior. These tests are normally carried out with the aid of tracers (e.g., NaCl) and are of the type stimulus response, that is, a known quantity of tracer is added to the influent of the bioreactor at the start of the test. The addition of the tracer can be carried out instantaneously (adding a small volume of tracer solution in a very short period of time) or continuously (feeding the reactor during a time period equivalent to at least three times the hydraulic retention time). The concentration of tracer in the effluent is then measured continuously a function of the time elapsed or at predetermined time intervals. As the total mass of tracer added and its initial concentration are known, the theoretical concentration of tracer in the effluent as a function of the time elapsed since the beginning of the test can also be calculated. Finally, the curves for the distribution of the residence times obtained during the hydrodynamic tests using a saline tracer and those related to the perfect mixing model can then be compared. In the case of MBBR, in particular, these preliminary tests can be started with the reactor containing a certain media filling fraction which can be gradually increased in order to observe the influence of the filling ratio on the hydraulic regime of the reactor.

One of the factors which strongly affect the hydrodynamics of a bioreactor is scale-up. Although, in general, the liquid phase shows perfect mixing behavior, in most cases, due to the relatively high air flows which are employed, the turbulence and subsequent possibility of collusions between the particles can present different intensities on distinct scales. Nevertheless, regardless of the scale at which the reactor is operated (laboratory, pilot, or full scale), the more intense the turbulence applied to the system, the greater the biofilm sloughing will be, which leads to an increase in the concentration of suspended solids in the liquid phase (TAVARES et al. 1995).

3.4.3 Dissolved Oxygen (DO)

The dissolved oxygen content is a limiting variable in biological treatment processes. A concentration of 2 mg/L is generally adopted as the minimum concentration required for the operation of aerobic biological reactors aimed at the removal of organic matter (METCALF and EDDY 1991). However, systems with immobilized biomass may require a greater concentration of DO due to the diffusion limitation inherent to these biofilm processes, as described in the previous section (Sect. 3.4.2).

In the case of moving bed biofilm reactors, the supply of oxygen by air bubbling is responsible not only for providing air for the microorganisms but also for maintaining the carriers in suspension. In this regard, the air flow rate employed in MBBR systems to maintain the carriers in suspension is much greater than that necessary in suspended biomass reactors.

Taking into consideration the dual function of the aeration in MBBR systems, the design of the aerators must be carried out in such a way that the air bubbles generated should present an adequate size. The bubbles should not be very large, since this would lead to a substantial drop in the oxygen transfer coefficient (kLa). Small bubbles, on the other hand, favor oxygen transfer to the liquid medium but do not promote sufficient carrier movement. In this regard, the best air supply conditions should be established to enable proper oxygen transfer to the liquid medium and adequate carrier movement, preventing excessive biofilm sloughing from the carriers.

3.4.4 Formation of Biofilm on Moving Carriers in MBBR

The way in which the biofilm is formed in the carrier media of MBBR systems is very distinctive. Initially, the biofilm formation process is slow, particularly when the turbulence caused by the aeration is high, as it increases the shear rate and may hinder the adhesion of the microorganisms to the support medium. In general, the process start-up is carried out with the inoculation of a mixed culture of bacteria, often originating from activated sludge systems showing good and stable performance.

Over time, with the operation of the bioreactor, the biomass gradually adapts to the conditions imposed, particularly those related to the nature of the wastewater to be treated. In some cases, given the high heterogeneity of the substances present in a certain wastewater and their potentially recalcitrant or inhibitory nature, the development of biomass adhered to the moving carriers is hindered.

In general, the biofilm formed on the carriers in MBBRs designed for organic matter (COD) removal is thicker than that of systems aimed only at nitrification, given the high growth rate of heterotrophs. Autotrophic nitrifying bacteria are very sensitive to variations in the influent characteristics. Thus, when problems arise during the operation of nitrifying systems, irreversible consequences can ensue, even potentially resulting in a substantial loss of the biofilm.

In the case of biofilm reactors in which the simultaneous removal of organic matter and nitrification takes place, besides competing for oxygen, the heterotrophic and nitrifying bacteria compete for space, possibly generating stratification in the biofilm structure. The more rapid growth of the heterotrophic bacteria leads to this microbial community being located in the more external layers of the biofilm, where the concentration of substrate and the sloughing of the biomass are greater. On the other hand, nitrifying bacteria are present in the deeper layers of the biofilm. Thus, a heterotrophic layer can form over a nitrifying community, which is disadvantageous to the latter, particularly when the concentration of DO in the liquid medium is low. This limitation of DO, which is the result of consumption and resistance to mass transfer through the heterotrophic layer, negatively affects the performance of the nitrification and may hinder establishment of nitrifiers. In contrast, if the DO level is high or sufficient to overcome the diffusion limitations in the biofilm, the heterotrophic layer can have a positive effect on the nitrifying microbial community, protecting it from sloughing (FURUMAI & RITTMAN 1994). Regarding this issue, an interesting study was carried out by Bassin et al. (2012), which showed that the presence of organic carbon at the initial stage of reactor operation reduced the time required to form stable nitrifying biofilms in MBBR systems. Stepwise reduction of influent COD caused an enrichment of the biofilm with nitrifiers, as observed by fluorescent in situ hybridization (FISH) analysis.

Figure 3.10 shows the biofilm formed in different types of AnoxKaldnes® carriers used in MBBRs. Figure 3.10a in particular displays the biofilm grown on Kaldnes K1 carriers from an autotrophic laboratory-scale nitrifying MBBR reactor treating synthetic wastewater with no COD input (unpublished data from Bassin, J.P). The biofilm is very compact and dense as it is enriched with slow-growing nitrifiers.

3.4.5 Extracellular Polymeric Substances (Exopolymers)

Extracellular polymeric substances (EPS), produced by microorganisms, play an important role in biological wastewater treatment processes. In the activated sludge process, exopolymers are responsible for the mechanical stability of the flocs. In biofilm reactors, EPS (especially polysaccharides) are crucial for the adhesion of bacteria to the supports (CAMMAROTA and SANT'ANNA 1998). In attached growth processes, the exopolymers function as cementation agents ("glue"), aiding the fixation of the microorganisms to the support medium and to each other.

However, there are situations where the concentration of EPS is so high that it can lead to operational problems. The exopolymeric substances affect the total COD in the system and increase the turbidity of the treated effluent. Thus, the quality of the final effluent is adversely affected. REIS (2007), operating an MBBR submitted to high organic loads (4.4–8.6 kg COD/(m^3 day)), observed a substantial polysaccharide production. In many cases, in order to remove the excess exopolymer secreted by the microorganisms, it is necessary to add a physicochemical treatment

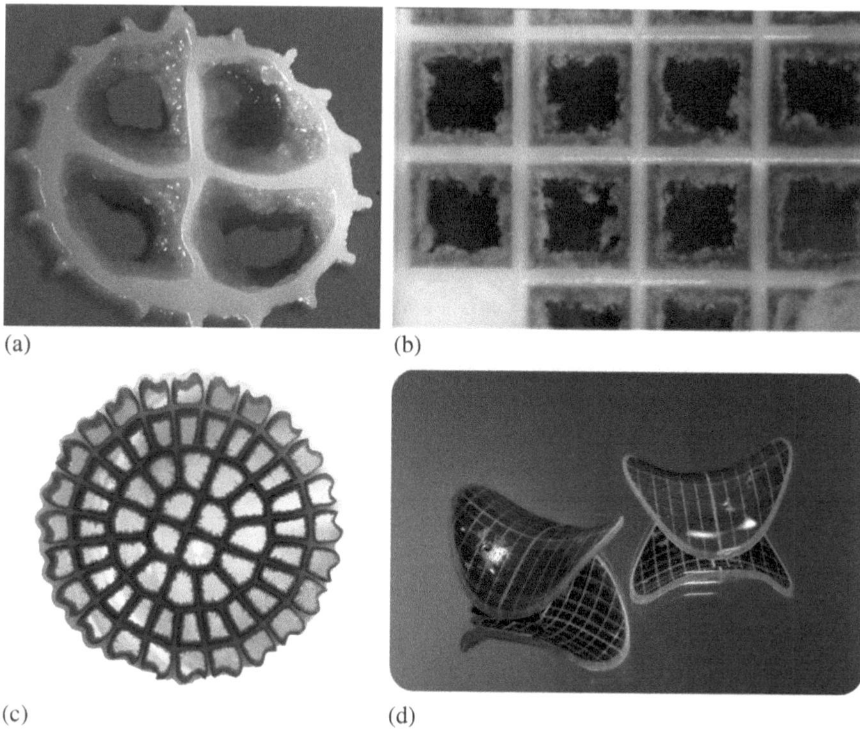

Fig. 3.10 Biofilm established in different MBBR carriers (Data from Bassin, J.P. (**a**) and Veolia Water Technologies—AnoxKaldnes (**b**, **c** and **d**)). (**a**) Biofilm on AnoxKaldnes K1. (**b**) Biofilm on AnoxKaldnes BiofilmChip P. (**c**) Biofilm on AnoxKaldnes K5. (**d**) Biofilm on AnoxKaldnes Z-400

stage, such as coagulation/flocculation, to the biological process. However, companies would clearly rather avoid this solution since it contributes to increasing the overall cost of the treatment.

Several studies have demonstrated that EPS are composed of many organic compounds. These include polysaccharides, proteins, lipids, and nucleic acids (FRØLUND et al. 1996). The complexity of the exopolymeric matrix hinders its characterization and requires not only some preliminary stages, involving the extraction and collection of these exopolymers, but also well-defined analytical procedures.

The complexity of the biofilm characteristics is related to various factors including the nature of the substrates, the diversity of the microbial species, and the characteristics of the support material (FLEMMING and WINGLINDER 2001). It is important to note that, although several studies have been carried out, a complete understanding of biofilm processes has not yet been established. There is a large number of variables to be investigated, seeking to expand our knowledge regarding these engineered bacterial aggregates. The same applies to the composition of exopolymeric substances which, due to their complexity, have become the focus of study for several researchers. The study of the production of these exopolymers and their characterization is essential for better understanding of the processes which occur in the biofilm.

The characterization of extracellular polymeric substances associated with biological wastewater treatment systems has shown a considerable variation in the composition and quantity of these substances. This variation can be attributed to the different responses obtained in the methods used for the extraction of the EPS components from microbial flocs and biofilms and to the different analytical methods used to quantify the different fractions (FRØLUND et al. 1996).

The influence of extraction methods on the composition of exopolymers is particularly notable, as evidenced in several results reported in the literature. Thus, on comparing the EPS composition in different bioreactors, it is important to remember that the distinct methods may lead to very different results. It is important to emphasize that the analytical methods used in these determinations can lead to different results regarding the recovery of extracellular polymeric substances. One example of this can be observed in the work carried out by FRØLUND et al. (1996). These authors evaluated different methods for the quantification of proteins. When the method described by LOWRY et al. (1951) was used, they obtained a concentration five times higher than that determined using the Bradford method (BRADFORD 1976). In general, the Bradford method rather than the Lowry method is recommended, since it has greater sensitivity and is less subject to interference from other components (DANIELS et al. 1994).

3.4.6 Microscopic Observation of the Biofilm

The biofilm formed on the carriers of the MBBR systems, besides containing bacterial cells responsible for the removal of organic and inorganic matter, may present a very diverse microfauna, characterized by the presence of a large number of micrometazoans and protozoans.

In many cases, due to their sensitivity to the variations in the process conditions (e.g., concentrations of substrate and dissolved oxygen and the presence of toxic substances), the protozoans and micrometazoans are used as indicators of the process treatment performance. As they feed on live bacteria, the turbidity and consequently the particulate organic matter content of the treated effluent can be reduced. Furthermore, pathogenic bacteria can be removed by these organisms.

Figure 3.11 shows some micrographs, obtained from optical microscopy, of the biofilm removed from the carriers of lab-scale nitrifying MBBR systems submitted to low organic loads and subjected to influent ammonium concentrations of 100–500 mg/L. As can be observed, the biofilm is colonized by fixed or stalked protozoans (Fig. 3.11a, e, and i), free ciliates (Fig. 3.11b), cylindrical worms (nematodes) (Fig. 3.11c), rotifers (Fig. 3.11d), thecamoebians (Fig. 3.11f, i), filamentous bacteria (Fig. 3.11g), and amoebas (Fig. 3.11h).

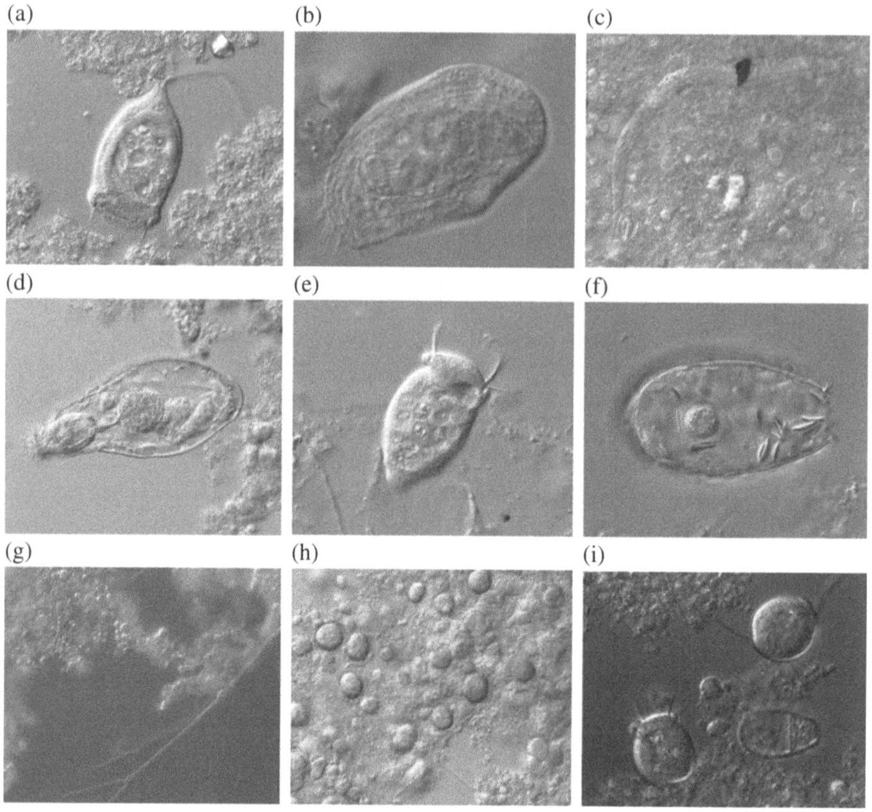

Fig. 3.11 Microphotographs of the biofilm adhered to the carriers from a nitrifying MBBR system as observed by scanning electron microscopy. fixed or stalked protozoans (**a**, **e**, and **i**), free ciliates (**b**), cylindrical worms (nematodes) (**c**), rotifers (**d**), thecamoebians (**f**, **i**), filamentous bacteria (**g**), and amoebas (**h**)

With the use of scanning electron microscopy (SEM), it was possible to observe the structure and composition of the biofilm adhered to the carriers taken from the same reactors from which the samples were analyzed by optical microscopy (Fig. 3.12). Figure 3.12a shows the biofilm structure characterized by microbial agglomerates arranged on the exopolysaccharide matrix. Figure 3.12b, c show the disordered arrangement of the bacterial cells around the extracellular polymeric material and the thecamoebians located between the bacteria arranged on the EPS matrix, respectively. Figure 3.12d highlights the organizational complexity of the biofilm, showing its different levels. Microalgae (diatoms) and cyanobacteria (cyanophyceans or blue algae) can be seen in Fig. 3.12e, f, respectively.

Figure 3.13a, b illustrate, respectively, a ciliated protozoan (family Euplotidae) and stalked protozoans (genus *Vorticella*). A cylindrical worm (nematode) can be observed in Fig. 3.13c, and several thecamoebians (genus *Euglypha*) can be seen in Fig. 3.13d. Figure 3.13e, f show, respectively, ciliated protozoans (family Euplotidae) in different positions.

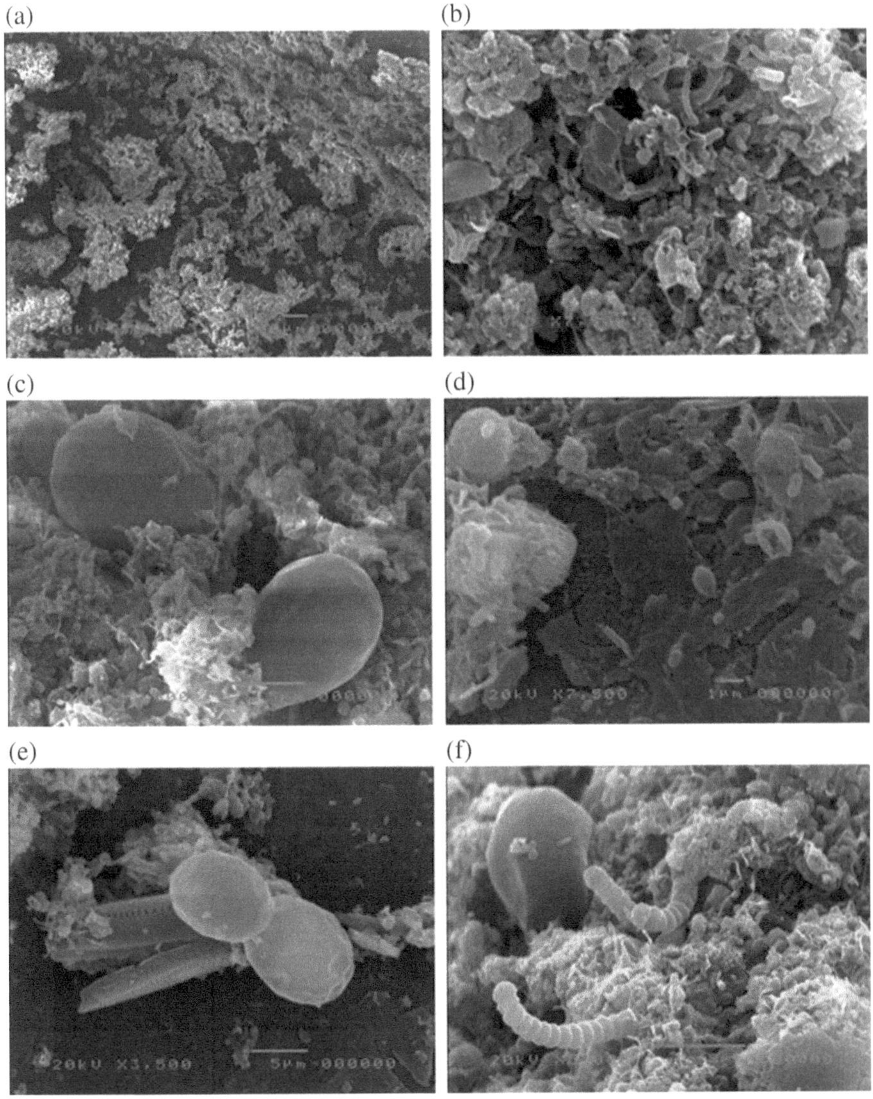

Fig. 3.12 Microphotographs of the biofilm adhered to the carriers collected in a nitrifying MBBR system as observed by scanning electron microscopy. (**a**) Magnification ×75. (**b**) Magnification ×7500. (**c**) Magnification ×5000. (**d**) Magnification ×5000. (**e**) Magnification ×3500. (**f**) Magnification ×5000

In Fig. 3.14a *Epistylis* protozoans can be observed, which are considered to be indicators of good quality effluent and which often occur in treatment systems operated under steady-state conditions and with permanent aeration (*apud*). A "pool" of bacterial cells can be observed in detail in Fig. 3.14b, and the stratification of the biofilm structure and arrangement of the bacteria in the exopolysaccharide matrix can be visualized in Fig. 3.14c, d.

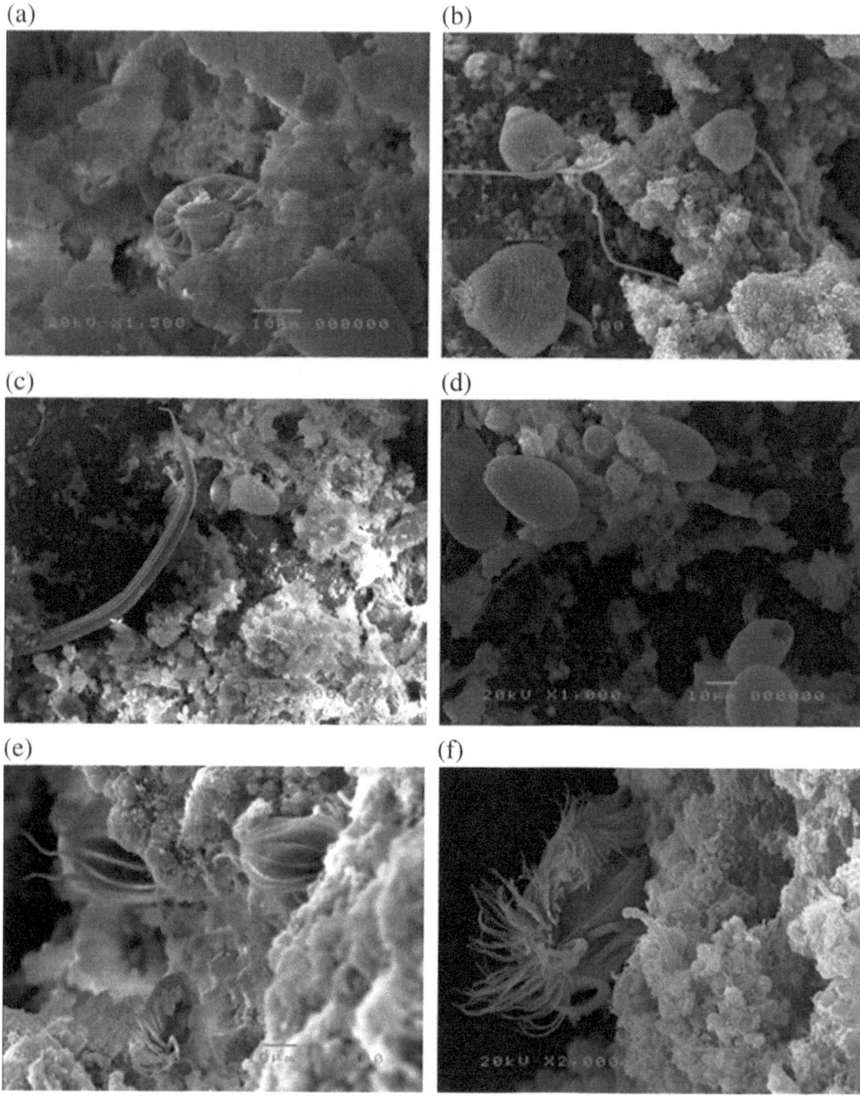

Fig. 3.13 Microphotographs of the biofilm adhered to the carriers in an MBBR system aimed at nitrification, obtained by scanning electron microscopy. (**a**) Magnification ×1500. (**b**) Magnification ×1000. (**c**) Magnification ×500. (**d**) Magnification ×1000. (**e**) Magnification ×1000. (**f**) Magnification ×2000

It is important to note that in many cases the micrometazoans and protozoans present in the biofilm and detected through optical microscopy are not observed by scanning electron microscopy. In fact, the sample preparation procedure carried out prior to observation by SEM involves several stages (washing, fixation, post-fixation, dehydration, and drying) which may remove these organisms from the

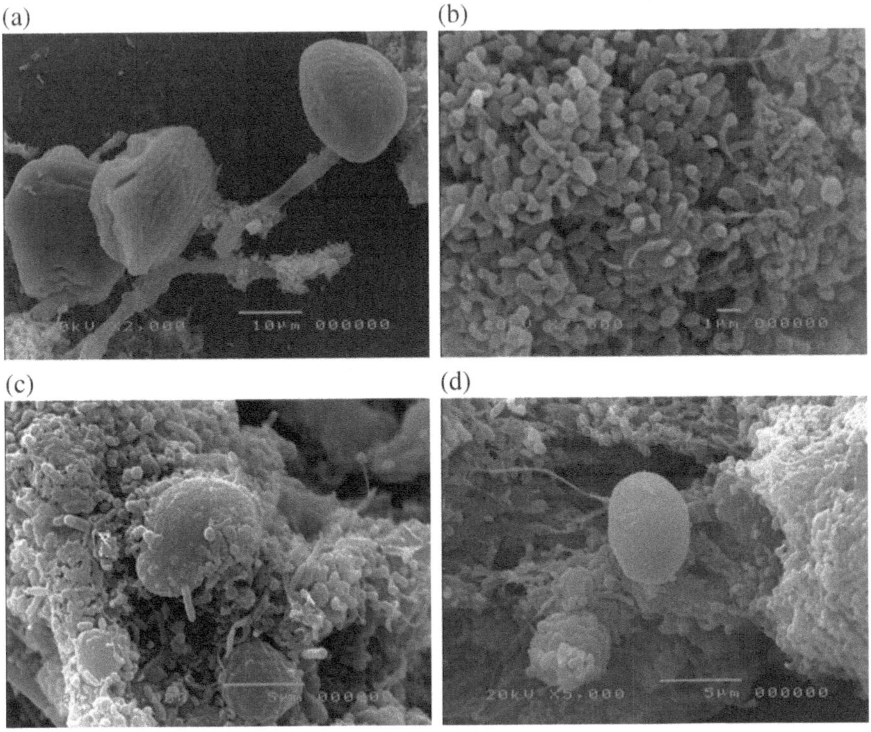

Fig. 3.14 SEM microphotographs of the biofilm adhered to the carriers in a lab-scale MBBR. (**a**) Magnification ×2000. (**b**) Magnification ×7500. (**c**) Magnification ×5000. (**d**) Magnification ×5000

biofilm. Therefore, combination of both techniques may give the complete panorama of the microfauna harbored in the biofilm: optical microscopy, for the almost immediate observation of the biofilm characteristics at a certain moment, and scanning electron microscopy, for the acquisition of high resolution images and observation of elements integrated with the biofilm which would not be identified by optical microscopy alone.

3.5 MBBR Applications

The moving bed biofilm reactor (MBBR) has been applied in the treatment of many kinds of wastewaters for the removal of organic matter, nitrogen, and, to a lesser extent, phosphorous. When strict environmental legislations are imposed on the discharge limits for phosphorus, as is the case in some European countries, MBBR represents an interesting alternative when combined with physicochemical processes (e.g., chemical precipitation).

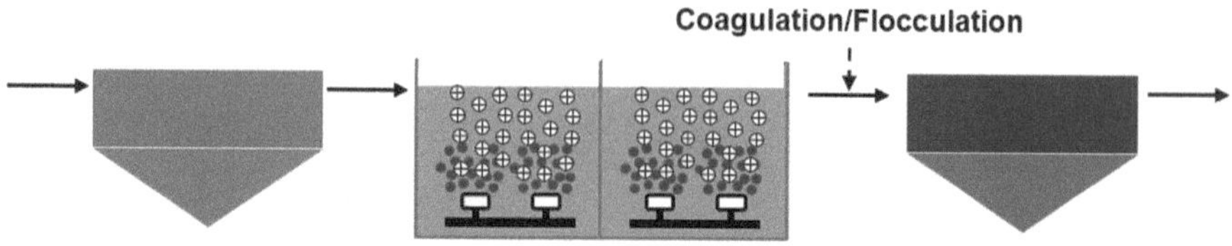

Fig. 3.15 MBBR followed by settling (biomass separation), with the inclusion of a physicochemical stage (coagulation/flocculation) for the removal of phosphorus (Adapted from ØDEGAARD 2006)

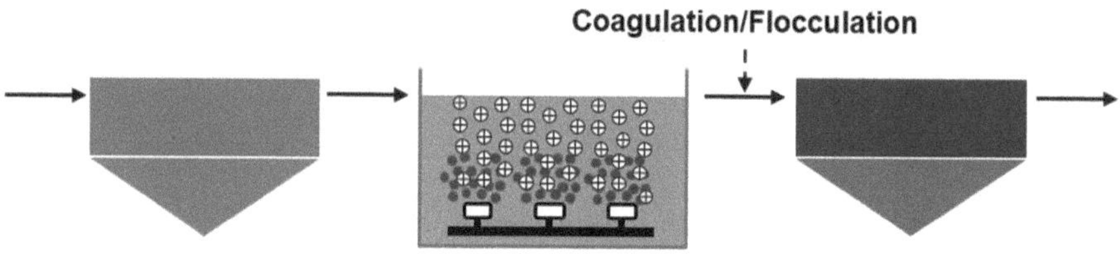

Fig. 3.16 High-rate MBBR followed by coagulation/flocculation and separation of biomass (Adapted from ØDEGAARD 2006)

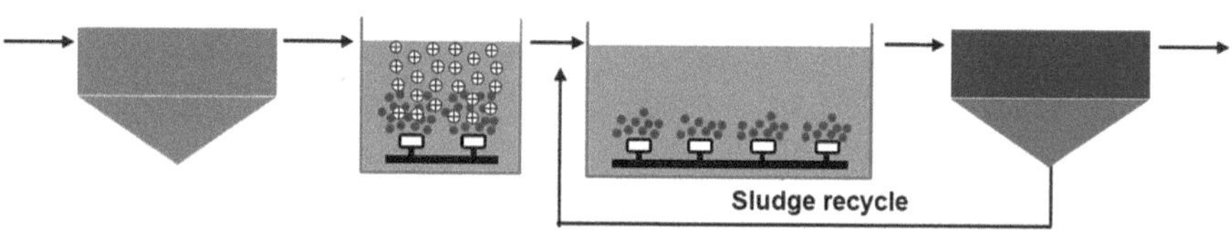

Fig. 3.17 MBBR applied as a pretreatment for existing activated sludge plants aimed at increasing the capacity of the plant or improving the overall efficiency of the process (Adapted from ØDEGAARD 2006)

3.5.1 Application of MBBR for Organic Matter Removal

Studies related to the removal of organic matter have generally aimed at improving the conditions in order to adapt the conventional activated sludge process to the MBBR configuration or to combine these and other processes in order to improve the performance of existing biological treatment plants. The application of the MBBR technology in combination with activated sludge systems helps to absorb the oscillations and shock loads in the feed stream, avoiding damage to the microbial community. In high-rate processes, the MBBR can be used as a stage following the biological treatment, together with the coagulation/flocculation process. In these cases, the hydraulic retention time is less than 1 h (ØDEGAARD et al. 2004).

In Figs. 3.15, 3.16, 3.17, and 3.18, some possibilities for the abovementioned combinations are illustrated. As regards the configuration shown in Fig. 3.15, the number of reactors in series required will be dependent on the pretreatment and the

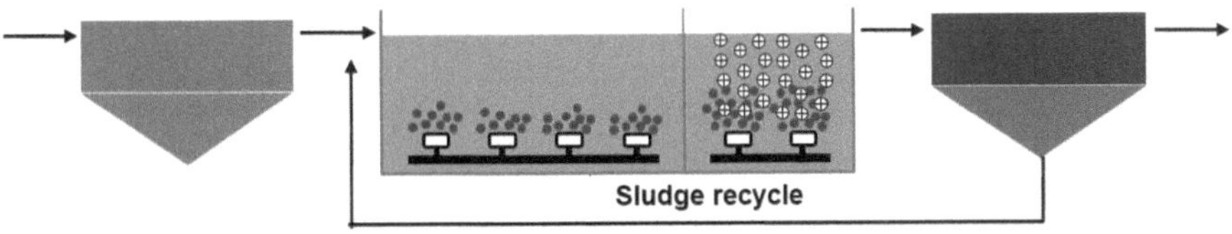

Fig. 3.18 Combination of activated sludge process and MBBR system where the support media are placed in the final part of the reactor in order to increase the reactor performance and/or enable/improve nitrification (Adapted from ØDEGAARD 2006)

characteristics of the influent wastewater. The use of coagulation/flocculation steps may be required in high-rate MBBRs where poor settling sludge may occur (Fig. 3.16). As the MBBR can withstand high organic loads, it can be used before the activated sludge plant, as shown in Fig. 3.17.

The introduction of supports in the final compartment of the aeration tank, as shown in Fig. 3.18, is aimed at obtaining a greater surface area for the development of the biomass. Thus, in the space destined for the moving bed reactor, the biological activity will be greater than that for the configuration without the inclusion of supports, although the total reactor volume remains unaltered. Consequently, the capacity for the removal of organic matter and nitrogen loads increases. The application of this configuration is convenient and is relevant to the reality of many existing treatment plants which are overloaded and lack space for their extension. Clearly, in this hybrid system (activated sludge + MBBR), conditions which allow the biofilm carriers to be retained in the aeration tank (installation of sieve at the outlet) need to be ensured, and the carriers need to be maintained in suspension without the occurrence of stagnant zones (use of aerators which provide a high air flow rate).

The combination of the MBBR technology with the activated sludge process in the same aeration tank also favors nitrification, since the solid retention time becomes partly uncoupled from the hydraulic retention time. This configuration is particularly interesting for plants operating at low temperatures, as under these conditions, the sludge age needed to support nitrification is relatively high due to low growth rate of nitrifying bacteria. The maintenance of high sludge age may not be achieved in a suspended biomass growth process. With the addition of plastic media in one compartment of the aeration tank, the overall solid retention time increases and favors the occurrence of nitrification in the biofilm, despite the unfavorable environment for this process in the bulk liquid phase.

It should be noted that not only can the MBBR technology be combined with the conventional activated sludge system, but it can also be applied as a posttreatment for ponds, functioning in this case as a polishing stage for the removal of nitrogen compounds or residual organic matter.

Another possibility is the operation of MBBRs in sequencing batch mode, configuration often referred to as sequencing batch biofilm reactors (SBBRs). Compared to conventional SBRs with suspended biomass, SBBRs ensure greater stability of the process due to partial immobilization of the bacteria, which may also lead to higher concentration of biomass within the reactor.

SBBR systems have been frequently used to remove nutrients, such as nitrogen. Besides promoting the development and maintenance of slow-growing nitrifying bacteria, the mode of operation of SBR allows the coexistence of aerobic and anaerobic/anoxic phases in the same tank, and therefore different processes can occur in well-defined sequential time intervals. The operation of an SBBR is similar to that of an SBR, with a filling phase, reaction (aerobic or anaerobic/anoxic), and discharge of the treated effluent. However, the traditional settling phase of the conventional SBR is not required, as the biomass is adhered to the carriers. Furthermore, cell loss arising from the occurrence of poor settling sludge is minimized. Similarly to the SBR, the duration of the phases is determined a priori, and the reactor performance, in terms of pollutant removal efficiency, can be improved when the reactor is fully automated, allowing the anaerobic/anoxic and aerobic periods to be controlled accurately.

The efficiency of the organic matter removal from domestic sewage in MBBR systems described in the literature is similar to those of the conventional activated sludge process, reaching values equivalent to 95% (ØDEGAARD and RUSTEN 1993; ØDEGAARD et al. 1994). The performance of the process is directly dependent on the biomass concentration within the reactor system, which is intrinsically related to the amount of support material available.

One of the parameters affecting the performance of biological treatment processes is the influent volumetric loading rate. In the case of biofilm reactors, such as the MBBR, the performance of the process depends on the surface loading rate, defined as the ratio between the volumetric load and the total surface area provided by the carrier material (RUSTEN et al. 1998). So, in the same reactor volume, the surface load can be varied by modifying the amount of support media available.

The residence time of MBBR reactors for the removal of organic matter is short (15–90 min), depending on the organic load applied. The soluble organic material is rapidly degraded, and the particulate matter is partially captured and then undergoes hydrolysis and degradation (ØDEGAARD 2006).

ØDEGAARD et al. (2004) evaluated the performance of some MBBR plants treating domestic and industrial wastewaters. These authors observed high COD removal, even at high organic loads. The air flow, besides maintaining the oxygen concentration required for the organic matter removal (around 3 mgO_2/L), was observed to prevent the accumulation of excessive biomass on the carriers, helping to maintain a thin biofilm.

Considering the good results obtained during the operation of MBBR systems in the treatment of domestic and food industry wastewaters (ØDEGAARD and RUSTEN 1993; RUSTEN et al. 1992), RUSTEN et al. (1994) evaluated the performance of these reactors in the treatment of wastewaters generated by the cellulose and paper industry. The authors used four wastewaters originating from different industrial plants, with total COD in the range of 360–1250 mg/L. In all cases, the cylindrical polyethylene biofilm carrier, Kaldnes® K1, was used, which shows a specific area of 500 m^2/m^3, density in the range of 0.92–0.96 g/cm^3, and dimensions of 10 mm diameter and 7 mm in height. The filling ratios (V_S/V_R) were 0.37 and 0.47 for the anaerobic and aerobic reactors, respectively. Several configurations were developed, including aerobic/anaerobic stages and inclusion of physicochemical stages (coagulation/flocculation, flotation, and sedimentation).

For the wastewater with a total COD of 360 mg/L, a configuration with three aerobic MBBR systems, arranged in series, followed by a coagulation/flocculation step was tested. The hydraulic retention time applied to the set of reactors was 40 min. In the physicochemical treatment, aluminum sulfate (7% m/v) and ferric chloride (3% m/v) were used together with 1 g/m^3 of anionic polymer. The removal of total COD by the three biological reactors was 45%. With the addition of the physicochemical coagulation/flocculation step, COD abatement increased to 78%. Based on these results, the Swedish company "Stora Papyrus Grycksbo AB" implemented this treatment system with an investment of the order of US$ 350,000 (RUSTEN et al. 1994).

In the case of the wastewater with COD of 550 mg/L, the configuration with three MBBRs in series followed by coagulation/flocculation and flotation was tested. The treatment started with one anaerobic reactor seeking to evaluate its capacity to remove chlorinated compounds. The hydraulic retention time (HRT) was 28 min per reactor, leading to an overall HRT of 1.4 h. The volumetric loads applied varied from 8 to 10 kg COD/(m^3 day). In the coagulation/flocculation step, only anionic polymer was used as flocculant, in the concentration range of 1.3–2.0 g/m^3. The removal percentages for total COD, soluble COD, and chlorinated compounds were 50%, 85–90%, and 90%, respectively. In addition, the effluent in the pilot unit did not show any turbidity, although its toxicity was high. Actually many compounds were identified as potential inhibitors of the biological process. The results indicated that a considerable part of the organic matter removal was obtained due to the incorporation of the coagulation/flocculation and flotation stages, which motivated the Swedish company *Stora Cell Industri AB*, in Skutskär, to continue their studies seeking at obtaining a process which could offer a better performance.

For the wastewater exhibiting a COD of 1250 mg/L, a configuration in which three aerobic MBBRs were arranged in series was tested. The HRT for this set of reactors ranged from 0.9 to 1.9 h. The volumetric loads applied varied from 17 to 28 kg COD/(m^3 day) and from 6 to 11 kgBOD$_7$/(m^3 day). The BOD removal was low (28–38%) for the range of HRTs adopted. The results were attributed to the high organic loads applied to the biological system. For the HRT of 1.2 h, the load in the first reactor was 75 kg COD/(m^3 day), while for a set comprised of three reactors, the load would be one third of this value.

3.5.2 MBBR Application for Nitrogen Removal

Prior to beginning a discussion regarding the application of MBBR technology for nitrogen removal, a brief description of the conventional nitrogen removal process, comprised of nitrification and denitrification steps, will be given.

Table 3.3 Important parameters in the nitrification process and their optimum ranges

Author	pH	*T* (°C)	DO
SURAMPALLI et al. (1997)	7.5–9.0	25–35 °C	>2 mg/L
EPA (1993)	7.5–8.5	35 °C for *Nitrosomonas* 35–42 °C for *Nitrobacter*	
HENZE et al. (1997)	7.5–8.0	30–35 °C	3–4 mg/L
METCALF and EDDY (1991)	7.5–8.0	>28 °C	

3.5.2.1 Nitrification

Nitrification is understood to be the limiting step in the conventional nitrogen removal process. This process is carried out through the action of two groups of bacteria: ammonium-oxidizing bacteria (AOB) and nitrite-oxidizing bacteria (NOB). The first group, belonging mainly to the genus *Nitrosomonas*, is responsible for the oxidation of ammonium to nitrite (nitritation). The second group, mostly belonging to the genera *Nitrobacter* and *Nitrospira*, promotes the conversion of nitrite to nitrate (nitratation), (RAMALHO 1983; METCALF and EDDY 1991). The steps involved in the nitrification process are described by Eqs. 3.1–3.3 (HENZE et al. 1997; MADIGAN et al. 1997).

1. Generation of nitrite

$$NH_4^+ + (3/2)O_2 \rightarrow NO_2^- + H_2O + 2H^+ \tag{3.1}$$

2. Generation of nitrate

$$NO_2^- + (1/2)O_2 \rightarrow NO_3^- \tag{3.2}$$

3. Global reaction

$$NH_4^+ + 2O_2 \rightarrow NO_3^- + 2H^+ + H_2O \tag{3.3}$$

The bacteria acting in the nitrification process are autotrophic and use inorganic carbon (CO_2) for cell synthesis. Thus, they are not dependent on organic compounds as carbon sources. They are also referred to as chemolithotrophic organisms, since they oxidize inorganic compounds in order to obtain energy (METCALF and EDDY 1991). Nitrifying bacteria have very low cell growth rates, which, under some circumstances, may complicate the operation of biological treatment systems. Since the biomass is produced in small quantities, the nitrification process becomes more susceptible to inhibition by a wide range of compounds (SORIA and CHAVARRIA 1978; HÄNEL 1988). Actually nitrification is among the most sensitive processes taking place in wastewater treatment bioreactors (JULIASTUTI et al. 2003).

The main factors which directly influence the nitrification activity include: pH, temperature, alkalinity, dissolved oxygen (DO) concentration, and organic carbon/nitrogen (C/N) ratio. Table 3.3 shows the values considered to be ideal for some important parameters in the nitrification process, according to some researchers.

Nitrification process can be inhibited by a wide range of wastewater components, including chemical compounds, heavy metals, high salt concentrations, and even high concentrations of ammonium and nitrite (substrates for AOB and NOB, respectively).

Factors such as pH, inhibitor concentration, suspended solid concentration, sludge age, and inhibitor solubility influence the magnitude of the inhibition. The fact that the bacteria responsible for the nitrification are restricted to only a few genera, associated with their slow cell growth, makes this process more susceptible to inhibition (GRUNDITZ and DALHAMMAR 2001).

3.5.2.2 Denitrification

Denitrification is the step which follows nitrification in the conventional nitrogen removal process. In this process, facultative anaerobic heterotrophic bacteria reduce the nitrate generated in the nitrification to nitrogen gas (MADIGAN et al. 1997). Organic carbon compounds such as carbohydrates, organic alcohols, amino acids, and fatty acids act as electron donor, while nitrate is the electron acceptor for denitrification. Thus, the occurrence of this process is related to the presence of an oxidizable substrate and an adequate concentration of nitrate (VAN RIJN et al. 2006).

Denitrification is of great importance since it contributes to mitigating the various adverse effects caused by nitrate in water bodies. Moreover, it is particularly important for the treatment of wastewaters with a low natural alkalinity. The loss of alkalinity through the release of H^+ ions during nitrification (see Eqs. 3.1 and 3.3) can be partially (~50%) compensated by the denitrification process. In addition, high nitrate concentrations in the effluent of the bioreactor can induce denitrifying activity in the secondary clarifiers. Consequently, rising sludge may occur, adversely affecting the clarification process and therefore the quality of the treated effluent (VON SPERLING 1996).

The importance of denitrification is emphasized due to the fact that nitrate is one of the main factors which contributes to the acceleration of eutrophication in receiving water bodies. Nitrate is also considered a priority pollutant due to its toxicity in relation to methemoglobinemia (blue baby syndrome) and the possible formation of nitrosamines in the gastric system, known for their carcinogenic effect on the organism (WHO 2003 *apud* ROCCA et al. 2006; McADAM and JUDD 2007). Rural areas characterized by intense agricultural activity are the most susceptible locations to nitrate contamination. The high quantity of nitrogen-based fertilizers employed in these areas corresponds to the greatest source of contamination of underground water reservoirs (ASLAN and CAKICI 2007). Furthermore, fertilizers, explosives, metals, and nuclear industry may generate residues with high nitrate concentrations ($N\text{-}NO_3 > 1000$ mg L^{-1}) (GLASS and SILVERSTEIN 1998).

Simplistically, the reduction of nitrate occurs in sequential steps, according to Eq. 3.4. During this sequence involving the transformation of NO_3^- to N_2, passing through gaseous nitrogen oxides, the oxidation state of nitrogen changes from +5 to 0 (ASLAN and CAKICI 2007; SOUSA and FORESTI 1999).

$$NO_3^- \rightarrow NO_2^- \rightarrow NO \rightarrow N_2O \rightarrow N_2 \tag{3.4}$$

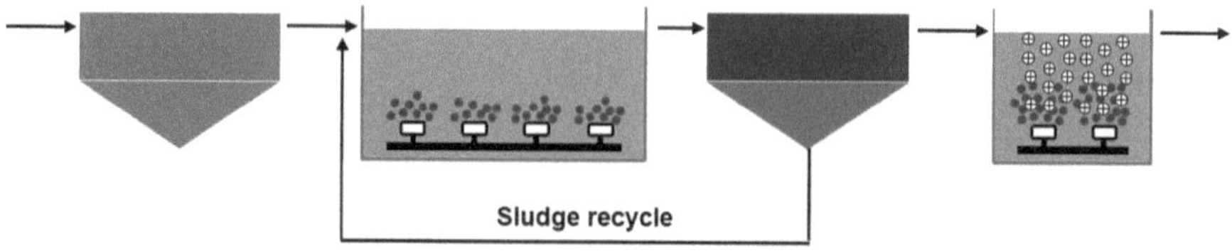

Fig. 3.19 MBBR placed after the conventional activated sludge plant to accomplish tertiary nitrification

Each of the reduction reactions and the catalytic enzymes are represented in Eqs. 3.5–3.8.

$$NO_3^- + 2e^- + 2H^+ \rightarrow NO_2^- + H_2O \quad \text{Nitrate reductase} \tag{3.5}$$

$$NO_2^- + e^- + 2H^+ \rightarrow NO + H_2O \quad \text{Nitrite reductase} \tag{3.6}$$

$$2NO + 2e^- + 2H^+ \rightarrow N_2O + H_2O \quad \text{Nitric oxide reductase} \tag{3.7}$$

$$N_2O + 2e^- + 2H^+ \rightarrow N_2(g) + H_2O \quad \text{Nitrous oxide reductase} \tag{3.8}$$

The first two steps, in which nitrate is reduced to nitrite and nitrite is reduced to nitrous oxide, are carried out, respectively, through the action of the enzyme nitrate reductase and nitrite reductase (SOUSA and FORESTI 1999). The determination of the enzymatic activity provides a good indicator of the denitrification rate, besides aiding an understanding of the overall process (NAIR et al. 2007).

3.5.2.3 Applications

Nitrification in MBBR systems has been studied using synthetic, domestic, and industrial wastewaters. As with any process, the nitrification rates are influenced by the organic load applied, DO concentration in the reactor, ammonium-nitrogen concentrations, temperature, pH, and alkalinity (RUSTEN et al. 2006).

Due to the diffusion effects inherent to biofilm systems, the concentrations of substrate and DO have an even greater effect on nitrification rate. In general, oxygen will be limiting at high ammonium concentrations. On the other hand, ammonium will be the limiting factor of the reaction when it is present in low concentrations (RUSTEN et al. 2006).

There are many configurations of MBBR systems (combined or not with activated sludge) designed for nitrification and/or complete nitrogen removal, as exemplified in Figs. 3.19, 3.20, and 3.21. Some of the flow schemes previously shown in Figs. 3.15, 3.16, 3.17, and 3.18 may also enable nitrification, although the removal of nitrate is conditioned to the presence of an anoxic zone.

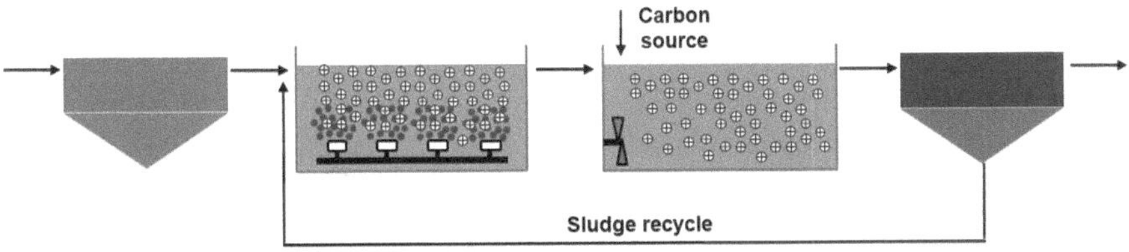

Fig. 3.20 Post-denitrification configuration with aerobic and anoxic MBBRs in series

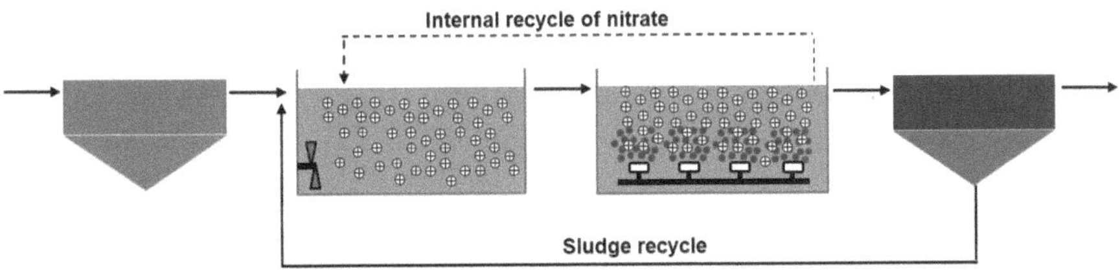

Fig. 3.21 Pre-denitrification MBBR system

In Fig. 3.19 the MBBR process is added downstream of the conventional activated sludge plant for tertiary nitrification. As most of the COD was removed in the activated sludge process, low amount of solids will be produced in the nitrifying MBBR. Therefore, it can be followed by a very compact a solid-liquid separation device (e.g., sand filter, disk filter).

In cases where denitrification is required, a second MBBR may be added after the nitrification reactor (post-denitrification configuration). In this reactor, anoxic conditions need to be assured in order to allow nitrate reduction. Instead of aeration, mechanical mixers are generally employed to promote the circulation of the biofilm carriers in the anoxic MBBR. It should be noted that the post-denitrification configuration requires the addition of an extra organic carbon source (organic acids, alcohols, etc.), since this is almost completely metabolized in the previous aerated step. This requirement increases the operational costs of the process. Figure 3.20 shows a schematic diagram of the post-denitrification configuration. No sludge recycle needed in pure MBBRs but will be required in hybrid biofilm-activated sludge processes (Fig. 3.20).

Alternatively, denitrification can be carried out before the nitrification (pre-denitrification configuration), thus making use of the organic carbon from the influent wastewater. In this case, internal recirculation of the nitrate formed in the nitrifying to the denitrifying reactor needs to be conducted, as shown in Fig. 3.21. In hybrid biofilm-activated sludge processes, sludge recycle needs to be implemented. In pure MBBR systems, however, biomass recirculation is not required.

A third operational possibility is to promote the nitrification and denitrification in a single reactor. In this case, sequencing batch biofilm reactors (SBBRs) are used, which is, as previously mentioned, a combination of MBBR technology and the working principle of the sequencing batch reactors (SBRs). The MBBR operating in a continuous regime, illustrated in the above flow diagrams, can be replaced with one or more SBBRs operating in series or in parallel. Their operation can be divided into operational cycles comprising filling, reaction, and withdrawal phases. In the reaction phase, aeration can be provided to allow nitrification, and, subsequently, an anoxic environment is assured to promote denitrification. In general, an addition of organic compounds is required in the anoxic stage of the SBBR. In the case of these reactors, a settling phase is not required, since the biomass is adhered to the carriers and not in suspension in the liquid medium. Strict control of the sequential steps of the SBBR is crucial in order to achieve proper functioning of the system.

HEM et al. (1994) demonstrated that the nitrification rate in MBBR systems is not only dependent on the concentrations and loads of the substrates present at a given time but is also influenced by the biofilm history, that is, the previous conditions to which the system was submitted. It was verified that when the biofilm was acclimatized with a high ammonium load, the nitrification rate obtained in the steady state was approximately twice than obtained when the biofilm was acclimatized with a low ammonium load.

The same authors applied an MBBR system to study nitrification, both in laboratory scale, using synthetic wastewater, and in pilot scale, using domestic sewage from primary or secondary treatment. The results showed that, when there is alkalinity in excess and no organic matter is present, both the ammonium and the concentration of DO can be limiting factors for the nitrification rate. The ammonium is no longer limiting when the oxygen concentration/ammonium concentration ratio is lower than approximately 3 gO_2/gNH_4-N. Under these conditions, oxygen becomes the limiting factor. The DO concentration has a strong influence on the nitrification process, particularly when it is the reaction limiting factor (HEM et al. 1994). It was observed that an increase in the organic load leads to a decrease in the nitrification rate. When the organic load exceeded 5 $gBOD_7/(m^2$ day), the nitrification was insignificant. When the system was fed with secondary effluent, nitrification rates of 0.7–1.0 gNO_x-N/(m^2 day) were reached at DO concentrations of between 4.5 and 5.0 gO_2/m^3 (HEM et al. 1994).

RUSTEN et al. (1995) studied nitrification in previously treated wastewater and raw wastewater, in a pre-denitrification system implemented in a pilot MBBR plant with carriers with a surface area of 310 m^2/m^3. The maximum nitrification rates achieved with the pretreated wastewater were 20–25% greater than those obtained with the raw wastewater in the system with recirculation.

RUSTEN et al. (1995) also observed that under oxygen-limited conditions, the effect of temperature on the nitrification process was insignificant within the temperature range of 7 and 18 °C. This was due to an increase in the DO concentrations at low temperatures, that is, the loss of the process efficiency with a decrease in temperature was counterbalanced by an increase in the DO content. Nitrification rates of 300–400 gNH_4-N/(m^3 day) were obtained at 10 °C, demonstrating the great efficacy of MBBRs systems in achieving good nitrification at harsh temperature conditions.

RUSTEN et al. (1995) also evaluated the transition from ammonium to DO as the limiting factor for the nitrification rate. When present in a concentration equivalent to 2 mg/L, oxygen became limiting for an ammonium concentration greater than 0.5 mgNH_4-N/L. On the other hand, when the oxygen level was 6 mg/L, the transition from ammonium-limited to oxygen-limited conditions occurred when the ammonium concentration reached 1.7 mgNH_4-N/L.

With the use of MBBR technology, WELANDER et al. (1997) evaluated the possibility of obtaining nitrification of a municipal landfill leachate, containing 460–600 mg/L of N-NH_4^+ and BOD in the range of 800–1300 mg/L. They operated three reactors in laboratory scale, each filled with a certain type of carrier: carrier A (extruded polyethylene tubes with significant rugosity with dimensions of 8 mm in height and 8 mm diameter), carrier B (polyethylene tubes with dimensions of 10 mm in height and 8 mm diameter), and carrier C (small cubes with 3 mm each side made of microporous cellulose). The surface areas of the carriers A, B, and C were 200, 390, and 1700 m^2/m^3, respectively. The filling fractions for these carriers in relation to the reactor volume were 60%, 60%, and 10%, respectively.

The effects of temperature and hydraulic retention time (HRT) on the volumetric nitrification rate were investigated. The nitrification rates in the steady state were obtained after approximately 1 month of operation. The values for this parameter showed a relatively weak dependence on temperature, the nitrification rate obtained at 5 °C being approximately 77% of that obtained at 20 °C. The relatively weak influence of the temperature is related to the fact that the nitrification is generally limited by the diffusion of oxygen to the biofilm. Thus, the reduction in the specific nitrification rate at low temperatures is counterbalanced by a greater penetration of oxygen into the biofilm (*apud* WELANDER et al. 1997; ØDEGAARD et al. 1994).

The HRT had a more pronounced effect on the nitrification rate, a considerable increase in this rate being observed with a decrease in the HRT. This strong effect is related to the greater ammonium load at short HRTs, leading to a higher nitrification rate. However, it should be noted that, although higher nitrification rates were reached during operation at shorter HRTs, the effluent ammonium concentration under these conditions did not reach the required discharge standards (<10 mgN-NH_4^+/L). Thus, in practice, the process should be operated at longer HRTs (WELANDER et al. 1997).

Despite the fact that carrier A had only half the surface area for microbial adhesion in relation to carrier B, the two carriers showed similar performance. This result is related to the fact that the surface of carrier A had greater rugosity in relation to carrier B, resulting in a broadening of the biofilm surface on a microscale. This effect may be strongly related to the very thin biofilm which develops on the carrier surface (not easily observed with the naked eye). Actually, it is unlikely that the surface area of the biofilm is greater for a rugose surface in relation to a smooth surface, particularly when the biofilm thickness is considerably greater than the unevenness of the surface (WELANDER et al. 1997).

The use of carrier C allowed higher volumetric nitrification rates to be obtained. For the other two carriers, the shortest HRT which allowed the discharge levels for ammonium (10 mgN-NH_4^+/L) to be reached at 20 °C was 4–5 days.

Conversely, with carrier C, even lower concentrations of ammonium were attained applying an HRT of only 22 h. In addition, the greatest efficiency was obtained with the use of this carrier, the maximum nitrification rate of 40 N-NH_4^+/(m^3 h) being obtained at 20 °C with an HRT of 14 h (WELANDER et al. 1997).

The better performance of carrier C was clearly due to its porous structure and the fact that the very thin biofilm did not block the pores of this support. Actually the effective surface area of the biofilm was greater than the macro-surface of the carrier, which explains the much higher volumetric nitrification rates achieved with carrier C compared with carrier B, despite the macro-surface available per unit volume being greater in the reactor with carriers of type B (234 m^2/m^3 at 60% of filling) than in that with carriers of type C (170 m^2/m^3 at 10% of filling) (WELANDER et al. 1997).

According to PASTORELLI et al. (1997), nitrification in MBBR reactors can be operated in multistage processes in order to achieve progressive enrichment in autotrophic microorganisms. In the initial stages, nitrification may become oxygen limited, mostly due to the competition between heterotrophic and autotrophic microorganisms. The linear relation between the nitrification rates and dissolved oxygen consumption leads to a high energy consumption associated with the aeration of the process. However, the nitrification can be controlled through an increase in the DO concentration at high ammonium loads or low temperatures. In the final stages, the ammonium concentration probably becomes the limiting step in the process rather than the DO. Therefore, lower nitrification rates are obtained in these stages.

Following this line of research, PASTORELLI et al. (1997) studied nitrification in a continuous flow MBBR, comprised of three stages. The nitrification tests were carried out at concentrations below 2.6 mgNH_4^+-N/L and also at concentrations above 4.0 mgNH_4^+-N/L. In the first case, two different situations could be observed. The first situation, corresponding to ammonium loads below 0.8 gNH_4^+-N/(m^2 day), allowed to obtain nitrification rate of 0.47 gNH_4^+-N/(m^2 day). The second, corresponding to ammonium loads of 0.8–1.4 gNH_4^+-N/(m^2 day) led to a nitrification rate of 1.07 gNH_4^+-N/(m^2 day) (PASTORELLI et al. 1997).

Under nitrogen loads of less than 0.8 gNH_4^+-N/m^2 (low loads), there was no increase in the nitrification rate when the DO concentration was increased. This implies that DO limitation did not occur. In the case of loads between 0.8 and 1.4 gNH_4^+-N/(m^2 day) (high loads), in contrast to the situation with low ammonium loads, the penetration of oxygen into the biofilm was not complete, and the nitrification rate was influenced by the oxygen concentration. For ammonium concentrations above 4.0 gNH_4^+-N/L, the diffusion in the liquid film also represented the limiting step in the process. It should be noted that, for DO concentrations below 2–3 mgO_2/L, no nitrification occurred (PASTORELLI et al. 1997).

Following the same trend of applying multistage processes, YU et al. (2007) operated an MBBR system with two stages in laboratory scale in order to remove ammonium from a synthetic wastewater at moderate concentrations (50–400 mg/L). The DO concentration was maintained at 2–3 mg/L, and the ambient temperature varied between 23 and 29 °C. The removal of ammonium reached 99% when the influent ammonium concentration was less than 300 mg/L. Effluent ammonium concentration was lower than 1 mg/L in the process effluent. It was observed that, even with an increase in influent ammonium concentration to 400 mg/L, which corresponds to a volumetric nitrogen load of 6 kgNH_4-N/(m^3 day), the removal efficiency remained at over 95%. Under these conditions, the ammonium concentration in the effluent was lower than 15 mg/L. The same authors noted that the composition of oxidized nitrogen (NO_x) was strongly influenced by the free ammonia, and the intermediate compound nitrite was present in significant concentrations when the concentration of free ammonia was greater than 2.5–3.0 mg/L.

RODGERS and XIN-MIN (2004) evaluated the performance of a system comprised of six vertical moving bed biofilm reactors for nitrogen removal. The reactor system was operated at high recirculation rates. Synthetic effluent was used, with an ammonium concentration which varied between 75 and 136 mgN/L and COD of around 600 mg/L. The last four tanks of the series were aerobic and the first two were anoxic. The removal of organic matter, evaluated in terms of raw BOD, reached 94–96%, and the total nitrogen removal varied from 77 to 88%. In anoxic reactors, the denitrification efficiency varied from 94 to 98%, and the denitrification rate per area of support varied within the range of 2.9–3.8 gNO_3^--N/(m^2 day). The nitrification efficiency in the aerobic tanks was over 95%. In this case, the maximum nitrification rates per unit area of support varied from 1.3 to 1.8 gNH_4^+-N/(m^2 day).

LUOSTARINEN et al. (2006) investigated the removal of residual COD and nitrogen from wastewater originating from an aerobic reactor operating at low temperatures (effluent 1) and a chicken farm wastewater (effluent 2). The system was comprised of four MBBRs in series, operating in sequencing batch regime, with and without aeration. During the periods without aeration (anoxic regime), mechanical mixers were used to homogenize the system. The anoxic phase was reached within 30–40 min after the interruption of aeration when the DO concentration reached 1.0 mg/L. The average operating temperature was 20 °C for effluent 1 and 10 °C for effluent 2. The operating cycles for all stages of the system lasted 1.8–2.2 days. In relation to the combined system (UASB + MBBRs), the removal efficiency of organic matter, total nitrogen, and phosphorus were 90%, 60–70%, and 80%, respectively. For the chicken farm wastewater, there was complete removal of organic matter, and for all of the reactors the nitrogen removal varied from 50 to 60%. Complete nitrification was achieved, although denitrification was hindered by the lack of a carbon source in the medium.

BASSIN et al. (2011) investigated the tertiary nitrification of treated domestic sewage with high levels of salinity in a laboratory-scale MBBR. The salinity of the treated sewage was increased from 1000 to 8000 mgCl^-/L. The results showed a good adaptation of the biomass adhered to the carriers, reaching over 90% removal for ammonium even at the highest salinity tested. The same authors, after investigating a gradual increase in the salinity of the treated sewage, fed the reactor with a mixture of sewage (8000 mgCl^-/L) and a treated industrial wastewater. The industrial wastewater had a very complex composition, since it originated from a chemical plant which produces agricultural pesticides, such as herbicides, insecticides, fungicides, and acaricides. Although this wastewater had passed through a conventional activated sludge system, it still contained many substances which are potential nitrification inhibitors. The percentage of this wastewater fed to the MBBR was gradually increased from 10 to 100%. The results showed that although the nitrification efficiency was observed to decrease, notably when there was an increase in the percentage of industrial wastewater mixed with the treated sewage, the biological system adapted to the new conditions, achieving a nitrification efficiency higher than 85% for all conditions tested. It can be concluded that the MBBR system played an important role in the development and maintenance of the nitrifying microbial community, even under inhibitory conditions.

LABELLE et al. (2005) investigated denitrification using MBBR technology with spherical polyethylene carriers, which provide a biofilm specific area equivalent to 100 m^2/m^3. Methanol was used as the carbon source, and the C/N ratio was varied. The concentration of nitrate was reduced from 53 mg/L to 1.7 ± 0.7 mgN/L, and the maximum denitrification rate (17.7 ± 1.4 gN/(m^2 day)) was obtained when a COD/N ratio of 4.2 was employed.

Several authors have reported the removal of phosphorus combining biological (MBBR) and physicochemical treatment. WANG et al. (2006) evaluated a combination of chemical precipitation (coagulation/flocculation) and biological processes for phosphorus and nitrogen removal in an MBBR. The process was stabilized with a DO concentration of over 2 mg L^{-1}, and the nitrogen removal (achieved by the simultaneous nitrification and denitrification (SND)) was around 90%. In the chemical precipitation, aimed at the removal of phosphorus (P), iron (II) was used in a ratio of 1:1.3 (P:Fe). The phosphorus removal efficiency was significant, indicating that the system is viable for applications on an industrial scale.

KERMANI et al. (2009) evaluated the removal of nutrients from synthetic wastewater in a moving bed biofilm reactor (MBBR) in laboratory scale. The experimental unit was comprised of four reactors in series (one anaerobic, two anoxic, and one aerobic) with no recirculation of sludge through the system. The reactors were operated in a continuous regime and submitted to different nitrogen and phosphorus loads as well as different HRTs. The anaerobic reactor (R1) aimed at biological removal of phosphorus, while the first anoxic reactor (R2) was responsible for minimizing the effect of the nitrate (coming from the second anoxic tank—R3) on phosphorus release taking place in anaerobic reactor. Part of the volume of R2 was returned to the anaerobic reactor in order to increase the COD removal and promote ideal conditions for phosphorus release. The second anoxic reactor (R3), placed after the first anoxic reactor (R2), received the recirculation of the nitrate from the aerobic reactor (R4) to allow the removal of most of the nitrate via denitrification. The reactor R4, in turn, was responsible for the nitrification. All of the reactors were maintained at a constant temperature of 28 ± 1 °C.

The concentrations of dissolved organic carbon (DOC), NH_4^+-N, and PO_4^--P were kept at 500–2000 mg/L, 25–125 mg/L, and 5–25 mg/L, respectively. The HRT varied between 8 and 48 h, depending on the inflow applied (10–60 L/day). Under ideal conditions (500 mgDOC/L and 62.6 N-NH_4^+/L), almost complete nitrification occurred with an NH_4^- N removal efficiency of 99.72% in the aerobic reactor. In this nitrifying tank, the average specific nitrification rate was 1.92 gN-NO_xproduced/(kg VSS h).

The authors also observed that the denitrification rate increased with an increase in the NO_x-N load in the second anoxic reactor. Under ideal conditions (500 mgDOC/L and 12.5 P-PO_4^-/L), the overall nitrogen and phosphorus removal were 80.9% and 95.8%, respectively (KERMANI et al. 2009).

3.6 Final Considerations

The implementation of domestic sewage and industrial wastewater treatment plants, which use the MBBR technology, is increasing worldwide. The MBBR process offers some advantages in relation to conventional activated sludge processes or even to other fixed biofilm-based systems. It requires less space for its implementation, and it is very suitable to accomplish nitrification and denitrification processes without the need for sludge recycle between different redox zones. In addition, some configurations of the MBBR systems which combine physicochemical stages allow the removal of phosphorus.

The biofilm processes, in general, present good efficiency in the removal of the organic matter present in a given wastewater. However, when the organic load is very high, a loss of efficiency can be expected. This also applies to the MBBR process, whose performance is adversely affected at high loads, under which frequent sloughing of the biofilm is observed. At high organic loads, a substantial amount of suspended biomass will be present. In this case, the MBBR functions more as a hybrid biofilm-activated sludge process than a pure biofilm reactor. Therefore, the design of pure MBBRs should consider operation at low or moderate loads.

For a complete overview of the MBBR process, further studies need to be conducted to better understand the relation between the support material, the aeration/mixing device, the reactor configuration, and the biomass immobilized on the moving carriers. This would aid the establishment of the most appropriate conditions for the development of different types of bacteria engaged in the treatment of wide range of wastewaters.

References

ASLAN, S., CAKICI, H. Biological denitrification of drinking water in a slow sand filter, *Journal of Hazardous Materials*, v. 148, n. 1-2, p. 253-258, 2007.

BASSIN, J.P; DEZOTTI, M. Primary, secondary and tertiary treatment of wastewaters. In: DEZOTTI, M. Processes and Techniques for Environmental Control of Liquid Wastewaters. Rio de Janeiro: e-papers, p. 53-241, 2008 (In Portuguese).

BASSIN, J.P., DEZOTTI, M., SANT'ANNA Jr., G.L. Nitrification of industrial and domestic saline wastewaters in moving bed biofilm reactor and sequencing batch reactor. *Journal of Hazardous Materials*, v. 185, n. 1, p. 242-248, 2011.

BASSIN, J.P., KLEEREBEZEM, R., ROSADO, A.S., VAN LOOSDRECHT, M.C.M., DEZOTTI, M. Effect of different operational conditions on biofilm development, nitrification, and nitrifying microbial population in moving-bed biofilm reactors. *Environmental Science and Technology*, v. 46, n. 3, p. 1546-1555, 2012.

BRADFORD, M. M. A Rapid and Sensitive Method for the Quantitation of Microgram Quantities of Protein Utilizing the Principle of Protein-Dye Binding. *Analytical Biochemistry*, v. 72, p. 248-254, 1976.

CAMMAROTA, M. C., SANT'ANNA JR., G. L. Metabolic blocking of exopolysaccharides synthesis: effects on microbial adhesion and biofilm accumulation. *Biotechnology Letters*, v. 20, n. 1, p. 1-4, 1998.

DANIELS, L., HANSON, R. S., PHILLIPS, J. A. *Chemical Analysis*. In: Washington, USA: ASM press, 1994. In: Gerhart, P., Murray, R. G. E., Wood, W. A., et al. (eds), *Methods for General and Molecular Bacteriology*, Cap. 4.

EPA - Environmental Protection Agency – USA, *Nitrogen Control*, EPA/625/R-93/010, 1993.

FLEMMING, H. C., WINGLINDER, J. Relevance of microbial extracellular polymeric substances (EPS) – Part I: Structural and ecological aspects. *Water Science and Technology*, v. 43, n. 6, p. 1-8, 2001.

FRØLUND, B., PALMGREN, R., KEIDING, K. Extraction of Extracellular Polymers from Activated Sludge using a Cation Exchange Resin. *Water Research*, v. 30, n. 8, p. 1749-1758, 1996.

FURUMAI, H., RITTMAN, B.E. Evaluation of multiple-species biofilm and floc processes using a simplified aggregate model. *Water Science and Technology*, v. 29, n. 10-11, p. 439-446, 1994.

GLASS, C., SILVERSTEIN, J. Denitrification kinetics of high nitrate concentration water: pH effect on inhibition and nitrite accumulation. *Water Research*, v. 32, n. 3, p. 831-839, 1998.

GONÇALVES, R.F.; CHERNICHARO, C.A.L.; ANDRADE NETO, C.O.; SOBRINHO, P.A.; KATO, M.T.; COSTA, R.H.R.; AISSE, M.M., ZAIAT, M. Post-treatment of wastewaters from anaerobic reactor in biofilm systems. In: Chernicharo, C.A.L. Post-treatment of wastewaters from anaerobic reactors. Belo Horizonte: PROSAB, p. 171-278, 2001 (In Portuguese).

GRUNDITZ, C., DALHAMMAR, G. Development of nitrification inhibition assays using pure cultures of *Nitrosomonas* and *Nitrobacter*, *Water Research*, v. 35, n. 2, p. 433-440, 2001.

HÄNEL, K. *Ellis Horwood series in water and wastewater technology biological treatment of sewage by the activated sludge process*. New York: Ellis Horwood Limited, 1988.

HEM, L.J., RUSTEN, B., ØDEGAARD, H. Nitrification in a moving bed biofilm reactor. *Water Research*, v. 28, n. 6, p. 1425-1433, 1994.

HENZE, M., HARRE, P., JANSEN, J.C., ARQIN, E. *Wastewater Treatment – Biological and chemical process*. Springer Verlag, 2nd edition, Germany, 1997.

JULIASTUTI, S.R., BAEYENS, J., CREEMERS, C. Inhibition of Nitrification by Heavy Metals and Organic Compounds: The ISO 9509 Test, *Environmental Engineering Science*, v. 20, n. 2, p. 79-90, 2003.

KERMANI, M.; BINA, B.; MOVAHEDIAN, H.; AMIN, M.M.; NIKAEEN, M. Biological phosphorous and nitrogen removal from wastewater using moving bed biofilm process, *Iranian Journal of Biotechnology*, v. 7, n. 1, p. 19-27, 2009.

LABELLE, M. A., JUTEAU, P., JOLICOEUR, M., VILLEMUR, R., PARENT, S., COMEAU, Y. Seawater denitrification in a closed mesocosm by a submerged moving bed biofilm reactor. *Water Research*, v. 39, n. 14, p. 3409-3417, 2005.

LOWRY, O. H., ROSENBROUGH, N. J., FARR, A. L. Protein Measurement with the Folin Phenol Reagent. *Journal of Biological Chemistry*, v. 193, p. 265-275, 1951.

LUOSTARINEN, S., LUSTE, S., VALENTIN, L., RINTALA, J. Nitrogen removal from on-site treated anaerobic effluents using intermittently aerated moving bed biofilm reactors at low temperatures. *Water Research*, v. 40, n. 8, p. 1607-1615. 2006.

MADIGAN, T. M.; MARTINKO, J. M.; PARKER, J. Brock biology of microorganisms. 8th ed. New York: Prentice Hall, 1997.

METCALF and EDDY. *Wastewater Engineering – Treatment, Disposal and Reuse*, USA: McGraw-Hill, 3rd edition, 1991.

McADAM, E. J.; JUDD, S. J. Denitrification from drinking water using a membrane bioreactor: Chemical and biochemical feasibility. *Water Research*, v. 41, p. 4.242-4.250, 2007.

NAIR, R. R.; DHAMOLE, P. B.; LELE, S. S.; D'SOUZA, S. F. Biological denitrification of high strength nitrate waste using preadapted denitrifying sludge. *Chemosphere*, v. 67, p. 1.612-1.617, 2007.

NICOLELLA, C.; VAN LOOSDRECHT, M.C.M; HEIJNEN, J.J. Wastewater treatment with particulate biofilm reactors. Journal of Biotechnology, v. 80, p. 1-33, 2000.

ØDEGAARD, H. Innovations in wastewater treatment: The moving bed biofilm process. *Water Science and Technology*, v. 53, n. 9, p. 17-33, 2006.

ØDEGAARD, H., RUSTEN, B. Wastewater treatment with aerated submerged biological filters. *Journal Water Pollution Control Federation*, v. 56, n. 5, p. 424-431, 1993.

ØDEGAARD, H., RUSTEN, B., WESSMAN, F. State of the art in Europe of the moving bed biofilm reactor (MBBR) process. Paper presented in WEFTEC'04 in New Orleans, 4th October, 2004.

ØDEGAARD, H., RUSTEN, B., WESTRUM, T. A new moving bed biofilm reactor – applications and results. *Water Science and Technology*, v. 29, n. 10-11, p. 157-165, 1994.

PASTORELLI, G., ANDREOTTOLA, G., CANZIANI, R., FRANGIPANE, E. F., PASCALIS, F., GURRIERI, G., ROZZI, A. Pilot-Plant experiments with moving-bed biofilm reactors. *Water Science and Technology*, v. 36, n. 1, 1997.

RAMALHO, R. S. *Introduction to wastewater treatment processes*. 2nd ed. New York: Academic Press, 1983.

REIS, Gelma Gonçalves. Influence of organic load on the performance of moving-bed biofilm reactors (MBBR). 134 f. Master's thesis in Chemical Engineering – Coppe. Universidade Federal do Rio de Janeiro, 2007 (In Portuguese).

ROCCA, C.D., BELGIORNO, V., MERIC, S. An heterotrophic/autotrophic denitrification (HAD) approach for nitrate removal from drinking water. Process Biochemistry, v. 41 p. 1022-1028, 2006.

RODGERS, M., XIN-MIN, Z. Biological nitrogen removal using a vertically moving biofilm system. *Bioresource Technology*, v.93, n. 3, p. 313-319, 2004.

RUSTEN, B., EIKEBROKK, B., ULGENES, Y., LYGREN, E. Design and operations of the Kaldnes moving bed biofilm reactors. *Aquacultural Engineering*, v. 34, n. 3, p. 322-331, 2006.

RUSTEN, B., HEM, L.J., ØDEGAARD, H. Nitrification of municipal wastewater in moving-bed biofilm reactors. *Water Environment Research*, v. 67, n. 1, p. 75-86, 1995.

RUSTEN, B., MCCOY, M., PROCTOR, R., SILJUDALEN, J.G. The innovative moving bed biofilm reactor/solids contact reaeration process for secondary treatment of municipal wastewater. *Water Environment Research*, v. 70, n. 5, 1998.

RUSTEN, B., MATSSON, E., BROCH-DUE, A., WESTRUM, T. Treatment of pulp and paper industry wastewater in novel moving bed biofilm reactors. *Water Science and Technology*, v. 30, n. 3, p. 161-171, 1994.

RUSTEN, B., ØDEGAARD, H., LUNDAR, A. Treatment of dairy wastewater in a novel moving bed biofilm reactor. Water Science and Technology, v. 26, n. 3/4, p. 703, 1992.

SALVETTI, R.; AZZELLINO, A.; CANZIANI, R.; BONOMO, L. Effects of temperature on tertiary nitrification in moving-bed biofilm reactors. *Water Research*, v. 40, p. 2.981-2.993, 2006.

SOKÓL, W., 2003, "Treatment of refinery wastewater in a three-phase fluidised bed bioreactor with a low density biomass support", *Biochemistry Engineering Journal*, v. 15, p. 1-10.

SORIA, F.L.; CHAVARRIA, J.M. Techniques for environmental protection, v. 2, Editorial Labor S/A, p. 235-239, 1978 (In Spanish).

SOUSA, J. T.; FORESTI, E. Environmental management and control. Use of anaerobic sludge as external carbon source of denitrification of wastewaters. *Revista Brasileira de Engenharia Agrícola e Ambiental*, v. 3, p. 69-73. Campina Grande/Paraíba: DEAg/UFPB, 1999 (In Portuguese).

STEWART, P.S. Diffusion in biofilms. *Journal of Bacteriology*, v. 185, n. 5, p. 1485-1491, 2003.

SURAMPALLI, R.Y.; TYAGI, R.D.; SCHEIBLE, O.K.; HEIDMANN, J.A. Nitrification, denitrification and phosphorus removal in sequential batch reactors. *Bioresource Technology*, v. 61, p. 151-157, 1997.

TAVARES, C.R.G., SANT'ANNA JR., G. L., CAPDEVILLE, B. The effect of air superficial velocity on biofilm accumulation in a three-phase fluidized-bed reactor. *Water Research*, v. 29, n. 10, p. 2293-2298, 1995.

VAN RIJN, J.V.; TAL, Y; Schreier, H.J. Denitrification in recirculating system: Theory and applications. Aquacultural Engineering, v. 34, p. 364-376, 2006.

VON SPERLING, M. Introduction to water quality and sewage treatment. 2nd ed. Belo Horizonte: Environmental and Sanitary Engineering Department, Universidade Federal de Minas Gerais, 1996 (In Portuguese).

WANG, X. J., XIA, S. Q., CHEN, L., ZHAO, J. F., RENAULT, N. J., CHOVELON, J. M. Nutrients removal from municipal wastewater by chemical precipitation in a moving bed biofilm reactor. *Process Biochemistry*, v. 41, n. 4, p. 824-828, 2006.

WELANDER, U.; HENRYSSON, T.; WELANDER, T. Nitrification of landfill leachate using suspended-carrier biofilm technology, *Water Research*, v. 31, 2351-2355, 1997.

WHO, Guidelines for drinking water quality, 3rd ed. Geneva (CH): World Health Organization, 2003.

XAVIER, J. B.; PICIOREANU, C.; VAN LOOSDRECHT, M.C.M. A framework for multidimensional modelling of activity and structure of multispecies biofilms. Environmental Microbiology, v. 7, n. 8, p. 1085-1103, 2005.

YU, H.B.; QUAN, X.; DING, Y.Z., Medium-strength ammonium removal using a two-stage moving bed biofilm reactor system, *Environmental Engineering Science*, v. 24, n. 5, p. 595-601, 2007.

http://www.environmental-expert.com

http://www.anoxkaldnes.com

http://www.aquamerik.com

Chapter 4
Aerobic Granular Sludge Technology

João Paulo Bassin

4.1 Introduction

Most wastewater treatment systems, such as the traditional activated sludge process, require large areas for their installation, which is mainly due to the need for large settling tanks and the low concentration of solids in the aeration basins. Considerable excess sludge production, high sensitivity to fluctuations in the applied load, and relatively low volumetric conversion capacities (0.5–2.0 kgCOD/m^3 day) are some disadvantages of the conventional processes.

The performance of the activated sludge-based processes is strongly dependent on the settling characteristics of the biomass present in the reactor. The microbial flocs must be separated from the treated wastewater in settling tanks and one part returned to the reactor. However, in some cases, poor settling flocs may exist, resulting in the washout of the sludge from the clarifiers. This of course reduces the quality of the treated effluent. In most cases, this loss of sludge is caused by the excessive growth of filamentous microorganisms, which may adversely affect the sludge sedimentation properties. The reactor configuration and the operation strategy can also influence the sludge characteristics. In the specific case of certain wastewaters, the formation of flocs with good sedimentation properties may be difficult to be accomplished.

The substantial increase in the number of inhabitants, in most cases concentrated in densely populated urban areas, has increased the need to improve the existing wastewater treatment plants or to construct new systems which are compatible with

J.P. Bassin (✉)
Chemical Engineering Program, COPPE, Federal University of Rio de Janeiro, Rio de Janeiro, Rio de Janeiro, Brazil
e-mail: jbassin@peq.coppe.ufrj.br

M. Dezotti et al., *Advanced Biological Processes for Wastewater Treatment*, DOI 10.1007/978-3-319-58835-3_4

the increasing amount of wastewater generated. In general, the availability of space for these constructions is limited, and thus it is imperative that these new installations occupy the smallest area possible.

In this context, there has been significant evolution in the wastewater treatment sector in response to greater demands regarding the quality of the treated effluent and the need to minimize the space used, to simplify the operation and to reduce the investment costs. Meanwhile, research on the development of new forms of biomass agglomerates has intensified, aiming to facilitate and improve the biomass retention in biological treatment processes, which is crucial to reduce the footprint of the treatment plants. As the biomass acts as a biocatalyst for the degradation of a wide range of pollutants, retaining a high solids concentration in the reactor leads to a greater treatment capacity. Thus, it is very important that the biomass has good sedimentation properties to ensure proper functioning of the biological wastewater treatment process.

In recent years several biofilm reactors in which the biomass is immobilized in fixed or moving carrier materials have been developed. Examples of such reactors include the trickling filter or biofilter, rotating biological contactor (RBC), and moving bed biofilm reactor (MBBR) (described in Chap. 3). As advantages, these processes exhibit a reduced area for plant installation, in some cases dispensing with the need for settling tanks (or allowing a reduction in their dimensions), and the ability to withstand high volumetric and organic loads.

Aerobic granulation, also known as aerobic granular sludge technology, is an even more recent innovation within biofilm reactors for wastewater treatment. Considered as a particular type of biofilm comprised of self-immobilized cells, aerobic granular sludge constitutes a promising technology which does not require a material support for biofilm growth. The first patent was filed by HEIJNEN and VAN LOOSDRECHT (1998).

It is interesting to point out that granular sludge technology was initially developed for strictly anaerobic systems in 1980. The formation of anaerobic granules was most frequently carried out in upflow anaerobic sludge blanket (UASB) reactors, in which high biodegradable organic matter removal from domestic and industrial wastewaters can be obtained (LETTINGA et al. 1980). The limitations associated with anaerobic granulation, which include a long reactor start-up, relatively high operation temperature requirement, strong dependence on the concentration of organic matter, and low efficiency for the removal of nutrients (nitrogen and phosphorus), motivated the development of aerobic granulation. Due to the numerous advantages of this process, it has become a focus for discussion among engineers dealing with the environmental issues (ADAV et al. 2008a, b).

In the past 10 years, many studies have been carried out to investigate aerobic granules, both to better understand the formation of these compact microbial structures and investigate the possibility of making aerobic granulation technology a new compact system for wastewater treatment.

Aerobic granulation could represent a solution for the operation of some reactors where flocculent sludge with poor settling features predominates. As no material support is required, the initial investment may be reduced. In addition, the formation of an oxygen gradient and the presence of a wide range of microorganisms within the aerobic granules allow the simultaneous removal of organic matter, nitrogen, and phosphorus. In this context, aerobic granulation has been the subject of several studies, mostly conducted in laboratory scale, with the first applications in pilot and industrial scale appearing recently.

4.2 General Characterization of Aerobic Granular Sludge Technology

Aerobic granules consist of highly compacted microbial aggregates containing millions of microorganisms per gram of biomass. Many different bacterial species with specific function in the degradation of a variety of pollutants present in complex wastewaters are harbored in such compact microbial structures (LIU and TAY 2004). Aerobic granules can also be considered as "mini-ecosystems," comprised of a mixed microbial population among which the desirable organisms can be selected through the application of specific operational conditions (DE KREUK et al. 2005a).

At the IWA workshop called "Aerobic Granular Sludge," carried out in Munich, Germany, in 2004, it was established that aerobic granules should be considered as aggregates of microbial origin, which do not coagulate under reduced hydrodynamic forces and which have a faster settling velocity than activated sludge flocs (DE KREUK et al. 2005a). At this workshop it was also determined that, for an aggregate to be considered an aerobic granule, it must have a structure in which the position of the microorganisms is not rapidly changed, as in the case of activated sludge flocs. In addition, besides settling rapidly, the aggregate comprised of biomass and extracellular polymers must be formed without the need for a material support and must have a diameter of at least 0.2 mm. The classification of granules should be carried out using sieves, which enable expressing the amount of granules present within the overall biomass. The granulation process can be considered complete when the amount of granules corresponds to 80% of the solids present in the reactor (DE KREUK et al. 2005a).

With an external spherical shape and a diameter which can vary between 0.2 and 5.0 mm, the granules density is much higher than that of activated sludge flocs (ADAV et al. 2008a). This property shown by the granules leads to other several interesting characteristics of these microbial aggregates (ADAV et al. 2008a; LIU and TAY 2004; BEUN et al. 1999; DE KREUK and VAN LOOSDRECHT 2004):

- Excellent sedimentation ability, facilitating the separation of the treated effluent from the granular sludge.
- Regular shape, smooth and almost round.
- They are visible and form a separate phase in the liquid during the aeration and sedimentation phases.
- They enable high biomass retention in the reactor, increasing the capacity to deal with high organic loads.

- The microbial structure is dense and strong.
- Both aerobic and anoxic (anaerobic) zones are present within the granules, which allows different biological processes to be carried out in the same system.
- They can operate with high flow rates.
- They are less vulnerable to the toxicity of chemical compounds and heavy metals compared with suspended sludge.
- There is no need for a support material (required in other biofilm systems), reducing the investment costs.
- They allow a reduction in the running cost of the treatment plant of at least 20% and in the space required of 75%.

The attractiveness of granular sludge reactors is highlighted by the fact that these systems can retain a large quantity of microorganisms, allowing the rapid transformation of pollutants and leading to improvements in the performance and stability of the reactor. Consequently, large volumes of wastewaters can be treated in compact reactors. Since they are larger and have a greater density compared with activated sludge flocs, the granules rapidly settle which facilitates the separation of the treated effluent from the biomass.

The aerobic granules are preferentially cultivated in sequencing batch reactors (SBRs), where the operation can be divided into temporal cycles. These, in turn, are characterized by the phases of filling, reaction (aerobic, anoxic, or anaerobic), sedimentation, and treated effluent (supernatant) discharge. The working principle of SBRs facilities the retention of high concentrations of granular sludge in the bulk and dispenses with the need for settling tanks to separate the biomass from the treated effluent as well as sludge return to the biological reactor.

In discontinuous systems, the excellent capacity for the sedimentation of the aerobic granular sludge, besides improving the separation of the biomass from the treated wastewater, allows that a longer period of each operation cycle can be allocated to the reaction phase. In SBRs used to cultivate aerobic granules, the cycle periods are generally a few hours (3–6 h). In each cycle a certain quantity of influent wastewater is added to the reactor, and the reaction phase is then started, whether it is aerobic, anoxic, or anaerobic, during which conversions occur. The end of each cycle is characterized by rapid settling of the granules (retaining them in the reactor), this step representing a primary criterion of the design (BEUN et al. 1999). Only particles of a certain size and density, able to rapidly sediment out, are retained in the reactor. On the other hand, particles with slower settling velocities, such as microbial flocs, are washed out of the system, allowing only the development of aerobic granules (LIU and TAY 2002; BEUN et al. 1999). After the settling period, the drainage/emptying phase begins, in which the upper part of the reactor content is removed as clarified effluent. Thus, the removal of organic matter and nutrients (N and P) and the sludge sedimentation step occur within the same reactor, resulting in a single tank containing a high concentration of granular sludge, which, consequently, allows high volumetric conversion rates (BEUN et al. 1999).

The physical step of settling is responsible for biomass selection, exerting a "selection pressure" (LIU and TAY 2002). Thus, the time allocated to sedimentation

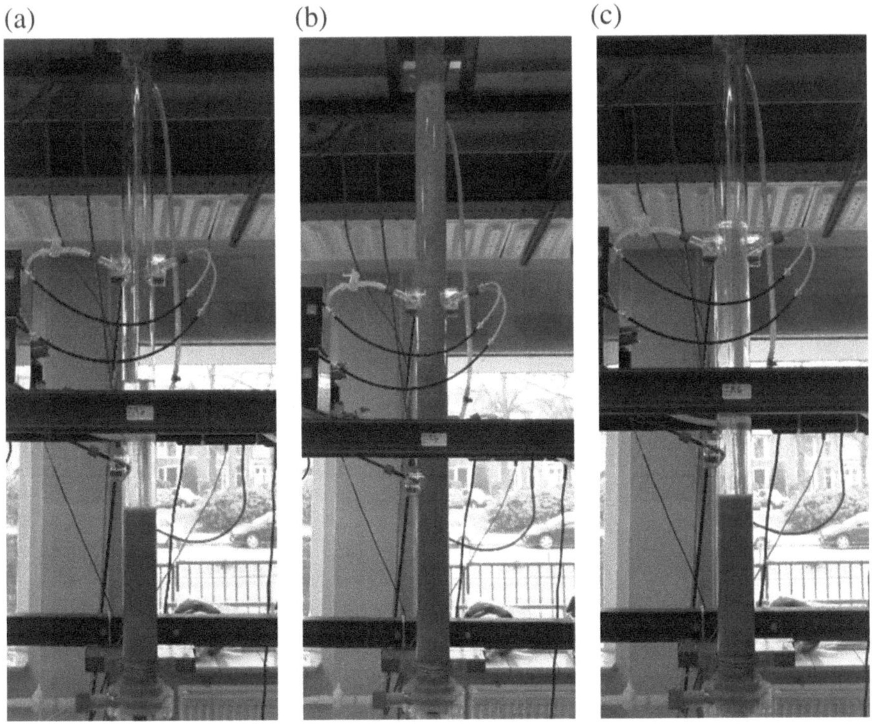

Fig. 4.1 Bubble column sequencing batch reactor used to cultivate aerobic granules. (**a**) Filling, (**b**) reaction, (**c**) sedimentation and effluent withdrawal

has a considerable influence on the aerobic granulation process and represents a key criterion of the design (BEUN et al. 1999). Short settling times lead to suspended biomass being washed out, and only the denser granules are retained (QIN et al. 2004). Figure 4.1 shows the different phases of the cycle in a lab-scale bubble column sequencing batch reactor, used to cultivate aerobic granules. As can be observed, the height/diameter ratio of the reactor is high (120 cm/6.5 cm = 18.5). As will be discussed later, this characteristic is important for the application of the aerobic granulation technology, considering that the settling velocity is one of the criteria for the selection of granules and to minimize the presence of poor settling microbial flocs.

In Fig. 4.1c, it can be observed that the treated effluent (supernatant) which is being drained is almost free from suspended solids, confirming the rapid and efficient sedimentation of aerobic granular sludge. In the case of the reactor shown in Fig. 4.1, the time allocated to the sedimentation of the granules is only 3 min. Small flocs are washed out of the reactor in every cycle, and thus the concentration of solids in the reactor, represented mainly by granules, increases during the operation.

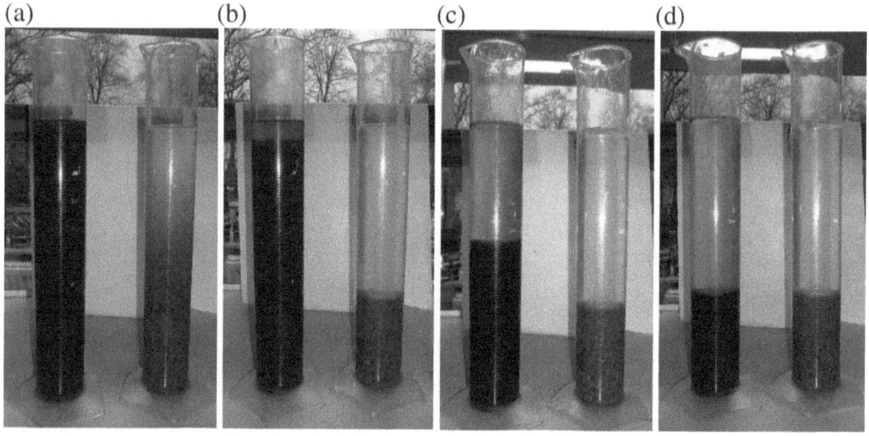

Fig. 4.2 Comparison of sedimentation over time of flocculent sludge and aerobic granules. (**a**) Start, (**b**) 30 s, (**c**) 12 min, (**d**) 20 min

The settling characteristics of activated sludge flocs and aerobic granular sludge is compared in Fig. 4.2, which illustrates their sedimentation at different time intervals. As can be observed, after only 30 s, the granules had completely settled out. The activated sludge flocs, however, required 20 min to reach the same level of sedimentation as the granular sludge.

Considering that the diffusion limitation increases when the biomass growth is no longer in suspension but instead in the form of biofilms/granules, the latter type of biomass always grows more slowly compared with biomass in suspension. Therefore, it is plausible to assume that bacteria preferentially grow in suspension rather than in the form of biofilms or granules since in the latter cases there are diffusion limitations associated with all of the components involved. Thus, only by avoiding the accumulation of suspended cells or flocs, through the application of reduced sedimentation times, will suitable granules be formed (BEUN et al. 1999, 2002). It is important to remember that slow-growing bacteria with a low substrate to cell yield coefficient ($Y_{x/s}$), such as nitrifying bacteria, show a greater tendency to grow in the form of biofilms/granules as compared to fast-growing aerobic heterotrophic bacteria (BEUN et al. 2002).

One parameter which can be varied in order to control the sedimentation step is the minimum settling velocity (v_{min}), which can be obtained by dividing the sedimentation height by the sedimentation time, the latter being selected and fixed as required. Bearing in mind that in most cases the granules have high settling velocities, the time allocated to the settling step is short, allowing more time for the degradation processes in the reaction period. This characteristic certainly makes the discontinuous operation mode (e.g., in SBRs) more attractive.

One factor to be highlighted relates to the size of the granules. As expected, the larger the particles, the greater their settling velocity will be. Thus, in this case, in particular, the cultivation of larger granules is favorable. However, it should be taken into consideration that diffusion effects may be present depending on the diameter of the granule. When the aim is, for instance, to create anoxic conditions inside the granules in order to allow denitrification, larger granules tend to be recommended. In larger granules the diffusion of oxygen to the interior of the granule is limited, especially when microbial populations situated in the outer layers of the granular biomass reduce the dissolved oxygen concentration. However, the control of the granular particles size is not trivial and depends on many operating parameters.

As previously noted, the granular sludge settling step in sequencing batch reactors is not troublesome since the granule settling velocity is much higher than that presented by flocculent biomass. However, as in any treatment process, there are obstacles to be overcome. In the case of the operation of granular sludge reactors, the main technical drawback is the instability of the aerobic granules, which can originate from the growth of filamentous bacteria, reducing the sedimentation capacity and leading to biomass washout from the reactor. This of course may limit the application of this technology to wastewater treatment. However, as will be observed in some studies involving aerobic granules reported in the literature, the growth of filamentous organisms at low or moderate levels does not lead to problems. In fact, these organisms may contribute to the stabilization of the granular structure, serving as a support for the formation of these microbial agglomerates.

While the settling velocity is one of the key operational strategies for granule selection and reduction of the amount of smaller particles in suspension, the growth rate of the microorganisms represents one of the main factors responsible for the density of the granules or biofilms. Organisms with high growth rates form less dense granules compared to those with low growth rates (VILLASEÑOR et al. 2000). Nitrifying microorganisms, for instance, form much denser biofilms than heterotrophs under the same conditions. Thus, a decrease in the bacterial growth rate contributes to increase the density of the biofilms/granules (VAN LOOSDRECHT et al. 1995) and, as will be observed later in this chapter, leads to the development of stable granules even under low oxygen conditions (DE KREUK and VAN LOOSDRECHT 2004). To achieve these objectives, some operational conditions must be ensured. Greater details regarding this issue will be provided in the section describing the factors affecting aerobic granulation (Sect. 4.4.4).

The formation of aerobic granules is also strongly influenced by shear forces, as described in Sect. 4.4.4. The hydrodynamics of the system is directly related to the development of dense aggregates, such as biofilms. From the microbiological point of view, biofilms and granular sludge can be considered to be the same, although they have obvious differences from the technical point of view. Thus, hypotheses created in an attempt to characterize biofilms are useful to gain a better understanding of the conditions required for the formation of stable aerobic granules with the desired characteristics.

Due to its distinct and special characteristics, aerobic granulation technology has been recently applied to the treatment of highly concentrated wastewaters containing organic compounds, nitrogen, phosphorus, toxic substances, and xenobiotics (JIANG et al. 2002; MOY et al. 2002; TAY et al. 2002b; LIN et al. 2003; ADAV et al. 2007a, b; ADAV and LEE 2008).

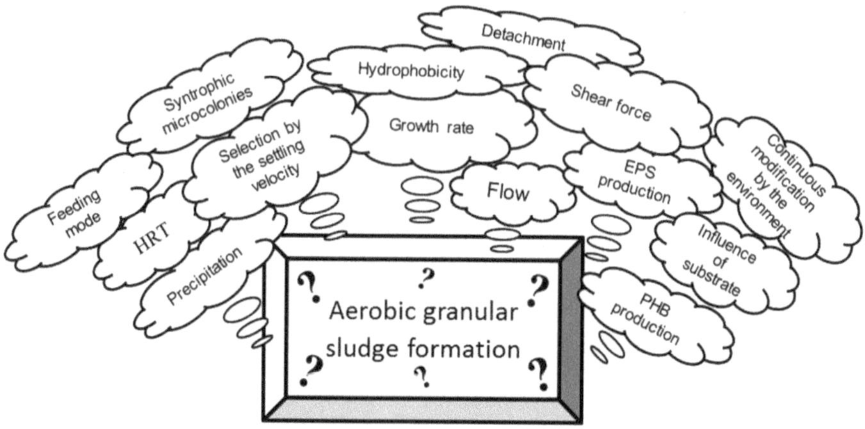

Fig. 4.3 Different ideas and hypotheses related to the formation of aerobic granular sludge (adapted from DE KREUK 2006)

Before presenting some studies involving the various applications of aerobic granules (Sect. 4.4.7), it is important to describe the aerobic granulation process and the different phases involved in the gradual transformation of flocculent sludge into granules (Sect. 4.4.3). The main factors which influence aerobic granulation, some case studies involving the formation of aerobic granular sludge, and the conversion processes which occur within the aerobic granules are addressed in Sects. 4.4.4, 4.4.5, and 4.4.6, respectively.

4.3 Formation of Aerobic Granules

Many different ideas and hypotheses were suggested to explain the formation of aerobic granules, as illustrated in Fig. 4.3. Several studies have proposed mechanisms to describe the granulation process, although there is still no consensus with respect to the process by which flocculent sludge is converted into granules.

Biogranulation basically involves interactions between cells, associated with biological, physical, and chemical phenomena which are related to the formation of stable and contiguous multicellular associations. In order for the bacterial cells present in a culture to aggregate, several conditions must be fulfilled (LIU and TAY 2004).

TAY et al. (2001a) described aerobic granulation as a self-immobilization process which does not require the use of a support material. Later LIU and TAY (2002) proposed the following stages for the aerobic granulation process:

1. Contact between microorganisms with the formation of aggregates via hydrodynamic, diffusion, gravitational, and/or thermodynamic forces
2. Stabilization of multicellular contact resulting from initial attraction, comprised of physical (van der Waals, attraction of opposite charges, thermodynamic

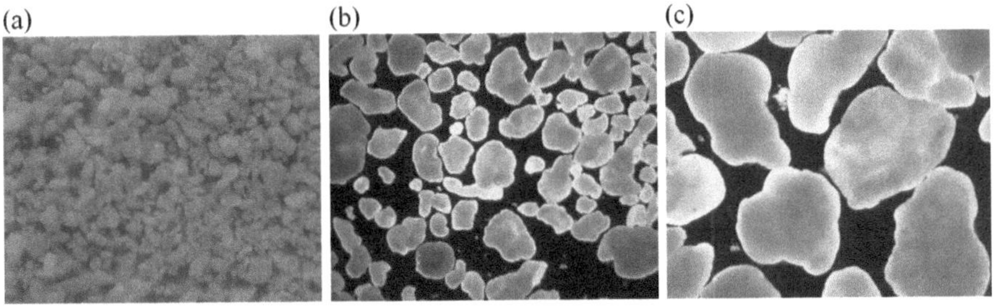

Fig. 4.4 Aerobic granules formed in a bubble column SBR. (**a**) Original size, (**b**) magnification 7.5×, (**c**) magnification 20×

forces, surface tension, hydrophobicity), chemical (ionic coupling, interparticle bonding), and biochemical interactions (fusion of cell membrane, cell receptor attraction, dehydration of cell surface)
3. Maturation of the cellular aggregates through the production of extracellular polymers and increase in cell groupings and metabolic changes, which facilitate interaction between cells and result in an organized microbial structure
4. Formation and stabilization of the tridimensional structure of the microbial aggregate in the steady state by way of hydrodynamic shear forces

Thus, aerobic granulation consists of a gradual process involving the transformation of the activated sludge into compact aggregates, which later acquire the form of granular sludge and finally mature granules (TAY et al. 2001a). In this context, rather than the simple random agglomeration of bacteria, the formation of biogranules consists of a microbial evolution process. In Fig. 4.4 the structure of mature granules can be observed, at different magnifications, cultivated in a lab-scale bubble column SBR, fed with synthetic wastewater containing acetate as a carbon source.

In general, the structures of biofilms/granules consist of highly complex ecosystems, in terms of their microbiology and morphology. In these systems, the structure influences the activity which, in turn, affects the structure. Due to the complexity of the structures of biofilms and granules, it is difficult to identify causes and effects of issues involving morphogenesis (VAN LOOSDRECHT et al. 1995). In addition, it should be noted that, from the engineering perspective, the events which occur during microbial granulation can be identified and, to a certain extent, understood. In contrast, the mechanisms associated with events which occur at the molecular or genetic level require a much more in-depth understanding.

Although several theories regarding the factors which are crucial for the development of granules have been extensively discussed, in most studies involving granular sludge, the possible contribution of protozoans and fungi in the formation of granules and their interactions with bacteria have been neglected. However, these eukaryotic organisms have a variety of important functions in activated sludge systems (conversion of biomass and water clarification) (FRIED and LEMMER 2003), and they are involved in the formation and structuring of the biofilms which play a role outside the field of wastewater treatment systems (HARTMANN et al. 2007).

In this context and taking into consideration preliminary observations of the granules in which stalked ciliates and fungi were present in significant numbers, WEBER et al. (2007) investigated the function of ciliated protozoans and fungi in the structural formation of microbial granules originating from conventional activated sludge. The authors analyzed the structure and development of different types of aerobic granules, cultivated in three laboratory-scale sequencing batch reactors (SBRs). These systems were fed, respectively, with wastewater rich in particulates containing malt and obtained from a mixture of barley powder and water (SBR_1), with wastewater from a brewery plant (SBR_2) and with a synthetic medium (SBR_3).

A detailed view of the architecture and composition of the granules has been obtained by applying scanning electron microscopy (SEM), optical microscopy, and confocal laser scanning microscopy (CLSM). The microscopic observations revealed that the granules consisted of bacteria, extracellular polymeric substances (EPS), protozoans, and, in some cases, fungi. The development of the granular sludge, from activated sludge until the obtainment of mature granules, was divided into three phases. During the first phase, stalked ciliated protozoans of the subclass Peritrichia started to be present in a great quantity in the activated sludge flocs, forming new stalks. These protozoans then began to proliferate and form large colonies while, at the same time, their stalks were colonized by bacteria. The colonization was intensified by the movement of their cilia, which provides a continuous flow of nutrients in the direction of the bacterial cells responsible for biofilm formation. After a few days, many ciliate cells got fixed to the surface of the microbial flocs. Most of the ciliate colonies formed had the shape of a tree and were comprised of the genera *Opercularia* and *Epistilys* (WEBER et al. 2007).

During the second phase, the sludge flocs became grouped, and a large growth of ciliates was observed. Concomitantly with the formation of voluminous flocs, a central zone appeared which consisted of the remains of the ciliates stalks and bacteria responsible for the production of EPS. The ciliate stalks served as a support for the development of the granules, since the bacteria used them as a substrate for their growth. The flocs which aggregated were considered to be precursors to the granules (WEBER et al. 2007).

Subsequently, with the start of the third phase, the ciliates were also colonized by bacterial cells and became embedded in the expanding biofilm. After a certain period, they were completely covered by bacteria and could no longer survive. Some cells of stalkless free-swimming ciliates appeared and left the biofilm. In this phase, compact bacterial granules were formed, which were gradually colonized by the free-swimming ciliates which managed to survive. These formed new stalks and colonies, which were used as a substrate for microbial growth. This study in particular emphasized the importance of ciliates in the formation of the structure of the granules, since they serve as a base for the growth of the microbial biofilm (WEBER et al. 2007).

As regards the fungi, their role in the granulation process is related to the fact that their filaments (hyphae), together with the ciliate stalks, act as a support for the bacterial growth, thus increasing the area available for bacterial colonization (WEBER et al. 2007). BEUN et al. (1999), as will be mentioned in Sect. 4.4.5, also noted the possible contribution of fungi as a support for the formation of granules, particularly during the initial phases of the granulation process.

WEBER et al. (2007) also reported that the mature granules formed were composed of several microbial layers, comprised of distinct bacterial species responsible for different processes, such as nitrification and denitrification. A dense central region was observed within these granules containing bacterial cells and EPS and a freely structured external portion consisting of ciliates and bacteria or fungal filaments (WEBER et al. 2007).

The exopolymeric material excreted by the microorganisms has been detected in significant quantities both in aerobic and anaerobic granules, forming a tridimensional matrix in which bacteria and other particulate material are jointly present (GROTENHUIS et al. 1991). These extracellular polymeric substances (EPS), which are comprised of polysaccharides, proteins, nucleic acids, humic acids, and lipids, contribute to the cellular adhesion and the formation of bioaggregates (biofilms and sludge flocs) and play an important role in the aerobic granulation process. They are advantageous in several aspects related to the survival of microbial cells under a great variety of circumstances. Furthermore, they provide stable operation of the aerobic granular sludge processes as they are strongly related to the stability of the granules (SCHMIDT and AHRING 1994; LIU et al. 2004b; NIELSEN and JAHN 1999). The importance of EPS is also highlighted considering that the blocking of its synthesis could hinder or even totally prevent microbial aggregation, as reported by CAMMAROTA and SANT'ANNA (1998) and by HWANG et al. (2006).

It should be noted that the polysaccharides are the only components of the exopolymeric material which are synthesized extracellularly for a certain specific function, while the proteins, lipids, and nucleic acids are present in the extracellular polymeric network due to the excretion of intracellular polymers or as a result of cell lysis (DURMAZ and SANIN 2001). Although present in all types of microbial aggregates, such as flocs and biofilms, exopolymeric materials are present in anaerobic and aerobic granules in greater quantity (TAY et al. 2001a). The composition of EPS is variable and is related to the microbial species present and their complexity, to the physiological state of the bacteria, and to the operational conditions in which the development of the granules occurs.

From the microbiological point of view, the extracellular polymeric substances can help to stabilize the membrane structure and serve as a protective barrier. LIU et al. (2004a) proposed the hypothesis that these exopolymers unite the cells and other particulate material in aggregates, facilitating their interaction and providing an extensive surface area for microbial binding, thus acting as biological "glues." Exopolymers are strongly associated with the cohesion and adhesion of cells, performing a crucial function in the microbial physiology and in maintaining the integrity of the structure of communities of immobilized cells. In addition, they can be involved in the regulation of the energy metabolism of bacteria. High shear forces stimulate bacteria to secrete more exopolymeric material, thus contributing to the formation of granules with strong compact structures. Under normal operating conditions, there is no need for the microorganisms to secrete excessive quantities of EPS. The increase in the production of EPS observed in biogranules is

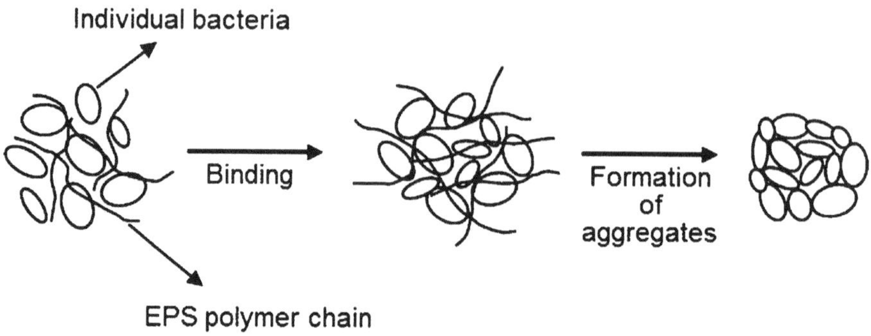

Fig. 4.5 Schematic illustration of the biogranulation process aided by the presence of EPS, evidenced through the formation of points of binding between microflocs and filaments (adapted from LIU et al. 2004a)

induced by the stressing operating conditions. In this context, several operational parameters can stimulate the bacteria to secrete greater amounts of these substances. These include the reactor configuration, the substrate composition, the substrate loading rate, the hydraulic retention time, hydrodynamic forces, the settling time in sequencing batch reactors, and the temperature (LIU and TAY 2004; LIU et al. 2004a, b).

Since EPS accumulate on the cell surface as capsule material, some cell characteristics, such as the hydrophobicity, charge density, binding site, and morphology, can be affected. It has been proposed that the extracellular polymeric material can reduce the negative charge of the cell surfaces, reducing the repulsion between neighboring microbial cells, favoring binding between them and with other inert particulate matter, and promoting their sedimentation in aggregates rather than as individual cells (SHEN et al. 1993; SCHMIDT and AHRING 1994). The hydrophobic interaction is recognized as being important to the biogranulation process, considering that most bacteria maintain contact with hydrophobic surfaces and with other cells, but do not bind to hydrophilic surfaces (GHIGO 2003). Figure 4.5 shows a schematic diagram of the biogranulation aided by the presence of extracellular polymeric substances, which act as agents for binding between neighboring cells and between cells and inert particulate matter.

Since it is responsible for the structural integrity of the aerobic granules, the exopolymeric material has been considered as not being easily biodegraded by the microorganisms which produce it, even in the absence of a substrate (SUTHERLAND 1999). Other studies have shown that during the period when an external substrate was not available, there was the induction of EPS degradation by the producer microorganisms, which resulted in the detachment of bacteria (ZHANG and BISHOP 2003; RUIJSSENAARS et al. 2000).

Although there is no consensus regarding the role of the different components which comprise the EPS matrix in the structural stability of the aerobic granules and despite the controversy regarding this subject, an interesting study was carried out by ADAV et al. (2008b). These authors sought to identify the individual contribution of the different components of EPS in terms of the structural stability of aerobic granules fed with wastewater containing phenol as the only carbon source. The granules were cultivated in a sequencing batch reactor, with an operation cycle of 6 h (5 min of filling, 320 min of aeration, 30 min of settling, and 5 min of effluent withdrawal) and with a high height/diameter ratio (*H*/*D* = 24). The granules had an average diameter of 900 μm (ADAV et al. 2008b).

The functions of the proteins, α- and β-polysaccharides, and lipids were evaluated via their selective hydrolysis using specific enzymes for each component of the EPS (proteinase K for proteins, lipase for lipids, α-amylase for α-polysaccharides, and β-amylase for β-polysaccharides), and the structural changes in the granules were observed using in situ fluorescence labeling and confocal laser scanning microscopy (CLSM). The concentration of proteins, carbohydrates, and lipids in the control granules (without the addition of enzymes) were 240.0 ± 13, 61.0 ± 9.4, and 51.1 ± 7.8 mg/gVSS, respectively. The protein/carbohydrate ratio of the granules fed with phenol was approximately 3.9 (ADAV et al. 2008b).

The authors observed from the results obtained in the fluorescence labeling of the control granules with different labels specific for each EPS component that the proteins and dead cells were distributed mainly in the center of the granule, while the active cells and the α-polysaccharides were located in the external regions of the granular biomass. β-Polysaccharides were located both in the center and in the external regions of the granules (ADAV et al. 2008b).

The selective enzymatic hydrolysis of the proteins, lipids, and α-polysaccharides had a minimal effect on the tridimensional structural integrity of the granules and did not lead to their disintegration. However, the selective hydrolysis of β-polysaccharides caused the fragmentation of the granules, generating small particles of 10–80 μm. These results reinforced the conclusions proposed by WANG et al. (2005), who stated that β-polysaccharides comprise the most important components in terms of the granule stability. These components of the extracellular polymeric matrix were attributed the function of acting as the main support of the granule structure, in which proteins, lipids, α-polysaccharides, and cells are interconnected, ensuring the mechanical stability and structural integrity of the granules. ADAV et al. (2008b) also stated that adequate mechanical resistance can be obtained using limited quantities of β-polysaccharides, a statement supported by the fact that the granules are stable even when the protein/polysaccharide ratio is high (around 4).

4.4 Factors Affecting Aerobic Granulation

The aerobic granulation process is affected by several operational parameters, notably the time allocated to the settling of the granules in sequencing batch reactors, growth rate of microorganisms, feed strategy, dissolved oxygen concentration, aeration intensity, reactor configuration, substrate composition/concentration, temperature, and pH. Thus, it should be possible to successfully cultivate aerobic granules if certain favorable conditions are established (LIU and TAY 2004).

The main factors which affect the formation of aerobic granules will be described below. The sequence in which these factors are presented represents, to a certain extent, the order of their importance in the formation of the aerobic granular sludge. It should be noted that in some cases the factors which significantly influence the aerobic granulation may be interrelated, with one factor affecting another. Thus, a global analysis of all of the factors should be carried out without separating them as if they act independently in the granular sludge formation process. In fact, their combination appears to be a crucial factor in the obtainment of stable granules.

4.4.1 Settling Time

The mechanism involved in the formation of aerobic granules in sequencing batch systems is not completely understood; however, most studies reported in the literature indicate that the settling time is one of the main factors related to the formation of dense stable granules. As mentioned in Sect. 4.4.2, reduced settling times lead to a strong "selection pressure" on the biomass present in the reactor. In fact, when long sedimentation periods are applied, flocs with poor settling characteristics are not effectively removed from the reactors. In this condition they may become dominant and therefore hinder the aerobic granulation. Thus, it can be considered that aerobic granules are formed when the settling times are below a certain critical value.

Some studies have indicated that reduced settling times, as selection pressure, contribute to enhancing aerobic granulation (JIANG et al. 2002; LIN et al. 2003; HU et al. 2005). TAY et al. (2002b) evaluated the granulation of nitrifying bacteria under different settling times and observed the importance of keeping them short in order to obtain the granules. WANG et al. (2007) observed that the granule stability can be improved with a gradual increase in the selection pressure.

Seeking to demonstrate the importance of the selection pressure and the effect of the settling time on the mechanism involved in the formation of aerobic granules, LIU et al. (2004b) operated four SBRs (SBR_1–SBR_4) of 2.5 L, with a height and diameter of 127 and 5 cm, respectively. The operation cycles of each reactor were similar, lasting for 4 h. Nevertheless, the settling time was different in each system. In the first phase, in order to evaluate the effect of the settling time on the formation of granules, this time was maintained at 20, 15, 10, and 5 min in SBR_1–SBR_4, respectively. The biomass concentrations were 5.3, 4.9, 5.5, and 5.4 g/L in SBR_1–SBR_4, respectively. In the reactor with the shortest settling time (SBR_4), excellent formation of granules was observed, with the absence of sludge flocs. However, in the other reactors, the percentages of granules formed were only 10% in SBR_1, 15% in SBR_2, and 35% in SBR_3 (LIU et al. 2004b).

After reaching a period of granule stabilization, the settling time was reduced from 20 to 5 min in SBR_1, 15 to 2 min in SBR_2, and 10 to 1 min in SBR_3. This procedure led to a high loss of biomass (washout). However, after reestablishing equilibrium in the reactors, the sludge flocs were completely replaced by aerobic granules. The authors also observed that the shorter the settling time applied, the lower the sludge volumetric index (SVI) obtained and the greater the hydrophobicity of the cell surface, contributing to the formation of stable granules (LIU et al. 2004b).

In the same context, QIN et al. (2004) evaluated the effect of the settling time on the granulation process in an SBR. The authors observed that the aerobic granules were successfully cultivated and became dominant only when the reactors were submitted to settling times of less than 5 min. On the other hand, at longer settling times, a mixture of granules and flocculent sludge was obtained.

4.4.2 Bacterial Growth Rate

As briefly mentioned in Sect. 4.4.2, the formation of dense stable aerobic granules is based, among other factors, on a reduction in the bacterial growth rate. One way to obtain this characteristic in systems fed with easily biodegradable substrates is to convert them into intracellular polymers, such as polyhydroxyalkanoates (PHA). One of the best ways to achieve this conversion is to submit the granular sludge reactors to two characteristic regimes: feast (with substrate available) and famine (with substrate completely consumed) (DE KREUK and VAN LOOSDRECHT 2004). The combination of these regimes is known as feast-famine. The feast-famine regime, which, in fact, represents a type of microbial selection, is easily obtained in reactors which operate in discontinuous mode, such as SBRs. This characteristic explains the preference for the use of these systems in the cultivation of aerobic granules, in agreement with the information given in Sect. 4.4.2.

During the feast phase, high concentrations of easily biodegradable substrates are present in the liquid medium, and they are stored as intracellular compounds by the microorganisms. When the feast phase is anaerobic, polyphosphate-accumulating organisms (PAOs), which are responsible for the biological removal of phosphate, or glycogen-accumulating organisms (GAOs) are selected depending on the availability of phosphorus in the influent (BRDJANOVIC et al. 1998; DE KREUK and VAN LOOSDRECHT 2004). On the other hand, when the feast regime is aerobic the growth of other heterotrophic microorganisms is favored. The stored substrates can be used by the microorganisms during the famine period (when no external substrate is available) for growth and maintenance. The growth rate during the famine regime is generally lower than during periods when highly biodegradable external substrates are present. Thus, the formation of intracellular polymers favors the development of dense stable aerobic granular sludge (BEUN et al. 2002), although this stability is not guaranteed under reduced oxygen concentration conditions (MOSQUERA-CORRAL et al. 2005). To ensure this stability, other operational conditions need to be ensured, as evidenced by DE KREUK and VAN LOOSDRECHT (2004). It is important to note that when the selection of bacteria such as PAOs and GAOs occurs in anaerobic-aerobic systems, the settling time (described in Sect. 4.4.1) becomes a less important factor, due to the fact that these bacteria have an inherent tendency to form microbial aggregates (DE KREUK et al. 2005a).

PICIOREANU et al. (1998) postulated that stable biofilms are formed when the ratio between the biomass growth rate and the diffusion transport is low (weak substrate and oxygen gradients). When this ratio is high, that is, when the gradients are strong, structures in the form of irregular flocs are predominant. This supposition was described through a characteristic G (growth) number, as shown in Eq. (4.1):

$$G = L_{\mathrm{Y}}^{2} \cdot \frac{\mu_{\mathrm{m}} C_{\mathrm{xm}}}{D_{\mathrm{S}} C_{\mathrm{So}}} \qquad (4.1)$$

where G represents parameters which considerably affect the structures of the biofilms and granules, C_{So} is the concentration of soluble nutrients in the bulk, D_{S} corresponds to the diffusion coefficient, C_{xm} represents the maximum biomass density in the biofilm, μ_{m} is the maximum specific microbial growth rate, and L_{Y} corresponds to the biofilm thickness or granule radius.

A low G value results in more stable and homogeneous biofilms or better granule formation. Therefore, stable formation of granules at low oxygen concentrations should occur when the microorganisms have a low growth rate (PICIOREANU et al. 1998). This hypothesis has been demonstrated experimentally in the work of DE KREUk and VAN LOOSDRECHT (2004), as described in Sect. 4.4.5.

4.4.3 Feeding Strategy

As mentioned in Sect. 4.4.2, the adoption of the feast-famine regime is directly related to the selection of microorganisms with a lower growth rate, which are able to convert easily biodegradable substrates into intracellular polymers. These microorganisms lead to the development of dense stable aerobic granules, even at low dissolved oxygen concentrations (DE KREUK and VAN LOOSDRECHT 2004).

Although some studies have shown that periods without substrate are not a prerequisite for aerobic granulation (LIU and TAY 2008; LIU and TAY 2007), increases in the hydrophobicity due to a lack of carbonaceous material have been reported (CHEN and STREVETT 2003; KJELLEBERG and HERMANSSON 1984). During the operation of an SBR, periods of aeration with no substrate available lead to the development of more hydrophobic bacteria, facilitating microbial adhesion. This appears to be a strategy of the microorganisms in response to the lack of substrate, and under these conditions their surface characteristics change (KJELLEBERG and HERMANSSON 1984; TAY et al. 2001a). From the thermodynamic point of view, an increase in the hydrophobicity of the cell surfaces leads to a reduction in the excess Gibbs energy at the surface, which results in greater interaction between cells and dense stable structure. In addition, cells present in colonies submitted to periods without feed can form fibrils which strengthen the intercellular interaction and communication (VARON and CHODER 2000). In this context, it is clear that periods in which substrate is not available favor the formation of granules with a high capacity for microbial aggregation, resulting in more dense, compact, and resistant granular biomass structures.

In contrast to the strategy of adopting the feast-famine regime, there is evidence in the literature that a long period of aeration with no substrate available can lead to a reduction in the granule stability (WANG et al. 2006). McSWAIN et al. (2004) obtained an improvement in the aerobic granulation performance when using an intermittent feeding system, with several filling phases. Under these conditions, the formation of dense compact aerobic granules was favored.

4.4.4 Dissolved Oxygen and Aeration Intensity

Aerobic granules can be formed under reduced dissolved oxygen conditions, such as 0.7–1.0 mg/L (DANGCONG et al. 1999), or at above 2 mg/L (TAY et al. 2002a; BEUN et al. 1999, 2002; DE KREUK and VAN LOOSDRECHT 2004). In some cases it has been reported that the stability of aerobic granules is negatively affected by a reduction in the dissolved oxygen concentration (DE KREUK and VAN LOOSDRECHT 2004). Greater details regarding the influence of this parameter are given in the description of case studies involving the formation of aerobic granular sludge (Sect. 4.4.5).

The aeration intensity and the upflow air velocity are related to the hydrodynamics and the shear forces (ADAV et al. 2008a; BEUN et al. 1999). These, in turn, can stimulate the production of extracellular polysaccharides (TAY et al. 2001b), which, as mentioned in Sect. 4.4.3, are related to the cohesion and adhesion of the cells. Consequently, they are important agents responsible for the structural integrity of communities of immobilized cells, as observed by TAY et al. (2001c).

The formation of aerobic granules can therefore be favored when they are submitted to high rupture forces, represented, in this case, by high aeration intensities and upflow air velocity. In addition, the development of filamentous bacteria able to reduce the excellent sedimentation capacity of the granular sludge is minimized.

ADAV et al. (2007b) compared the granulation processes in three identical reactors fed with wastewater containing phenol and submitted to different aeration intensities (1–3 L air/min). At low air flows (1 L/min), granules were not formed. On the other hand, at the highest air flow tested (3 L/min), mature stable granules were obtained (1.0–1.5 mm) with compact inner core. At intermediate air flows (2 L/min), large granules developed (3.0–3.5 mm) with abundant filaments. The same authors stated that intermediate aeration levels are not able to provide sufficient oxygen or to break the filaments present in abundance, leading to possible decrease in the performance of SBRs (ADAV et al. 2007b).

In a study carried out by TAY et al. (2001a), the effect of shear forces on the aerobic granulation process was evaluated. The authors began the operation of an SBR fed with glucose as the carbon source submitted to two surface air velocities (0.008 and 0.025 m/s). At the lowest air velocity applied, only soft flocs were formed and granules were not observed. On the other hand, when the system was submitted to an air velocity of 0.025 m/s, granules with a regular shape were obtained. The authors also added that, according to the laws of thermodynamics, shear forces caused by upflow aeration can force the dense aggregates periodically submitted to conditions where substrate is lacking (considered to be an important factor in aerobic granulation as discussed in Sect. 4.4.3) to form regular-shaped granules, which have a minimum of free energy at the surface.

LIU et al. (2005) submitted two airlift sequencing batch reactors to different upflow air velocities (2.2 and 3.3 cm/s, respectively). In the reactor in which the lower velocity was used, there was a decrease in the SVI value from 103.5 to 47.2 mL/g after the appearance of the first granules. However, a few days later, filamentous organisms appeared which led to an increase in the SVI to 170 mL/g and, consequently, biomass washout from the reactor. On the other hand, in the reactor in which an upflow velocity of 3.3 cm/s was applied, the reduction in the SVI was greater, dropping to 26.5 mL/g. In this reactor the quantity of filamentous bacteria was very low, and the granules formed had a well-defined structure.

4.4.5 Reactor Configuration

Another factor which influences the formation of aerobic granules is the reactor configuration, which affects the flow of liquid and the microbial aggregation inside the reactor (BEUN et al. 1999). Aerobic granulation, when carried out in reactors with the format of a column and upward flow, differs from the process carried out in perfectly mixed reactors, mainly due to the hydrodynamic properties of each system, which modify the interactions between the flow and the microbial aggregates. The reactors in the format of a column, with a high height/diameter ratio, promote a long circular flow trajectory, allowing the microbial aggregates to be constantly subjected to hydraulic friction and high turbulence (also provided by air upflow), and regular-shaped granules tend to be formed. On the other hand, in perfectly mixed reactors, the microbial aggregates move with flows dispersed in all directions and are subjected to variable shear forces, upflow trajectories, and random collisions. These conditions favor the formation of granules with irregular shapes and sizes (LIU and TAY 2002).

Considering that the settling velocity is an important criterion for granule selection, the high height/diameter ratio of column reactors is advantageous. Firstly, this ensures a long circular trajectory which allows hydraulic friction between aggregates, and, secondly, when associated with the absence of a secondary settling tank, this characteristic results in a compact reactor.

WINKLER et al. (2011) and BASSIN et al. (2012a) observed segregation of the biomass along the sludge bed of granular sludge reactors with a high height/diameter ratio aimed at the simultaneous removal of nitrogen and phosphorus. Through fluorescence in situ hybridization (FISH) analysis, the authors observed that at the bottom of the sludge bed, there was a greater amount of polyphosphate-accumulating organisms (PAOs), while at the top, glycogen-accumulating organisms (GAOs) were dominant. These latter organisms only compete with the PAOs for organic matter and do not contribute to phosphorus removal. Thus, their presence in the reactor is undesirable. Through the selective removal of sludge from the top of the reactor to control the solids retention time, it was possible to reduce the quantity of GAOs and allow PAOs to become dominant. With this approach it was possible to obtain good phosphorus removal efficiencies even with the system operating at 30 °C, which is favorable for the development of GAOs. Both studies highlight that the biomass segregation along the sludge bed, which apparently is more commonly observed in reactors with high height-diameter ratio, offers an additional possibility for influencing the competition between different microorganisms with the aim of obtaining the desired microbial population.

DE KREUK and VAN LOOSDRECHT (2004) compared two granular sludge reactor configurations (sequencing batch airlift reactor, SBAR, and sequencing batch bubble column, SBBC reactor), which are the most commonly used systems for the cultivation of aerobic granules. In SBBC reactors combined with a pulse feeding regime under aerobic conditions, instable granules were formed (BEUN et al. 2000; LIU and TAY 2002). The stability achieved by the SBAR is probably due to the fact that in this type of reactor, the local shear forces are stronger. When the feeding was carried out under anaerobic conditions in an SBBC reactor, stable sludge granules were formed, although the reactor start-up period was longer. These observations indicate that on reducing the maximum growth rate of the microorganisms through the selection of PAOs and GAOs, which induced the complete conversion of acetate into PHB, the effect of the shear force on the granule formation appears to become less important. Thus, the characteristics of the granules formed in the SBBC reactor and the SBAR were similar. However, the fact that it took longer to obtain granulation in the SBBC reactor compared with SBAR is an indication that the hydrodynamic forces in the latter reactor are more favorable. Further information regarding a comparison between these two reactor configurations is given in Sect. 4.4.5 (Table 4.1).

It should be emphasized that the possibility of using a bubble column reactor, rather than an airlift reactor, in large scale, has significant economic advantages, considering that the former is easier and cheaper to construct. In addition, the effluent discharge can be more easily carried out in an SBBC reactor, since in the case of SBAR, the presence of a riser requires more attention to allow the same discharge in the riser and the downcomer tubes. If this design detail is not taken into consideration, the granular sludge could be washed out of the reactor (DE KREUK and VAN LOOSDRECHT 2004).

4.4.6 Substrate Composition and Concentration

Various substrates have been used to cultivate aerobic granules, and the main ones include glucose, acetate, phenol, starch, ethanol, molasses, sugar cane, and other synthetic components (LIU and TAY 2004; TAY et al. 2002a; BEUN et al. 1999; ZHENG et al. 2005; ADAV et al. 2007a, b). Details regarding the cultivation of granules with real wastewater have also been reported (SCHWARZENBECK et al. 2004;

Table 4.1 Comparison of different reactor technologies (SBAR, SBBC, and BASR) used in different studies involving granular sludge (adapted from BEUN et al. 2000)

SBAR (Sequencing batch airlift reactor)	SBBC (Sequencing batch bubble column)	BASR (Biofilm airlift suspension reactor)
• Discontinuous system feeding	• Discontinuous system feeding	• Continuous system feeding
• Substrate: acetate	• Substrate: ethanol	• Substrate: acetate
• No secondary settling tank required	• No secondary settling tank required	• No secondary settling tank required
• Requires riser	• Does not require riser	• Requires riser and three-phase separator
• Does not require support	• Does not require support	• Requires support
• Selection variable: settling time	• Selection variable: settling time	• Selection variable: HRT
• Detachment mainly controlled by hydrodynamic conditions	• Detachment mainly controlled by hydrodynamic conditions	• Detachment mainly controlled by carrier concentration
• Nitrification and denitrification occur	• Nitrification and denitrification occur	• Denitrification does not occur
• Biomass concentration: 4 g/L	• Biomass concentration: 3 g/L	• Biomass concentration: 0.7 g/L
• Biomass density: 48 g/L	• Biomass density: 12 g/L	• Biomass density: 15 g/L
• Average granule diameter: 1.0 mm	• Average granule diameter: 2.0 mm	• Average diameter of biomass particles: 0.35 mm (carrier diameter: 0.26 mm)
• Specific surface area of granules in reactor: 785 m^2/m^3	• Specific surface area of granules in reactor: 340 m^2/m^3	• Specific surface area of granules in reactor: 1700 m^2/m^3

ARROJO et al. 2004; CASSIDY and BELIA 2005; INIZAN et al. 2005; TSUNEDA et al. 2006; FIGUEROA et al. 2008). Based on these studies, it can be observed that most wastewaters containing sufficient amounts of biodegradable organic matter are successfully treated by means of aerobic granular sludge.

TAY et al. (2001a) evaluated the characteristics of granules cultivated in two sequencing batch reactors, one fed with glucose (SBR_1) and the other with acetate (SBR_2) as carbon source. Both reactors were inoculated with activated sludge originating from a wastewater treatment plant. The SVI of the sludge used as the inoculum was high, corresponding to 230 mL/g. This value is associated with a significant presence of filamentous bacteria. After 1 week of operation, it was observed that the filaments gradually disappeared from the reactor fed with acetate, while they remained in that fed with glucose. The SVI values for the compact aggregates formed in SBR_1 and SBR_2 were 190 and 178 mL/g, respectively.

After 2 weeks of operation, the formation of round granules was observed in both reactors. Filamentous bacteria continued to predominate in the reactor fed with glucose. In the reactor fed with acetate, these bacteria disappeared completely. Nonetheless, the formation of aerobic granules had a similar pattern in the two reactors. The SVI of the granules in this stage of the aerobic granulation process decreased to 115 mL/g in SBR_1 and to 114 mL/g in SBR_2 (TAY et al. 2001a).

After 3 weeks of operation, mature granules were obtained in both reactors. The granules fed with glucose had a soft external surface, which was probably related to the predominance of filamentous bacteria. The SVI of the granules obtained in this stage of the granulation process lays within the range of 51–85 mL/g in SBR_1 and between 50 and 80 mL/g in SBR_2. The average diameters of the granules cultivated in SBR_1 and SBR_2 were 2.4 and 1.1 mm, respectively (TAY et al. 2001a).

Differences were observed in the microbial diversity of the granules depending on whether they were fed with glucose or acetate. In the former case, the granules consisted mainly of bacteria which had a round shape, particularly their inner core, and some rod-shaped bacteria with some filaments. In the latter case, the microbes were predominantly represented by rod-shaped bacteria. In relation to the settling velocity, an important parameter for aerobic granulation, values of 35 and 30 m/h were obtained for the granules cultivated in SBR_1 and SBR_2, respectively. The high settling velocities associated with granular sludge are strongly related to the dense microbial structure of these microbial aggregates.

The concentration of substrate may also affect the growth of filamentous bacteria. In general, granules cultivated in SBRs receive constant concentrations of organic substrates in terms of chemical oxygen demand (COD) (DANGCONG et al. 1999; BEUN et al. 1999; TAY et al. 2001a). After the formation of mature granules, the biomass concentration in the reactor is typically in the range of 10 to 20 g/L. In systems operating in batch regime, the relation between the initial substrate concentration and the initial biomass concentration (S_0/X_0) can be used to describe the substrate availability to the microorganisms (LIU and LIU 2006). LIU and LIU (2006) demonstrated that during the operation of an SBR with granular sludge, the concentration of biomass increased, resulting in a decrease in the S_0/X_0 ratio. Under these conditions of low substrate availability and a high biomass concentration, the authors observed a significant increase in filamentous microorganisms.

4.4.7 Sludge Used as Inoculum

The quality of the sludge inoculated into the reactors aimed at the formation of aerobic granules is related to its macroscopic characteristics, settleability, surface properties (hydrophobicity and charge density), and microbial activity (LIU and TAY 2004). In most studies, the aerobic granules were from inoculum originating from activated sludge systems. The characteristics of the bacterial community present in the inoculum is essential for the granulation process. Hydrophilic bacteria tend to bind less to the microbial flocs in comparison to hydrophobic bacteria, which constitute most of the free bacteria in the effluent of treatment plants (ZITA and HERMANSSON 1997). The hydrophobicity of cells influences the granulation process, as described in Sect. 4.4.3. With a higher number of hydrophobic bacteria in the inoculum, the granules form more rapidly and with better settling characteristics (WILEN et al. 2007).

4.4.8 Temperature

The formation of stable granules is affected by the temperature. In general, most studies on aerobic granular sludge technology have been carried out at ambient temperature (20–25 °C). DE KREUKI et al. (2005b, c) investigated the short- and long-term effect of temperature variations on the stability of aerobic granules and on the conversion processes which occur within the granular structure in a sequencing batch reactor. During the reactor start-up, when the temperature applied was 8 °C, granules with an irregular structure were formed, and growth of filamentous bacteria was observed, leading to a substantial washout of biomass and operational instability. DE KREUK et al. (2005c) also observed that when the reactor start-up was carried out at 20 °C, it was possible to operate the system in a stable manner, even when the temperature was further decreased to 15 °C and subsequently to 8 °C. The results obtained by these authors stress the importance of carrying out the start-up of granular sludge reactors, in pilot or full scale, in periods of high temperature, such as in summer. The denitrification capacity and the nutrient removal rate were reduced at low temperatures (DE KREUK et al. 2005c), as it will be detailed in the applications of aerobic granules (Sect. 4.4.7).

4.4.9 pH

The pH of the reactor significantly affects the microbial growth rate. Oxidation at high organic loads produces sufficient quantities of CO_2 to reduce the pH in non-buffered solutions (McSWAIN et al. 2004). Fungi grow well at low pH and can contribute significantly to the initial granulation, as noted in Sect. 4.4.3 (McSWAIN et al. 2004; BEUN et al. 1999). These organisms are able to release protons in exchange for NH_4^+ in solution, thus contributing to decreasing the pH value (*apud* ADAV et al. 2008a). YANG et al. (2008) observed that aerobic granulation at pH 4.0, in the presence of fungi, favored the formation of granules of approximately 7 mm. On the other hand, at pH 8.0, conditions under which the granulation was controlled by bacteria, the granule size reached only 4.8 mm. Despite these findings, the effects of pH on aerobic granulation are still not completely understood (ADAV et al. 2008a, b).

4.4.10 Addition of Divalent Cations

Aerobic granulation can be influenced by the addition of divalent cations, such as iron and calcium. According to some researchers, high concentrations of these cations can accelerate the formation of granules, increase their stability and resistance, and improve their settling characteristics (LIU and TAY 2004; DE KREUK et al. 2005a).

Divalent cations adhere to negatively charged groups on the surface of bacteria and molecules of extracellular polysaccharides, acting as binding points. These, in turn, increase the microbial agglomeration (LIU and TAY 2004).

JIANG et al. (2003) observed that the addition of calcium to the feed medium in concentrations of 100 mg Ca^{2+}/L enabled the formation of granules to occur more rapidly (16 days) in comparison with cases where this cation was not added (32 days). Furthermore, it was observed that granules fed with calcium had better settling characteristics and a higher content of polysaccharides (JIANG et al. 2003).

4.5 Case Studies Involving the Formation of Aerobic Granules

In this section, some studies related to the formation of aerobic granular sludge will be discussed. It should be noted that some of them represent the first investigations in this field of study, and thus they are described in greater detail. In addition, the studies described below demonstrate the influence of some factors which affect the aerobic granulation process, a topic discussed in Sect. 4.4.4.

BEUN et al. (1999) cultivated granules in a bubble column SBR, with a volume of 2.25 or 2.5 L. The internal diameter of the column was 5.6 cm and the total height was 150 cm, which provides a high height/diameter ratio. As mentioned in Sect. 4.4.5, high height/diameter ratios tend to favor the formation of granules with a regular shape, which is related to the hydrodynamic forces present in the reactor. The inoculum used was sludge from a conventional SBR accomplishing COD removal. The experiments were conducted at ambient temperature (20 ± 2 °C), and air was introduced by means of a small bubble diffuser located at the base of the column.

The reactor was fed with synthetic wastewater with a COD of 0.83 g/L, total Kjeldahl nitrogen (NKT) of 0.04 gN/L and total phosphorus of 0.16 gP/L. The cycle was 3 or 4 h. Of this total, periods of filling (with aeration) and settling were 2 min. The discharge period was 1 min, and the aeration period was the remaining time, that is, 177 min for the 3 h cycle and 237 min for the 4 h cycle. The main parameter chosen to select biomass with a high settling velocity was the sedimentation time. The settling velocity varied between 12 and 24 m/h (BEUN et al. 1999).

After the inoculation of the reactor with 10 mL of cells in suspension originating from the aforementioned SBR (start-up period), observations carried out directly in the reactor and by microscopy revealed that pellets of filamentous fungi (filamentous granules) predominated in the reactor, and these functioned as an immobilization matrix on which the bacteria could grow and form colonies. The fact is that the fungi easily formed mycelial filaments which settle easily and could be removed from the reactor. Bacteria do not have this property when in suspension, being washed out of the reactor, which explains the predominance of filamentous fungi during the reactor start-up period (BEUN et al. 1999).

The granules formed were not stable, due to the shear force (mainly caused by the relatively high surface air velocity of 0.041 m/s), and they broke into pieces within a few days, with the release of filaments from the surface of the pellets, which then became more compact. The pellets grew up to a diameter of 5–6 mm and then underwent lysis probably due to the oxygen limitation in their internal parts (BEUN et al. 1999).

Thus, washout of most of the biomass occurred (notably of the filaments which were not able to settle rapidly and stay in the reactor), leading to a new granulation stage. The granules formed in this second stage were not generally filamentous and consisted mainly of bacteria. In this stage, colonies of bacteria remained in the reactor since they were large enough to settle rapidly. These colonies, in turn, later grew forming granules (BEUN et al. 1999). Clearly, this proposed mechanism was based particularly on experiments with this reactor, which was inoculated with a small quantity of suspended sludge, only capable of settling slowly. If the inoculum had consisted of flocs and/or small granules, the mechanism would be different.

The morphology of the granules was analyzed through an evaluation of some parameters which are commonly used to characterize this type of microbial agglomeration, such as the form factor (0 = line, 1 = circular) and the aspect ratio, that is, how round a particle is (minimum Feret diameter/maximum Feret diameter; 0 = line, 1 = circular; Feret diameter, maximum distance between two points along the boundary selected). The values obtained for these two parameters during the whole operational regime were around 0.45 and 0.79, respectively, these being independent of the settling velocity, surface air velocity, organic load applied, and HRT (BEUN et al. 1999).

The pH value fluctuated around 6.5. The carbon source (ethanol) was consumed constantly at the maximum rate until its concentration was negligible, although nitrification did not occur. After the ethanol had been completely consumed, CO_2 continued to be produced due to the conversion of compounds stored within the cells (BEUN et al. 1999).

During the period considered ideal for the formation of stable granules, the reactor was submitted to an organic load of 5 kgCOD/(m^3 day), and the total cycle time was 3 h. The cell retention time increased during these periods from 1.8 to 3.4 days, mainly due to the increase in the sludge bed volume. Under these conditions, the formation of granules with an average diameter of 3.3 mm and a density of 11.9 $gSSV/L_{granules}$ was observed. At high organic loads, as expected, higher biomass growth was observed. When the other parameters (minimum settling velocity, surface air velocity and HRT) were ideal, biomass accumulated in the system and its concentration increased. The organic load did not directly affect the granulation under the conditions tested, although it did influence the final shape of the granules (BEUN et al. 1999).

It was observed that a short HRT (6.75 h) and a high shear rate were favorable for the granulation process. Also, it was verified that the minimum settling velocity (24 m/h) could be applied only when the organic load was low (2.5 COD/(m^3 day)). At this load, less biomass accumulated in the reactor, and thus a higher minimum settling velocity could be applied without the granules exerting an influence on each other during sedimentation. However, with an increase in the organic load to 5 kgCOD/(m^3 day), the quantity of granules increased, and the sedimentation was adversely affected. Under these conditions, the biomass was not able to settle below the treated effluent discharge level, and thus it was partially washed out of the reactor during the effluent withdrawal stage. Under these circumstances, reducing the minimum settling velocity, that is, increasing the settling time, led to improved sedimentation and to a greater accumulation of biomass in the reactor. According to the authors, another way to avoid washout of the biomass would be to counterbalance the effect of the increased organic load by applying a high shear rate, allowing the formation of more compact granules and stable operation (BEUN et al. 1999).

DANGCONG et al. (1999) observed the aerobic granulation process in a sequencing batch reactor (SBR) operating in laboratory scale fed with a synthetic medium containing sodium acetate as the organic substrate and ammonium chloride as the nitrogen source. The reactor was inoculated with activated sludge containing filamentous bacteria. The seed sludge SVI was in the range of 250–300 mL/g. Oxygen concentration was initially maintained at 3.5–4.0 mg/L. The HRT was 8 h and the biomass retention time was 20 days. The total cycle was 4 h (0.5 h of filling, 0.75 h of reaction, 2.5 h of sedimentation, and 0.25 h of supernatant removal). The reactor was maintained at 25 °C under stirring at 400 rpm to ensure good mixing and oxygen transfer.

The authors observed that after 20 days, the filamentous bacteria disappeared and the SVI decreased to 100–150 mL/g. The dissolved oxygen concentration in the reactor was then maintained at lower values within the range of 0.7–1.0 mg/L. Under these conditions, microbial flocs were formed and the SVI increased to 150–200 mL/g. However, the flocculent sludge was gradually converted to granules. After 1 month of operation at low DO concentration, almost all of the biomass was in the form of granules. At this stage of the process, the removal efficiencies for COD, N-NH_4^+, and nitrogen were 95, 95, and 60%, respectively. Filamentous bacteria disappeared, and the SVI was 80–100 mL/g, even though the DO concentration was less than 1.0 mg/L. The reactor maintained the same characteristics for 3 months of operation (DANGCONG et al. 1999).

The granular morphology analyzed by microscopy revealed that the granules were spherical, in contrast to flocculent sludge. The diameter of the granules varied between 0.3 and 0.5 mm. These values are small compared with the diameter of the granules formed in reactors operating under anaerobic or anoxic conditions (2–3 mm). In the case of this research, in particular, the shear forces caused by the stirring and aeration did not allow the formation of granules with large diameters. Nonetheless, the structure of the granules, particularly their inner core, was very similar to that of granules cultivated under anoxic conditions. This finding suggests that the internal region of the granules might have been subjected to anoxic conditions due to the low DO concentration in the liquid. This hypothesis is reinforced by the fact that 60% of the influent ammonium nitrogen was denitrified to nitrogen gas (DANGCONG et al. 1999).

The authors also observed a high COD removal rate (2.16 gCOD/(gSS day)) and high nitrification activity (0.24 gN-NH_4^+/(gSS day)). The organic and nitrogen loads were 1.5 kgCOD/(m^3 day) and 0.18 kgN-NH_4^+/(m^3 day), respectively. Even operating under this organic load and DO concentration of 0.7–1.0 mg/L, the ammonium fed to the system was completely nitrified. The authors also observed that the COD removal or even the carbon oxidation activity increased with an increase in the air flow, although the DO concentration was maintained constant (less than 1% of air saturation). These results suggest that the granular sludge activity was strongly influenced by the oxygen supply, but not by the DO concentration of the bulk (DANGCONG et al. 1999).

BEUN et al. (2000) cultivated aerobic granular sludge in a 3 L sequencing batch airlift reactor (SBAR). The HRT applied was 5.6 h and the organic load was 2.3 kgCOD/(m^3 day). The duration of each cycle of the SBAR was 3 h. Of this total, 3 min corresponded to feeding, 173 min to aeration, 2 min to settling, and 5 min to the effluent withdrawal. The reactor was fed with a synthetic medium containing sodium acetate as carbon source and ammonium chloride as nitrogen source. The reactor was inoculated with sludge originating from an activated sludge system. The reactor operation was initiated twice using different inocula. The first operational period had a duration of 41 days (Experiment 1), while the other lasted 55 days (Experiment 2).

In Experiment 1, 30 days were required to reach the steady state, while in Experiment 2, the time necessary to reach stable operation was 10 days. The difference in the duration of the start-up periods is probably related to the inoculum used. It should be noted that Experiment 1 was carried out to describe the start-up period of the reactor and the initial development of the granules. The aim of Experiment 2 was to describe the stability of the granules under certain conditions since this reactor was operated for a longer time (55 days) (BEUN et al. 2000).

The sludge inoculated into the reactor in Experiment 1 consisted of a mixture of filaments and small particles. After 30 days of operation, the diameter of the granules was approximately 1.2 mm, and they had a smooth surface. During this period, the presence of flocculent biomass was not observed visually. From the start-up to 30 days of operation, the concentration of biomass in the reactor increased from 2 to 6 g/L, and the solids retention time (SRT) increased from 2 to 30 days, later reducing to 17 days. The increase in the SRT was caused by a decrease of filamentous and flocculent sludge. The selection of granules was achieved by applying a short settling time in the reactor with a high height/diameter ratio. Sludge particles with low settling velocities were washed out of the reactor with the treated effluent, and only granules with good settling characteristics (minimum settling velocities of 16.2 m/h) were retained in the system. This allowed a reduction in the effluent biomass concentration (BEUN et al. 2000).

During Experiment 2, the biomass concentration in the reactor remained stable at 4 g/L, and the solids retention time was, on average, 9 days. The average diameter of the granules obtained in the steady state was 1.0 mm. The biomass density had a stable value of 48 $g/L_{granules}$ (BEUN et al. 2000).

Through the determination of the acetate concentration during one representative SBAR operation cycle, it was observed that this organic substrate was consumed in 21 min (feast period). The concentration profiles of NH_4^+, NO_3^+, and NO_2^- showed that both nitrification and denitrification occurred in the reactor. During the feast period, the concentrations of NO_3^- and NO_2^- decreased, due to the occurrence of denitrification with acetate as carbon source. In the period without acetate available (famine period), the concentrations of NO_3^- and NO_2^- initially increased (nitrification was more rapid than denitrification) and later decreased again (denitrification with endogenous substrate as carbon source) (BEUN et al. 2000).

It should be noted that BEUN et al. (2000) did not expect denitrification to occur, since the reactor was operated only under aerobic conditions. During the famine period, the concentration of dissolved oxygen in the SBAR was almost 100% of air saturation. In general, in aerated and continuously fed airlift reactors, denitrification does not occur (TIJHUIS et al. 1994 *apud* BEUN et al. 2000). The reactors were fed with the same influent and submitted to the same loads as the SBAR used in this research, although they require the introduction of phases without aeration to allow the denitrification process to occur. In the SBAR, significant denitrification occurs even under conditions which are not ideal for this process. These observations are probably related to the fact that the feeding of the reactor was discontinuous rather than continuous (BEUN et al. 2000).

To exemplify the effect of the reactor configuration on the formation of granular sludge and highlight important design criteria, Table 4.1 shows a comparison between the SBAR described in BEUN et al. (2000), the bubble column reactor used by BEUN et al. (1999), and the biofilm airlift suspension reactor (BASR) operated by TIJHUIS et al. (1994). The values for some parameters obtained for each reactor in the steady state (submitted to an organic load of 2.5 kgCOD/(m^3 day) and a surface air velocity of 86.4 m/h) are given in Table 4.1.

In comparison with SBBC, much denser granules with a smaller diameter were obtained in SBAR when the same substrate load and aeration intensity were applied. In SBBC, the distribution of the hydrodynamic forces along the column was more homogeneous compared with SBAR. In the latter reactor, the hydrodynamic forces are greater at the bottom of the reactor where the liquid and particles change the direction of the flow. The stronger hydrodynamic forces concentrated in certain regions of the SBAR probably result in the formation of denser granules with smaller diameters. Also, in the SBBC the stratification of the granules was observed. In the upper part of the reactor, less biomass in the form of granules will be present in comparison with the lower part. This leads to more rapid growth of the granules at the top of the reactor which, in turn, promotes the formation of less dense granules with a larger diameter, and, subsequently, higher detachment and instability are observed (BEUN et al. 2000).

In the same way, on comparing the granules formed in the SBAR with those formed in the continuously fed BASR, denser granules with a larger diameter were obtained in the SBAR. In the case of the BASR, the development of granules is related to the biofilm fragments originating from the rupture of the biofilm particles. The reason for smaller diameters obtained in the BASR may be related to the presence of carriers, which promote a stronger shear force and consequently greater detachment. The granules of the BASR are probably less dense because this reactor was continuously fed, in contrast to the SBAR, which was operated discontinuously (BEUN et al. 2000). As mentioned in Sect. 4.4.2, the adoption of a feast-famine regime obtained through operation in discontinuous mode favors the development of microorganisms with a lower growth rate, which consequently promotes the development of dense granules. Further details regarding the comparison of these different reactor configurations will be provided below in the description of the study carried out by BEUN et al. (2002).

BEUN et al. (2002) cultivated aerobic granular sludge in a sequencing batch airlift reactor (SBAR), with a volume of 3 L, for a period of 140 days. The ratio between the height of the riser and the diameter was high (90 cm/6.25 cm = 14.4), as commonly observed in reactors aimed at the cultivation of aerobic granules. The synthetic medium fed to the reactor had an acetate concentration of 18.3 Cmmol/L, which corresponds to an organic load of 2.5 kgCOD/(m^3 day), and the HRT was 5.6 h. The temperature of the reactor was maintained at 20 °C, and the pH was fixed at 7.0 ± 0.2. The reactor operating cycle comprised 2 min of feeding, 170 min of aeration, 3 min of settling, and 5 min of discharge/drainage of supernatant (treated effluent). The inoculum used was activated sludge originating from a wastewater treatment plant with nutrient (N and P) removal.

In general, it was observed that the acetate (source of organic carbon) was completely consumed within a few minutes (feast period). When no substrate remained, endogenous respiration occurred (famine period). The period of transition from the feast to famine phase was observed through a rapid increase in the dissolved oxygen (DO) concentration in the reactor. During the feast period, the DO concentration in the reactor was relatively low (75% of air saturation) due to the consumption of oxygen during acetate metabolism. After the complete consumption of the substrate, the DO concentration became almost 100% of air saturation (BEUN et al. 2002).

In the first 12 days of operation, the settling period was gradually decreased from 5 to 4 min and then to 3 min, the value applied for the rest of the reactor operation. On the first day of operation, the feast period was approximately 100 min, decreasing to 60 min on the third day. From the third to the fourth day, the feast period increased again due to the washout of large quantities of biomass. The reactor walls became completely covered with biomass during the first 4 days. On the sixth day, the biomass fixed on the walls was removed (BEUN et al. 2002).

Small granules were formed in the reactor within 1 week after the inoculation with activated sludge originating from a conventional wastewater treatment plant. The granulation process then began, and the growth of biomass on the reactor walls decreased. The concentration of biomass in the reactor began to increase considerably, and thus the duration of the feast period decreased once again (BEUN et al. 2002).

It should be noted that in attempting to explain the formation of sludge granules at the beginning of the reactor operation, one possibility is that they originate from the growth of biofilm on the reactor walls, although during the reactor cleaning, the biofilm was completely removed. Considering the high liquid circulation rate, small pieces of the biofilm could be broken off, constituting the initial material for the formation of granules (BEUN et al. 2002).

The selection of only dense granules from a mixture of granules, filaments, and flocs was achieved due to the different settling velocities of the different biomass agglomerates: granules (biomass with high settling velocity) and filaments and flocs (biomass with low settling velocity). This strategy was successful considering that a rapid transition from the presence of sludge in the form of flocs to granular sludge occurred. Another hypothesis which could explain the formation of granules is that flocs with high settling velocity originating from the inoculum may represent the beginning of the granulation process (BEUN et al. 2002). It is interesting to note that in this study, there was no intermediate phase with the predomination of fungi, as observed in a previous study carried out by BEUN et al. (1999).

When the steady state was reached (at around day 37 of operation), the average granule diameter was 2.5 mm, biomass density was 60 gVSS/L, and the settling velocity was greater than 10 m/h. Also, the settling time was chosen so that only particles with a settling velocity greater than 10 m/h would be retained in the reactor. The rest of the biomass which did not settle sufficiently rapidly was washed out with the treated effluent (BEUN et al. 2002).

The biomass consisted of both heterotrophic and nitrifying microorganisms, which were able to withstand some disturbances, such as sudden increases in pH and fluctuations in the DO concentration. After these operational instabilities, the granules generally partially broke up. However, they rapidly returned to their state before the disturbance (BEUN et al. 2002).

The authors also observed that during the operational period (at around day 80), the biomass concentration and density in the reactor decreased due to the deterioration of the granules caused by a decrease in the DO concentration to 50% of air saturation between days 66 and 71 of operation, and thus smaller granules were washed out of the reactor. A low sludge production was observed which was associated with the sludge age of 50 days (BEUN et al. 2002).

BEUN et al. (2002) also cut the granules with the aid of a cutting blade in order to examine the internal structure under a light microscope, and they observed the presence of two layers in the granule: one central layer, with a diameter of around 1.7 mm, and one external layer approximately 0.4 mm thick. The latter had a denser structure in relation to the central part of the granule, which had a softer, more gelatinous structure, besides being transparent. Large empty cavities resulting from complete lysis inside the biomass were not observed in the center of the granules. The depth of oxygen penetration was calculated as between 17 μm (during the feast period) and 20 μm (during the famine period) (BEUN et al. 2002).

The authors also carried out a comparison between the SBAR aimed at the formation of granules and a continuously fed turbulent system (BASR), operated by TIJHUIS et al. (1994). This comparison had been previously described in BEUN et al. (2000), however, in less detail. In the BASR, granules (which, in principle, are biofilms without a material support) were formed from biofilm fragments originating from the rupture of the biofilm particles. The most notable difference between the two systems is related to the biomass density, which was much higher in the SBAR compared with the BASR, as previously highlighted in relation to the work reported in BEUN et al. (2000). As mentioned above, the reason for this considerable difference in the density may be attributed to the way in which the reactors were fed, with continuous feeding in the BASR and intermittent feeding in the SBAR. One point to be taken into consideration is that when there is substrate present in the SBAR (feast period), the acetate penetration into the granules is greater than 500 μm, bearing in mind that the concentration of this substrate in the liquid is high due to the pulse feeding. On the other hand, in the BASR with continuous feeding, the concentration of acetate in the liquid is always low (<0.1 Cmmol/L), which does not allow complete penetration of acetate to the interior of the biofilm (penetration < 20 μm). Thus, the microorganisms present deep within the biofilm will be deprived of acetate, which does not occur in the SBAR, where the cells present in the center of the granules can assimilate acetate and grow (using both oxygen and nitrate as electron acceptors). It is this growth inside the granules of the SBAR which makes the biomass increase, in contrast to the situation in the BASR (BEUN et al. 2002).

As previously mentioned, the electron acceptor required for the conversion of acetate can be O_2 or NO_3^-. In both reactors (SBAR and BASR), O_2 does not completely penetrate the biofilm. In the BASR, the O_2 penetration depth is around 80 μm, which is greater than the penetration of acetate. The O_2 penetration depth in the SBAR during the feast period is around 20 μm, which is less than the depth of acetate penetration. This value can be even lower in cases where the DO concentration in the bulk is decreased. The microorganisms located deeper within the particle than this depth can use NO_3^- as an electron acceptor. When this is not available, cell lysis occurs in the central region of the biofilm giving it a less dense structure, and it finally breaks into pieces (as observed when applying a DO concentration of 50% of air saturation) (BEUN et al. 2002).

The authors also highlight the importance of the influence of the shear force on the biofilm structure, which is determined by the balance between the biomass growth and the shear force (VAN LOOSDRECHT et al. 1995). Considering the total number of particles in the reactor, detachment should be, in principle, greater in the BASR (filled with basalt as the support material, probably with a high frequency of collisions between the particles), since the number of particles in this reactor is ten times higher than that in the SBAR, which would lead to the formation of a less dense biofilm than in the BASR. However, as previously mentioned, the specific growth rate of the biomass in the SBAR is lower than in the BASR, which leads to the formation of a denser biofilm. The balance between these two factors is obviously favored in the SBAR compared with the BASR, since the biomass density in the former system is much higher than that in the latter (BEUN et al. 2002).

GARRIDO et al. (2001) observed that the settling properties of the sludge obtained from an industrial-scale sequencing batch reactor (SBR_{ind}) showing high COD and nitrogen removal from wastewater originating from a laboratory which analyzes industrial dairy products were unsatisfactory. This conclusion was based on the fact that the sludge volume index (SVI) did not fall below 100 mL/gTSS and the zone settling velocity (ZSV) was around 0.3 m/h. Based on the results obtained by GARRIDO et al. (2001) and ARROJO et al. (2004) studied the possibility for the formation of aerobic granular sludge in two sequencing batch reactors (SBR_1 and SBR_2), in order to improve the biomass settling characteristics. Both reactors SBR_1 and SBR_2 were fed with the same industrial wastewater fed to SBR_{ind} and inoculated with the sludge from this latter reactor.

The two reactors were operated under similar conditions during most of the experimental period. However, in the case of SBR_1, an anoxic phase with a duration of 10–30 min was included at the beginning of each cycle, by sparging with nitrogen. The temperature varied between 15 and 20 °C, and the dissolved oxygen was between 0 and 8 mg/L. With no automatic control of the pH, this parameter varied between 7.4 and 8.5. SBR_1 was initially fed with a synthetic medium up to day 27 of operation. This synthetic medium was then gradually replaced with industrial effluent, maintaining a reasonably constant COD. After 48 days of operation, this reactor was fed only with industrial effluent. The operation of SBR_2 began 50 days later, and this reactor was fed only with industrial effluent. The duration of the operational cycle was 3 h (3 min of filling, 171 min of anoxic and aerobic reaction, 1 min of settling, 0.5–3 min of drainage, and a rest period of 2–4.5 min) for both reactors, with a volume exchange ratio of 50% (ARROJO et al. 2004).

In both systems, the poor settling flocculent sludge was almost completely washed out in the first 7 days, since the operational strategy adopted included a short settling period and rapid effluent removal. Only microbial agglomerates with settling velocities higher than 9 m/h were retained in the reactors. Three weeks after the reactor start-up, the formation of small aggregates with an average diameter of 1.05 mm was observed. Flocs in suspension gradually disappeared from the reactor, and the settling properties of the aggregates obtained were good, as were the SVI (60 mL/gSSV) and ZSV (20 m/h). Granules with similar morphology (distinct from the flocculent sludge used as the inoculum) developed in the two systems, and the size distribution was 0.25 and 4.0 mm. The biomass concentration was around 0.2 gTSS/L at the beginning of the experiments and increased to 3 gTSS/L after 50 days of operation. This parameter later reached a stable concentration of 5–6 gTSS/L. The percentage of volatile suspended solids (VSS) in relation to the total suspended solids (TSS) varied from 87 to 95% (ARROJO et al. 2004).

The experimental results showed that the use of industrial wastewater did not appear to adversely affect the development of granules or the accumulation of biomass. The biomass density was around 10–15 $gVSS/L_{granules}$ in both reactors, values similar to those reported by BEUN et al. (1999) (a study discussed earlier in this section), that is, 11.9 $gVSS/L_{granules}$.

MOSQUERA-CORRAL et al. (2005), seeking to find better conditions for the removal of nitrogen in an aerobic granular sludge system, studied the short- and long-term effects of a reduction in the oxygen concentration on the performance of a sequencing batch airlift reactor (SBAR). The reactor was operated in successive cycles of 3 h, with 3 min of filling, 169 min of aeration, 3 min of settling, and 5 min of effluent withdrawal. The short period of settling combined with the height of the reactor contributed to the selection of particles with a settling velocity higher than 12 m/h.

For a period of 150 days, the SBAR was operated under the same conditions as those described in BEUN et al. (2000), the only difference being the COD/N ratio, which was changed from 14 to 8. The oxygen concentration employed was 100% of air saturation. The granules, with an average diameter of 0.6 mm, were formed after 30 days of operation. The biomass concentration in the reactor stabilized at around

5 gVSS/L, the solids retention time reached a stable value of around 25 days, and the maximum density of the granules was 53 $gTSS/L_{granule}$. A high density combined with a regular granule shape resulted in a low value for SVI_5, which fluctuated between 48 and 75 mL/gTSS (MOSQUERA-CORRAL et al. 2005).

After 150 days of operation, the oxygen concentration was reduced to 40% of air saturation aiming to increase the nitrogen removal efficiency, as indicated by the model developed by BEUN et al. (2001) and discussed in Sect. 4.4.7. Ten days after the reduction in the O_2 concentration, filamentous structures were observed on the surface of the granules. Starting from day 174 of operation, the granules began to disintegrate, resulting in a reduction in their size and form factor (<0.52). The settling of the granular sludge was considerably affected by its disintegration, with substantial biomass washout from the reactor occurring. The biomass concentration decreases from 5 to 3.5 gVSS/L, and the solids retention time decreased to 8 days. The SVI_5 increased to 100 mL/gTSS, a value considered to be high for the settling conditions applied in the SBAR, and this resulted in the washout of most of the biomass. The lower oxygen penetration depth of the granule when the DO concentration was reduced from 100 to 40% of air saturation may have led to a reduction in the production of extracellular polymeric substances (EPS) in the inner region of the granules, weakening and rupturing their structure and leading to a decrease in their density. Under these circumstances, the settling properties of the granular sludge were adversely affected, leading to biomass washout from the reactor (MOSQUERA-CORRAL et al. 2005).

The SBAR operation was then restarted with new inoculum applying an oxygen concentration of 40% saturation. The first granules were formed after 10 days of operation, and they were small and unstable, being frequently washed out with the effluent. The average biomass concentration was 0.144 gTSS/L, and the SVI_5 was 200 mL/gTSS. Since it was not possible to obtain stable granules under these conditions, the reactor operation was stopped.

The results obtained by MOSQUERA-CORRAL et al. (2005) highlighted the difficulty associated with obtaining stable granules in the presence of low oxygen concentrations, and to achieve this objective, the concentration of this element needs to be increased. However, it is clear that in order to save energy and achieve good denitrification efficiencies, a low O_2 concentration needs to be applied.

The hypothesis raised by PICIOREANU et al. (1998) (discussed in Sect. 4.4.2) is in agreement with the results obtained by MOSQUERA-CORRAL et al. (2005), since low oxygen concentrations lead to an increase in the diffusion limitation and, consequently, to a stronger concentration gradient in the granule. If the proposed explanation discussed by these authors, in which stable biofilms are formed when the ratio between the biomass growth rate and diffusive transport is low (weak substrate and oxygen gradients), is correct in relation to the factors responsible for controlling the granule structure, then the selection of organisms with low growth rates would lead to the formation of stable granular sludge even in the presence of low dissolved oxygen concentrations. In order to reduce the growth rate of the organisms in aerobic granules, easily biodegradable substrates, such as acetate, should be converted into substrates which are slowly degraded, for instance, intracellular stored polymers (e.g., polyhydroxyalkanoates—PHA).

In the studies previously described in this section, such as that reported in BEUN et al. (2002), a period of feeding in pulse form was applied, in which the substrate was able to penetrate the granule completely, being partially used for rapid growth (40%) and partially stored as intracellular polymer (30–70%). This period was followed by a long phase of aeration in which the organisms grow using the polymers stored intracellularly at a low growth rate. As noted in Sect. 4.4.2, the period with substrate available for growth is called the feast phase, and the period in which the organisms use internally stored substrate is known as the famine phase. A reduction in the oxygen concentration leads to an increase in the duration of the feast phase (substrate present) and, consequently, to a longer period in which the growth rates are higher. In addition, it promotes a stronger O_2 gradient inside the granules which is not desirable when the formation of stable granules is required, as mentioned by PICIOREANU et al. (1998). Both the increase in the period with higher growth rates and the stronger gradient inside the granules lead to the growth of filaments and unstable granular sludge (DE KREUK and VAN LOOSDRECHT 2004).

According to DE KREUK and VAN LOOSDRECHT (2004), in order to obtain stable aerobic granules even under conditions of low oxygen concentrations, the bacterial growth rate needs to be reduced during the total cycle of sequencing batch reactors. This can be achieved through the selection of different types of bacteria which completely convert the easily biodegradable substrates into intracellular stored polymers which are slowly degraded, rather than only 60%, as in the study carried out by BEUN et al. (2002).

This conversion can be carried out by polyphosphate-accumulating organisms (PAO) and glycogen-accumulating organisms (GAO), which are able to convert all of the acetate into PHA (such as polyhydroxybutyrate, PHB) under anaerobic conditions. It is in this context that DE KREUK and VAN LOOSDRECHT (2004), based on previous studies involving theoretical concepts of biofilm morphology which indicate that the biofilm structure is dependent on the shear forces, microorganism growth rate, and substrate diffusion into the biofilm (VAN LOOSDRECHT et al. 1995; PICIOREANU et al. 1998), demonstrated that the selection of PAOs and GAOs in aerobic granules leads to the formation of stable granular sludge, even under low oxygen conditions. To this aim, the authors modified the feeding phase, which was no longer aerobic and in the pulse form, but instead a long period under anaerobic conditions followed by the aerobic reaction phase. In this way, the conditions required for the selected bacteria (PAOs and GAOs) to store all of the substrate inside the cells, without the occurrence of growth, were created, which consequently allowed the formation of stable granules.

Two 3 L reactors were operated, one being an SBAR and the other an SBBC. Both were operated in sequencing batch mode with the only difference being the presence of a riser in the former. In the first experiment, the oxygen concentration was not controlled, and thus it had values close to 100% of air saturation. In the second experiment, the dissolved oxygen concentration used was 40 or 20% of air saturation. Both reactors were submitted to successive cycles of 3 h. Of this total, 60 min was allocated to filling from the reactor base (plug-flow regime through the granule bed under anaerobic conditions), 112 min to aeration, 3 min to settling (to maintain particles with settling velocity greater than 12 m/h in the reactor), and 5 min to the

Table 4.2 Main characteristics of the granules obtained during the different stages of the experiment (DE KREUK and VAN LOOSDRECHT 2004)

	Long feeding period[a]					Pulse feeding regime[b]	
	SBAR				SBBC	SBAR	
Granule characteristics	DO 100%	DO 40%	DO 20%	DO 20%	DO 40%	DO 100%	DO 40%
Dominant organisms	PAO	PAO	PAO	GAO	PAO	Heterotrophs	
Stability	Stable	Stable	Stable	Stable	Stable	Stable	Stable
Local shear	Yes	Yes	Yes	Yes	No	Yes	Yes
Average diameter (mm)	1.3	1.1	1.3	1.2	1.1	1.6	5.0
Density ($gSSV/L_{biomass}$)	89	87	78	108	90	53	13
Solids concentration ($gVSS/L_{reactor}$)	8.5	12	16.5	15	13	5.1	0.9
SVI_8	24	20	14	17	19	50	200
Cell retention (days)	40	67	70	71	63	8	<5

[a]DE KREUK and VAN LOOSDRECHT (2004)
[b]MOSQUERA-CORRAL et al. (2005)

effluent discharge. The HRT and the applied organic load were 5.6 h and 1.6 kgCOD/(m^3 day), respectively. The feeding medium was composed of sodium acetate as a carbon source, NH_4Cl as a source of ammonium nitrogen, and KH_2PO_4/K_2HPO_4 as a source of phosphorous (DE KREUK and VAN LOOSDRECHT 2004).

Table 4.2 summarizes the main characteristics of the granules obtained during the different stages of the experiment carried out by DE KREUK and VAN LOOSDRECHT (2004) employing a long anaerobic feeding period. In addition, for comparison purposes, the properties of the granules obtained by MOSQUERA-CORRAL et al. (2005), who applied pulse feeding, are described.

As can be observed in Table 4.2, in contrast to the results obtained in previous studies in which a completely aerobic feast/famine regime was employed, DE KREUK and VAN LOOSDRECHT (2004) obtained stable granules even at low oxygen concentrations (40 and 20% of air saturation). Furthermore, the characteristics of these granules were similar to those formed at high dissolved oxygen concentrations (100% of air saturation). Filaments were not observed on the surface, and the sludge settleability was not affected, as shown by the low SVI_8 values. The solids retention time and the solids concentration increased with a decrease in the oxygen concentration (DE KREUK and VAN LOOSDRECHT 2004).

Due to the alternating periods of anaerobic feeding (presence of acetate) and aerobic reaction and the availability of phosphorus in the feeding influent, the selection of polyphosphate-accumulating organisms (PAOs) occurred. The acetate was completely converted into intracellular stored polymers (PHB) during the anaerobic feeding period, and the phosphate was released to the liquid medium. During the aerobic period, growth occurred through the use of stored PHB, while the phosphate in the liquid medium was converted to polyphosphate inside the cells.

Therefore, another advantage associated with the selection of PAOs is the possibility to remove phosphate from the wastewater. Ammonium was nitrified during the aeration period, and the nitrate formed was used as an electron acceptor in the anoxic zones located inside the granules. Thus, under DO conditions of 20% of air saturation, removal efficiencies of 100% for COD, 99% for phosphorus, and 90% for nitrogen were observed.

DE KREUK and VAN LOOSDRECHT (2004) also observed, through FISH analysis, that the biomass was mainly composed of PAOs. In contrast, when phosphate was no longer added to the feeding medium, the PAO population disappeared and was replaced by GAOs. These organisms, as previously mentioned, are also able to store acetate under anaerobic conditions. The energy required by GAOs to consume acetate is produced by glycogen hydrolysis. Glycogen reserves are replenished during the aerobic phase from stored PHB, which is also used for cell growth and maintenance. Despite the changes which occurred in the dominant populations, the characteristics of the granules did not show significant modifications.

Another point of interest in relation to the work of DE KREUK and VAN LOOSDRECHT (2004) should be mentioned here. The long feeding period under anaerobic conditions, besides allowing the complete conversion of the substrate (acetate) into intracellular stored polymers (PHB), improving the granule stability under low oxygen conditions, permitting the removal of phosphorus, and promoting better efficiencies for simultaneous nitrification/denitrification (SND), can simplify the operation of sequencing batch aerobic granular sludge reactors. When the duration of the feeding phase of N reactors represents 1/N of the cycle time, a continuous influent flow can be obtained. The plug-flow regime through the granular sludge bed allows a simultaneous feeding/discharge period. Approximately 90% of the void volume of the settled bed can be simultaneously replaced with the addition of influent, and thus the effluent discharge phase becomes unnecessary. In this operation mode, the influent and effluent flows become continuous, which greatly facilitates the design of the pre- and posttreatment installations (DE KREUK and VAN LOOSDRECHT 2004). It also resolves some difficulties encountered on scaling up the operation, such as the need for excessively large pumps or equalization tanks.

4.6 Conversion Processes in Aerobic Granules

Granular sludge technology has the great advantage that it allows the removal of COD and nutrients (nitrogen and phosphorous) in a single reactor. The operation of this reactor, as described in Sect. 4.4.2, is preferentially carried out in the sequencing batch mode. In order to obtain the simultaneous removal of these pollutants, some operational conditions must be ensured.

The mechanisms involved in nutrient removal with aerobic granules are basically the same of that observed for the conventional activated sludge process. The main difference is that, in the former technology, biological processes do not occur in different tanks but instead simultaneously in different regions inside the granules.

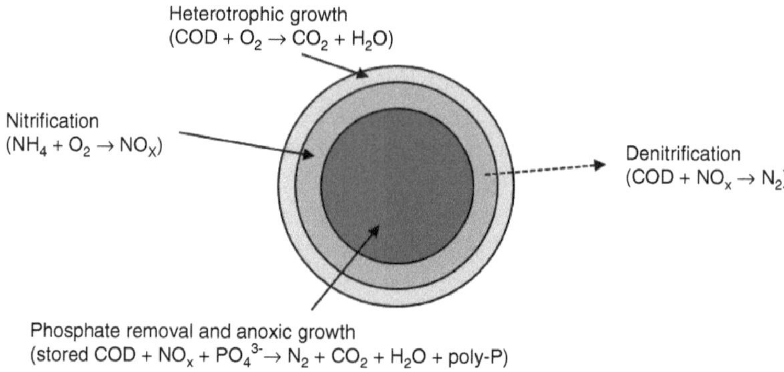

Fig. 4.6 Schematic representation of the different layers of the aerobic granules (adapted from DE KREUK 2006)

In this regard, the measurement and estimation of the capacity of specific conversions is complicated compared to suspended biomass. The most appropriate experimental conditions and methods to determine specific ammonium, nitrite, and phosphate uptake rates under normal operation of aerobic granular sludge reactors have been previously evaluated and are described by Bassin et al. (2012b). These authors proposed methodologies that may serve as an experimental frame of reference for investigating the metabolic capacities of microbial functional groups in aerobic granular sludge processes.

The simultaneous nitrification/denitrification process (SND) is an important mechanism which occurs in aerobic granules. The occurrence of this process is related to the presence of an external aerobic zone in the biofilm/granule for nitrification and an internal anoxic zone for denitrification (POCHANA and KELLER 1999). The distribution of heterotrophic microorganisms in the granules has a strong influence on the process. Since the removal of organic matter and nitrification/denitrification occur simultaneously in aerobic granular systems, there is competition between the heterotrophic and autotrophic microorganisms for space and oxygen. The former are located in the external layers since they have high growth rates, while the latter, mostly slow-growing organisms, are confined to deeper regions, where oxygen availability is limited (VAN LOOSDRECHT et al. 1995). Figure 4.6 shows a schematic representation of the structure of an aerobic granule comprised of an aerobic layer and an anaerobic/anoxic layer, as well as the different processes which occur in these layers.

During the feast regime, the concentration of the external carbon source is high, and this substrate will be completely dispersed in the interior of the granules and will be stored anaerobically (by PAOs or GAOs), aerobically, or anoxically (by other heterotrophic organisms) in the form of intracellular polymers (e.g., polyhydroxyalkanoates—PHA). The dissolved oxygen, used for growth and substrate storage, is rapidly consumed by the heterotrophic microorganisms located in the external layers of the granules. The dissolved oxygen is also consumed by autotrophic microorganisms which, being dependent on this element for nitrification, are

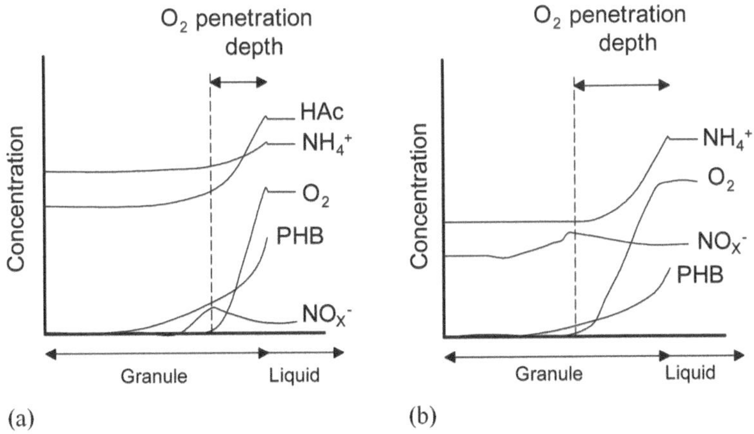

Fig. 4.7 Concentration profiles for acetate (HAc), PHB, NH_4^+, NO_x^- ($NO_2 + NO_3$), and O_2 inside the granules during the feast (**a**) and famine (**b**) periods of a sequencing batch reactor (SBR)

located in the aerobic layers of the granules, converting ammonium into nitrite and nitrate. During the feast period, the penetration of oxygen into the granules is limited. The storage of substrate (acetate) in the form of PHA and the growth occur aerobically in the external layer or anaerobically/anoxically inside the granules. It should be noted that during the feast phase of the operation cycle in an SBR, denitrification of NO_x (NO_2 and NO_3) compounds remaining from the last cycle can occur through the use of the external carbon available during this period. This process is indicated by the dotted line in Fig. 4.6.

During the period when the external carbon source is exhausted (famine phase), only the substrate stored inside the cells, in the form of PHA, is available to the granules. This substrate is used by the microorganisms for growth, which occurs much more slowly compared with the feast period (BEUN et al. 2001). The penetration of oxygen during the famine phase is greater, due to the reduction in the respiration rate of heterotrophic organisms during this period, although it continues to be limited due to the increase in the nitrification rate (THIRD et al. 2003). Heterotrophs have a lower requirement for oxygen in the famine phase, during which they grow using intracellularly stored PHA. Consequently, oxygen is no longer limiting for nitrifying bacteria (BEUN et al. 2001), and nitrification rate increases. The nitrate produced during nitrification can be simultaneously denitrified inside the granules using the stored substrate (PHA) as a carbon source (electron donor). Good nitrogen removal efficiencies are achieved when the aerobic and anoxic regions are well balanced during the aerobic phase (BEUN et al. 2001). The volume of these regions is influenced by the oxygen concentration in the bulk and by the granule size. Fig. 4.7 shows the theoretical concentration profiles for acetate (HAc), PHB, NH_4^+, NO_x^- ($NO_2 + NO_3$), and O_2 inside the granules during the feast (a) and famine (b) periods of a sequencing batch reactor (SBR).

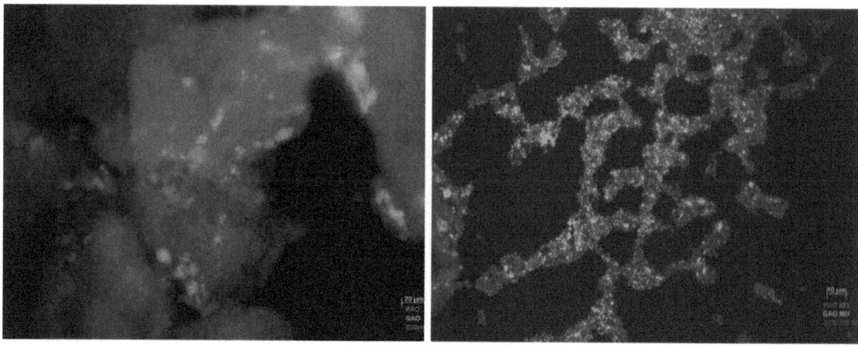

Fig. 4.8 Predominance of PAOs and GAOs in aerobic granules cultivated in a bubble column reactors. The *red* represents PAOs (PAOmix) and the *green* represents GAOs (GAOmix). The *red* and *green* are slightly modified due to the superposition of *blue*, which represents the entire bacterial population (EUBmix)

Studies on the formation of aerobic granules, described in Sect. 4.4.5, present differences regarding the operation mode of the reactors. This led to the formation of granules with different properties. Also, the operational conditions employed influenced the microbial conversion processes.

To exemplify this, when phosphorus removal is to be achieved, the discontinuous granular sludge reactor must be fed, under anaerobic conditions, for a relatively long period, favoring the development of PAOs. The adoption of a long anaerobic feeding regime, together with the creation of conditions for the growth of PAOs (and eventually GAOs), results in the formation of dense stable granules even in the presence of low dissolved oxygen concentrations (DE KREUK and VAN LOOSDRECHT 2004), as discussed in Sect. 4.4.5. In addition, regarding the application of aerobic granulation technology on a large scale, it can be assumed that sequencing batch reactors fed with a pulse regime in which the influent wastewater is added almost instantaneously, represent an economically and technically unrealistic option.

In order to illustrate the presence of the abovementioned organisms, Fig. 4.8 shows the presence of polyphosphate-accumulating organisms (PAOs) and glycogen-accumulating organisms (GAOs) in aerobic granules cultivated in different bubble column reactors, in laboratory scale, fed with acetate as carbon source, NH_4Cl as nitrogen source, and KH_2PO_4/K_2HPO_4 as phosphorus source. The detection of these microorganisms was carried out by fluorescence in situ hybridization (FISH), using the specific probes PAO 462/PAO 651/PAO 846 (represented by PAOmix in Fig. 4.8) and GAOQ431/GAOQ989 (represented by GAOmix in Fig. 4.8) (CROCETTI et al. 2000, 2002). Figure 4.8a shows the predominance of PAOs, as observed by FISH of biomass samples enriched with white colored granules. On the other hand, the dominance of GAOs in other granular sludge sample is illustrated in Fig. 4.8b. This sludge sample, in particular, was obtained in a reactor where phosphate was suppressed from the influent medium to allow the enrichment of GAOs.

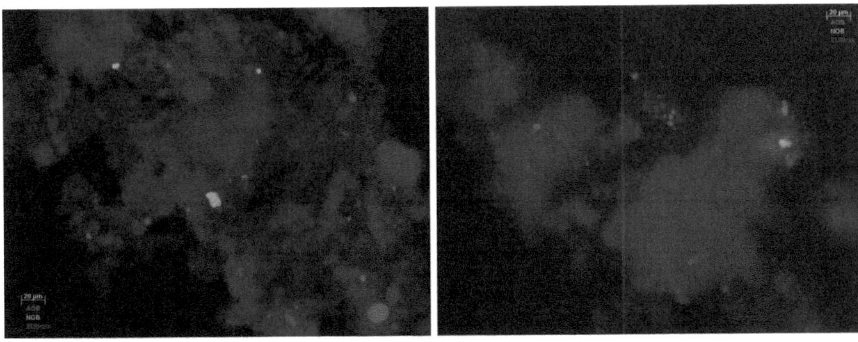

Fig. 4.9 Detection of AOB and NOB in aerobic granular sludge achieving simultaneous COD, nitrogen, and phosphate removal. The *red* indicates AOB, whereas the *green* indicates NOB. The *red* and *green colors* are slightly modified due to the superposition of *blue*, which represents the entire bacterial population (EUBmix)

The presence of GAOs in enhanced biological phosphorus removal (EBPR) systems is undesirable, since these microorganisms compete with PAOs for the substrate (volatile fatty acids) under anaerobic conditions but do not contribute to the removal of P-PO_4^{3-}. Operational strategies which favor the development of PAOs to the detriment of GAOs should be taken into consideration when the aim is to increase the phosphorous removal capacity (OEHMEN et al. 2006). Further details regarding the factors affecting the competition between these two microbial populations can be found elsewhere (LOPEZ-VAZQUEZ et al. 2008, 2009).

It should be noted that the application of the biological phosphorous removal can also simplify the simultaneous nitrification/denitrification (SND) process. In most biofilm systems consisting of heterotrophic and autotrophic organisms, the former dominate in the external layers since they compete more successfully with nitrifiers for dissolved oxygen and space in the biofilm. This distribution is not favorable for SND in continuous systems, bearing in mind that most of the organic components will be consumed in the external aerobic regions and thus cannot be used as electron donors for the denitrification in the interior of the biofilm. In addition, this type of system is more sensitive to oxygen concentrations, since heterotrophic organisms grow much more rapidly than nitrifying organisms. This does not occur when polyphosphate-accumulating organisms are selected, as these bacteria can coexist with nitrifiers in the same external layer of the granules due to their similar growth rates (BRDJANOVIC et al. 1998), as observed by DE KREUK et al. (2005b).

Figure 4.9 shows the occurrence of nitrifiers, both ammonium-oxidizing bacteria (AOB) and nitrite-oxidizing bacteria (NOB), in aerobic granules cultivated in a laboratory-scale bubble column reactor, as observed by FISH. In this case, the probes used were Nso1225/Nso190/Neu653 (specific for AOB), Ntspa662/Nit1035 (specific for NOB), and EUB 338mix for general bacteria. The references for the probes are given in Chap. 6.

In Fig. 4.9 it can be observed that population of nitrifying bacteria (AOB/NOB) represents a very small part of the total bacterial population. This result is consistent with the fact that in most wastewater treatment plants, the nitrifying autotrophic microorganisms correspond to a small part of the total bacterial population. However, despite their almost negligible contribution, these microorganisms are responsible for most of the dissolved oxygen consumption. In systems with on-line monitoring of the DO concentration, an increase in its concentration is clearly observed when the nitrification process ends, that is, when all of the ammonium has been oxidized to nitrate.

It should be noted that FISH analysis was carried out only in order to confirm the presence or absence of the abovementioned microbial groups and the position of the microorganisms illustrated in Fig. 4.9 does not correspond to their location inside the granules. In this case, a procedure in which the granules were crushed was adopted in order to facilitate the hybridization of cells to be later observed by FISH. The observation of the actual position occupied by the microorganisms in the granular sludge can be carried out with a confocal laser scanning microscope, which allows optical sections of the material to be analyzed, which are then grouped to construct tridimensional images. In the specific case of aerobic granules, successive optical sections allow the exact location of the microbial consortiums responsible for the different conversion processes to be determined.

In contrast to the nitrifiers and some heterotrophic bacteria, the capacity of PAOs to use nitrate as electron acceptor (in this case they are referred to as denitrifying polyphosphate-accumulating organisms, DPAOs), combined with their ability to store organic compounds (PHB) within the cell, enables their presence at the center of the granules (maintained under anoxic conditions).

It is also important to consider that to favor the simultaneous nitrification/denitrification process, the oxygen concentration must be selected so as to facilitate the occurrence of both processes. Low oxygen concentrations lead to an increase in the anoxic volume of the granules, rich in organisms able to store polymers, increasing the denitrifying capacity. Under these conditions, however, the aerobic external layer in which the autotrophic organisms are active is reduced, which may lead to a decrease in the nitrification efficiency. Also, in many cases, it is not possible to maintain stable operation when the oxygen concentration is lowered to below a certain critical value, notably due to the disintegration of the granules and their consequent washout from the reactor (MOSQUERA-CORRAL et al. 2005). On the other hand, high oxygen concentrations lead to a decrease in the anoxic regions inside the granules due to the complete penetration of oxygen, consequently reducing the denitrifying capacity.

Thus, it is clear that the ratio between the volumes of the aerobic layer and the anoxic central layer of the granules is important and determines the efficiency of the SND process. The greater the concentration of dissolved oxygen, the greater the oxygen penetration depth under the same oxygen consumption rate, which leads to a smaller anoxic volume inside the granules. Therefore, the size of the granules is an important variable in terms of the reactor operation, although it is difficult to control. Fig. 4.10 represents schematically the aerobic volume (external region) and

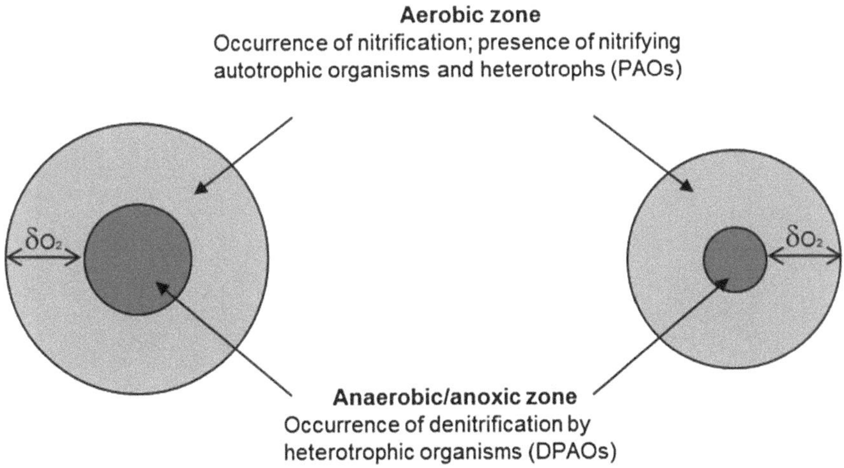

Fig. 4.10 Reduction in the anaerobic or anoxic zone with a decrease in the granule diameter at constant dissolved oxygen concentration (adapted from DE KREUK et al. 2005b)

the anoxic volume (internal region) of granules with different diameters submitted to the same oxygen concentration.

DE KREUK et al. (2005b) observed that granules with smaller diameters lead to a decrease in the nitrogen removal efficiency, indicated by an increase in the nitrate concentration in the effluent. On the other hand, ammonium oxidation was not affected by the different sizes of the granules. The highest nitrogen removal was obtained with granules larger than 1.3 mm. However, for diameters larger than 1.7 mm, the granules began to disintegrate, forming small irregular-shaped particles, with a small effective anoxic zone. In fact, the process conditions are not the only factors which influence the granule diameter. The interior of large granules tends to destabilize due to endogenous respiration, and the granules break into small fractions which can grow again to form new granules (DE KREUK et al. 2005b). In Sect. 4.4.7, the influence of the oxygen concentration on the SND process as assessed in some literature studies will be discussed.

Table 4.3 shows the nitrification and nitrogen removal efficiencies obtained by different researchers, who employed different operating conditions. It can be observed that the results obtained are closely related to parameters such as the granule size, oxygen concentration, and COD/N ratio. The low nitrogen removal efficiencies obtained by QIN and LIU (2006) and MOSQUERA and CORRAL et al. (2005) in comparison with those obtained by KIM et al. (2004) are associated with the fact that the former authors used higher dissolved oxygen concentrations, a condition which is not favorable for denitrification. Certain special circumstances, which are dependent on the granule diameter, DO concentration and COD/N ratio, can lead to the formation of microbial aggregates which are rich in substrate, with an interior region where no oxygen is available due to diffusion limitation. In this case, ideal conditions for denitrification are created.

Table 4.3 Nitrification and nitrogen removal efficiencies in granular sludge systems submitted to different operating conditions

Authors	COD/N	DO (mg/L)	*T* (°C)	Particle diameter (mm)	Organic load (gCOD/(L day))	Nitrogen load (gN/(L day))	Nitrification efficiency (%)	Nremoval efficiency (%)
KIM et al. (2004)	21	≥2	25	1.0–2.0	2.5	0.12	97	97
QIN and LIU (2006)	4.4–13.3	≥4	25	0.8–4.6	2	0.15–0.45	≥99	24–32
MOSQUERA and CORRAL et al. (2005)	8.3	8	20	1.6	1.6	0.19	100	16
TSUNEDA et al. (2003)	0	≥2	–	0.36	0	Up to 1.5	100	–
KIM and SEO (2006)	0	≥2	20–25	0.3–0.5	0	Up to 2.5	100	–

The aforementioned operational aspects are only some of the factors which strongly influence the treatment of wastewaters by aerobic granules. In fact, many other factors must be taken into consideration when aerobic granules are used for the treatment of a certain wastewater. The composition of the different substrates and their gradients along the structure of the granules, the operating temperature, the reactor configuration, and the pH comprise a significant group of factors which affect the diffusion coefficient, conversion rates, granule size, spatial distribution of the biomass, and density. These, in turn, strongly influence one another, and the effect of each one individually is difficult to be determined experimentally.

Computational models can be used to efficiently evaluate the relative importance of the individual parameters in the biofilms/granules and provide the information on the main factors affecting the treatment performance and the distribution of the different microbial populations in the granules. These models can also serve as a tool to improve the design of aerobic granular sludge reactors. The work by BEUN et al. (2001), which will be described in Sect. 4.4.7, provides an example of the use of models together with experimental data to investigate the effects of some operational parameters.

It should be mentioned here that the assessment of the conversions which occur in the granules by means of mass balances can be complicated, particularly due to the fact that the reactors used operates in sequencing batch mode. In these discontinuous systems, interpreting the results and reporting concentrations should be carried out with caution. For instance, the effluent which is not removed because it is below the effluent discharge port dilutes the influent. In addition, the nitrite or nitrate which remains from the previous cycle is firstly denitrified directly, using external organic carbon originating from the feed. The amount of nitrogen removed must therefore be related to the nitrogen concentration present in the influent, the nitrogen remaining from the previous cycle, and the quantity removed during the aeration phase. The nutrient balance of these reactors should ideally take into consideration the composition of the gas formed during the process.

Bassin et al. (2011) have studied an important factor which can influence the nitrogen mass balance calculations in aerobic granular sludge processes. These authors observed the occurrence of an ammonium adsorption phenomenon inside the granules in laboratory- and pilot-scale aerobic granular sludge reactors. Adsorption tests have shown that ammonium adsorption in aerobic granular sludge can be considerably higher than that occurring in activated sludge and Anammox granular sludge. For mass balancing over a treatment plant, this is no real problem. Under steady-state conditions, adsorption does not make a difference to the ammonium effluent concentrations. However, a significant error can be introduced in the calculations when ammonium adsorption is neglected in granular sludge bioreactor systems that are characterized by strongly variable ammonium concentrations as a function of place (plug-flow systems) or time (batch systems).

4.7 Application of Aerobic Granules in Wastewater Treatment: From Laboratory Studies to Full-Scale Experiences

Granular sludge technology can be applied for the removal of a wide variety of organic compounds and nutrients, such as nitrogen and phosphorus. The great advantage associated with the use of aerobic granules combined with operation of the reactors in sequencing batch mode is that the treatment can be carried out in a single tank. As mentioned in Sect. 4.4.6, the COD removal, nitrification, and denitrification processes can be carried out simultaneously due to the presence of aerobic, anoxic, and/or anaerobic regions in the granules.

BEUN et al. (2001) investigated the effect of the dissolved oxygen (DO) concentration on the performance of a sequencing batch airlift reactor (SBAR) containing granular sludge, particularly with regard to the nitrogen removal. The authors used both experimental data and results obtained from the model developed. The latter were used not only to describe the effect of low DO concentrations on nitrogen removal but also to predict the effect of several process conditions on the nitrogen removal efficiency.

The reactor was operated for 142 days. On day 42, a small quantity of nitrifying sludge was added to accelerate the accumulation of nitrifying microorganisms in the reactor. The feed consisted of sodium acetate as a carbon source (18.3 Cmmol/L) and ammonium chloride as a source of ammonium (40 mgN-NH_4^+/L). The performance of the SBAR was described by a model obtained in Aquasim (*apud* BEUN et al. 2001). The model was used to simulate the processes which occurred in the reactor under the same conditions applied in the laboratory experiments and to obtain some information on the physical distribution of the autotrophic and heterotrophic biomass within the granules (BEUN et al. 2001).

It was observed that the nitrification, denitrification, and COD removal can occur simultaneously in a granular sludge sequencing batch reactor. During the feast phase (period with acetate available), the concentration of NH_4^+ decreased slowly, and the amount of NO_x ($NO_2 + NO_3$) remaining from the previous cycle was denitrified with acetate. After the acetate had been completely consumed (beginning of the famine period), the NH_4^+ concentration immediately began to decrease more rapidly in comparison to the feast regime, due to the occurrence of the nitrification process. The NO_x concentration increased during this period. From the nitrogen mass balance, it was observed that denitrification occurred during the famine period. When all of the NH_4^+ had been converted, the NO_x concentration did not decrease. This implies that, after this period, denitrification no longer took place. According to the authors, denitrification did not continue due to the complete penetration of oxygen into the granules after the nitrification had been completed. At DO concentration corresponding to 100% of air saturation, it was observed that 87% of the NH_4^+ was removed via nitrification, and the rest was used for biomass growth. Around 70% of the NO_x formed by nitrification was denitrified (BEUN et al. 2001).

In order to improve the denitrification efficiency, from day 66 to 71 of reactor operation, the DO concentration was reduced to 50% of air saturation 40 min after the beginning of each cycle until its end. This concentration was reached by mixing air and nitrogen gas. With this procedure an increase in the denitrification was observed, although ammonium was detected in the effluent due to DO limitation (BEUN et al. 2001).

On days 79, 100 and 142, the DO concentration was decreased to a constant value of 20% saturation 40 min after the beginning of a certain cycle until its end. The NH_4^+ concentration in the effluent increased, and the NO_x concentration in the effluent decreased. Under these conditions, 83% of the NH_4^+ was converted to NO_x by nitrification, and 100% of the NO_x formed was denitrified. It was clearly demonstrated that on decreasing the DO concentration to 20% of air saturation, the nitrification rate decreased and the denitrification potential increased substantially (BEUN et al. 2001).

Bearing in mind that the model provided a good description of the experimental data obtained at DO of 100% and 20% of air saturation, it was used to evaluate the effect of DO on the nitrification and denitrification, varying the concentration of this parameter within this range (20–100%). The nitrogen balance during one cycle was used to calculate the nitrogen removal efficiency of the reactor. The authors clearly observed that the increase in the DO concentration contributed to improving the nitrification efficiency and reducing the NH_4^+ concentration, although it led to a decrease in the denitrification efficiency, increasing the NO_3^- concentration. The maximum nitrogen removal was obtained at DO of 40% air saturation (BEUN et al. 2001).

In another simulation, the performance of the SBAR at DO of 40% air saturation was studied in greater detail, since under this condition the maximum nitrogen removal percentage was observed. The feast period was more than twice as long as the period of operation with DO of 100% air saturation, and the oxygen penetration during this period was low. In the case of acetate, there was complete penetration in the granules, regardless of the DO concentration. Since oxygen was not present in the center of the granule during the feast period, and with the availability of NO_3^- originating from the previous cycle, the acetate could be stored in the central region of the granules, under anoxic conditions, in the form of PHB (BEUN et al. 2001). Although the concentration of PHB was low in the center of the granules compared with the external layers, the presence of NO_3^- in the famine period and the low oxygen concentration promoted the occurrence of denitrification in the central region of the granules, where stored polymers were used as carbon source (BEUN et al. 2001).

At DO of 40% air saturation, NH_4^+ was not completely removed during the cycle, the concentration of NO_3^- in the effluent was lower than during SBAR operation at 100% air saturation, and the nitrogen removal efficiency was 80%. At this DO concentration, the depth of oxygen penetration was small, and the occurrence of nitrification was limited to the external layers of the granules. An internal anoxic zone was established in the center of the granules, favoring denitrification. The authors clearly demonstrated that the DO concentration has a notable effect on the nitrogen removal. With an increase in the DO concentration, nitrification was favored. On the other hand, the same increase in DO had an adverse effect on denitrification. Thus, an ideal balance between nitrification and denitrification needs to be established through controlling the DO concentration in the reactor to ensure maximum nitrogen removal (BEUN et al. 2001).

In another simulation, the nitrification was considered to be inactive during the operational cycle, and the effect of nitrification on the oxygen penetration depth was observed. In the simulation without the occurrence of nitrification, the amount of nitrate remaining from the preceding cycle was denitrified during the feast period. It was also observed that, even during the feast period of the standard operational cycle (DO of 100% air saturation), the removal of NH_4^+ occurred mainly due to biomass growth, given that the ammonium removal rate was the same as that of the cycle in which nitrification was neglected by the simulation. This finding is associated with the fact that during the feast phase, the fast-growing heterotrophic microorganisms responsible for the degradation of acetate dominate the competition with nitrifiers for oxygen. From the beginning of the famine phase of the standard cycle, NH_4^+ removal by nitrification increased (BEUN et al. 2001).

During the feast period, the penetration depth was relatively small for the simulations both with and without nitrification. Once the acetate had been completely consumed in the simulation with nitrification, the depth of oxygen penetration was 350 μm while NH_4^+ was present in the liquid medium. Under these circumstances, the oxygen was used for nitrification. When ammonium had been totally consumed, the oxygen penetrated completely into the interior of the granules. In the simulation in which the occurrence of nitrification was neglected, when acetate was no longer available (end of feast period), the oxygen was able to completely penetrate into the granular biomass, since it was not used for nitrification. These results obtained by the model suggest that the activity of nitrifying bacteria is responsible for the fact that the oxygen does not reach the center of the granules, favoring the occurrence of denitrification in this region (BEUN et al. 2001).

The authors also investigated the effect of the storage and degradation of PHB on the denitrification, by performing a simulation in which the accumulation and degradation of this compound are considered to be inactive during the operation cycle of an SBAR. In the standard simulation (with PHB), growth of the granules occurred throughout the cycle. During the feast period, the biomass grew due to acetate degradation, while during the famine period, the growth was on PHB (BEUN et al. 2001). The duration of the feast period in the simulation without PHB was almost twice that of the standard simulation with PHB. The growth of the granules occurred only during the feast phase, although the quantity of biomass produced was much greater in this period than during the whole cycle of the standard simulation. The removal of NH_4^+ was greater under conditions without PHB compared with the feast period under conditions with PHB due to the higher biomass growth rate of heterotrophs in the former case, although nitrification did not occur during this period. After all of the acetate had been consumed (beginning of famine period), the removal of NH_4^+ was attributed only to nitrification, and from this period onward biomass growth did not occur (BEUN et al. 2001).

With regard to the role of PHB in denitrification during the famine period, in the simulation without this intracellular polymer, it was observed that all of the NH_4^+ removed was converted into NO_3^-, that is, denitrification did not occur. In the famine period of the simulation with PHB, 50% of the NH_4^+ was nitrified to NO_3^-, indicating that the denitrification process occurred. In the latter case, the PHB stored in the center of the granule can be used as a carbon source for denitrification when an external carbon source (acetate) is no longer available, so long as NO_3^- (originating from nitrification) is present as an electron donor and not oxygen (BEUN et al. 2001).

BEUN et al. (2001) also noted the fact that, in systems with continuous operation, no storage and degradation of organic substrate occurred, since its concentration in the reactor is always low, preventing the occurrence of feast and famine periods. According to the authors, in these systems the acetate penetration depth is always much lower than the respective depth for oxygen, and thus anoxic conditions need to be introduced to promote denitrification which, in turn, requires the reactor operation mode to be adjusted. In addition, an external carbon source should be added for the denitrification process, given that intracellular polymers are not available in these continuous systems (BEUN et al. 2001).

As regards the distribution of the microbial community in the structure of the granules, the authors also speculate that in a system fed in a discontinuous form, for instance, SBAR, the organic substrate (in this case acetate) completely penetrates the biofilm or granule due to the high concentration in the liquid medium, particularly at the beginning of the feast period. Since the oxygen is present only in the external layers of the biofilms, it is expected that nitrifying bacteria will be located in these regions where oxygen is available, and that the acetate is stored in the form of PHB by the heterotrophic population, under anoxic conditions, in the interior regions of the biofilms/granules (BEUN et al. 2001).

The stratification of the granule structure was also predicted by the simulation carried out by BEUN et al. (2001). The autotrophic biomass was located mainly in the external region of the granules. The location of the heterotrophic community was less well defined. These organisms were present both in the central regions of the granules (using acetate for growth during the feast period and using NO_3^- as an electron acceptor) and in the external layers of the granules (using the oxygen available as an electron acceptor).

The simulation at DO of 40% of air saturation showed that, in the long term, the DO influenced the microbial distribution in the granular sludge. Under these conditions, a greater fraction of the autotrophic biomass was located in the aerobic external region of the granules in comparison with condition in which DO corresponded to 100% air saturation. This situation is explained considering that the nitrifying biomass needs oxygen for the conversion of NH_4^+. The heterotrophic population can use either oxygen or NO_3^- as the electron acceptor and can thus be present also in the center of the granules. It was observed that the exact location of the autotrophic biomass influenced the nitrogen removal and the distribution of these organisms is, in turn, affected by the concentration of dissolved oxygen in the reactor (BEUN et al. 2001).

DE KREUK and VAN LOOSDRECHT (2004) showed that the selection of polyphosphate-accumulating organisms and glycogen-accumulating organisms (through adopting an anaerobic feeding regime) promoted an improvement in the stability of aerobic granules at low oxygen concentrations. Based on this previous study, DE KREUK et al. (2005b) studied the factors which are important in order to obtain the simultaneous removal of nitrogen and phosphorus in aerobic granular sludge reactors.

The operational conditions used by DE KREUK et al. (2005b) were the same as those applied in an earlier study (DE KREUK and VAN LOOSDRECHT 2004), as described in Sect. 4.4.5. In the search to identify the main conversion processes which occurred at different dissolved oxygen concentrations, the authors observed that after 52 days of operation under saturated oxygen conditions, complete consumption of acetate occurred during the anaerobic period. After 65 days, phosphate removal of 95% was achieved, considering that the initial concentration was 20 mgP/L. On days 65 and 233 of operation, the average concentration of phosphate released into the liquid medium during the anaerobic feeding phase was 86 mgP/L, while the effluent concentration was only 0.4 mgP/L. The ratio between the phosphate released and the acetate consumed was 0.44 P-mol/C-mol, which is close to that obtained at pH 7 for a culture highly enriched with PAOs, i.e., 0.55 (SMOLDERS et al. 1994 *apud*).

During the first 2 days after the reactor start-up, ammonium-oxidizing organisms were inhibited with allylthiourea (ATU) in order to suppress nitrification. As a consequence, the presence of nitrate was prevented during the anaerobic feeding period, and therefore the development of PAOs was not affected. The complete oxidation of ammonium was observed 39 days after the addition of ATU had been stopped, although it took around 100 days for all of the nitrite to be oxidized. After 154 days of operation, ammonium and nitrite were no longer detected in the effluent. Due to the operation at oxygen saturation of 100%, incomplete denitrification occurred, and total nitrogen removal amounted to 34% (27% due to the biomass growth) (DE KREUK et al. 2005b).

The concentration of oxygen was lowered to 40% of air saturation in order to improve the denitrification efficiency, decreasing the oxygen penetration depth and, as a consequence, the aerobic volume of the granules. This change did not affect the phosphate removal, which remained at around 97%. The acetate continued to be totally consumed during the anaerobic period, while the average ratio between the phosphate released and acetate consumed showed no variation (0.45 P-mol/C-mol). The denitrification efficiency increased slowly during this period, with concentrations below 5 mgN-NO_3/L being measured in the effluent 64 days after operation at DO of 40% air saturation. Ammonium and nitrite were completely oxidized, and the average nitrogen removal was 98% (DE KREUK et al. 2005b).

Subsequently, the authors observed that a change in the morphology of the granules occurred, and they changed from spherical to irregular-shaped particles, with fissures in the direction of their centers. One of the consequences was a reduction in the nitrogen removal efficiency, which dropped to values between 50 and 70%. In order to improve the removal of nitrogen, the dissolved oxygen concentration was further decreased, from 40 to 20% of air saturation. The phosphate removal remained high (94% on average) during the first 90 days of operation under these conditions. However, later the phosphate concentration began to increase, probably due to the low biomass yield resulting from the high solids retention time applied (71 days) (DE KREUK et al. 2005b). In order to completely remove the phosphate, the sludge age needs to be maintained at relatively low values. If this condition is not respected, there will not be efficient removal of phosphate, since the only way of removing this nutrient is by removing biomass rich in polyphosphate (i.e., polyphosphate-accumulating organisms) in appropriate proportions.

The concentrations of nitrogen compounds ($N\text{-}NH_4^+$, $N\text{-}NO_2^-$, and $N\text{-}NO_3^-$) did not change immediately after the decrease in oxygen concentration from 40 to 20% of air saturation. After 30 days of operation under this condition, a drop in the nitrogen efficiency was observed as a result of the reduction in the external aerobic layer where nitrifiers were present, although the capacity of these microorganisms was reestablished over time. During operation in the steady-state at DO of 20% of air saturation, in which the granule size was greater than 1.3 mm, the anoxic volume containing denitrifying polyphosphate-accumulating organisms (DPAOs) was large enough to allow denitrification. The highest nitrogen removal efficiency obtained was 94% (DE KREUK et al. 2005b).

DE KREUK et al. (2005b) also evaluated the short-term effect of both an increase and a decrease in the dissolved oxygen concentration on the performance of the reactor. The experiments were carried out during operation under steady-state conditions at DO of 40% of air saturation. This concentration was modified during the cycle to 100, 40, and 10% air saturation. The phosphate consumption rate decreased with a reduction in the oxygen concentration, due to the reduced oxygen penetration depth, resulting in an increase and decrease in the anoxic and aerobic volumes, respectively. It was also observed that, in the absence of nitrate (obtained through the inhibition of nitrification by allylthiourea), the phosphate consumption rate decreased by around 45% in relation to the maximum obtained during the operation at 100% air saturation with nitrification/denitrification. The highest nitrogen removal efficiency recorded in this experiment was obtained at 40% air saturation, which corresponds to the long-term DO concentration in which the granules were cultivated.

The same authors also observed that with high phosphate concentrations in the influent (19.6 $mgP\text{-}PO_4^{3-}/L$; COD/P = 20.2) and with a low substrate to cells conversion factor (0.25 gVSS/gCOD), the value calculated for the P content in the granules was 0.20 gP/gVSS. The amount of P in the sludge reported in the literature is usually lower, varying from 0.02 to 0.14 (FALKENTOFT 2000). Thus, it was assumed that the phosphate removal could be partially attributed to precipitation occurring inside the granules. This can be confirmed by the increase in the fixed solids (ash) content of the granules enriched with PAOs (representing 30–41% of the total solids) under anaerobic feeding conditions, compared to that observed for the granular biomass from the reactor operated under aerobic conditions and subjected to pulse feeding (where ash corresponded to 6% of the total solids). In addition, most granules were white, although some were light brown. These latter granules had a fixed solids content of 17.2%, while for the white granules, the corresponding value was 50.4%, suggesting the presence of precipitates in the white biomass (DE KREUK et al. 2005b). It should be noted that the precipitation inside the granules may be advantageous since, besides favoring the removal of phosphate, it makes the granules heavier, increasing their settling velocity.

MOSQUERA-CORRAL et al. (2005), as discussed in Sect. 4.4.5, evaluated the best operating conditions for the removal of nitrogen in aerobic granular systems. To this aim, they observed the short- and long-term effects of a decrease in the oxygen concentration on the performance of a sequencing batch airlift reactor (SBAR). The authors obtained complete removal of ammonium after 120 days of operation at DO of 100% air saturation. Under these conditions, the average nitrogen removal efficiency was 16%.

When the lowest oxygen concentration (40% air saturation) was applied, the denitrification efficiency increased, and the NO_x compounds (NO_2 and NO_3) were almost completely denitrified during the first days of operation under these conditions. These results are related to an increase in the anoxic volume of the granules with the decrease in the oxygen concentration, increasing the denitrification capacity. In contrast, the nitrification efficiency was reduced, and thus ammonium was present in the effluent. This decrease could be related to the competition, for oxygen, between the ammonium-oxidizing microorganisms and the heterotrophic microorganisms in the external layer of the granules. An interesting result obtained by MOSQUERA-CORRAL et al. (2005) was that the aerobic biomass volume in the external layer of the granules reduced to a lesser degree than the ammonium oxidation rate. When the oxygen concentration was reduced from 100 to 50% air saturation, for instance, the aerobic volume decreased by 27% while the ammonium oxidation rate dropped by 47%. These data indicate a structure composed of several layers within the aerobic zone with the heterotrophic organisms growing further toward the outside than the nitrifying bacteria. According to VAN LOOSDRECHT et al. (1995), this stratification is mainly caused by the differences between the growth of the two populations of microorganisms and due to the competition for oxygen. The decrease in the nitrification rate did not result in an overall reduction in nitrogen removal. The average nitrogen removal was 63% at an oxygen concentration of 40% air saturation, indicating that most of the nitrite and nitrate was denitrified under these conditions.

MOSQUERA-CORRAL et al. (2005), as also described in Sect. 4.4.5, restarted the operation of the system with new inoculum and applied an oxygen concentration of 40% of air saturation. Under these conditions, the volumetric acetate consumption rate was very low due to the long feast period of 160 min. Nitrification was not observed since the nitrifying bacteria were not able to develop in this unstable system. The formation of stable granules was not possible in this system, and therefore the reactor operation was stopped.

In experiments carried out to study the short-term effect of a decrease in the dissolved oxygen concentration on nitrogen removal, MOSQUERA-CORRAL et al. (2005) observed that reducing the DO from 100 to 50, 40, 20, or 10% of air saturation did not affect the acetate consumption rate, since the duration of the feast phase was similar at all DO concentrations tested. The ammonium oxidation decreased, and the nitrate production increased when the oxygen concentration was lowered. At oxygen concentrations below 20% of air saturation, the ammonium concentration in the effluent increased, indicating that complete nitrification was not achieved. NO_x compounds were denitrified during the feast phase using acetate as the carbon source. During the famine phase of the cycles with high oxygen concentration (40–100% of air saturation), denitrification did not occur. On the other hand, at low DO concentrations, this process did occur during the famine phase, using PHB stored intracellularly as electron donor. The nitrogen removal percentage increased with a decrease in the oxygen concentration, reaching a maximum value of 34.5% when employing a DO level of 10% air saturation.

The main biological conversions taking place in laboratory-scale aerobic granular sludge sequencing batch reactors have been evaluated by Bassin et al. (2012a). Two bioreactors carrying out simultaneous COD, nitrogen, and phosphate removal were operated at different temperatures (20 and 30 °C). In contrast to previous studies on this topic, Bassin et al. (2012a) not only described the general performance in terms of nitrogen and phosphate removal but also investigated the importance of specific subpopulations of PAOs (PAO clade I and II) on phosphate and nitrogen conversions. PAO clade I use both nitrite and nitrate as electron acceptors for denitrification, whereas PAO clade II is only able to use nitrite, since these microorganisms lack the nitrate reductase enzyme. This particular work aimed to link the nitrogen and phosphate conversions to the microbial community structure. Complete nitrification/denitrification and phosphate removal were achieved in both systems submitted to low dissolved oxygen concentration (less than 2 mg/L).

The authors observed that a considerable fraction of the phosphate removal was coupled to denitrification (denitrifying dephosphatation). This particular step is carried out by denitrifying polyphosphate-accumulating organisms (DPAOs), which use intracellular polymers stored under anaerobic conditions as electron donors for denitrification and do not require the addition of an external carbon substrate to carry out this process (Kuba et al. 1993; Kerrn-Jespersen and Henze 1993). Besides an efficient use of the incoming COD, the use of oxidized forms of nitrogen (e.g., nitrite or nitrate) rather than oxygen as electron acceptors for phosphate uptake also leads to energy savings due to the lower aeration requirement.

Denitrifying glycogen-accumulating organisms (DGAOs) were observed to be the main microbes responsible for the reduction of nitrate to nitrite. A significant fraction of the nitrite was further reduced to nitrogen gas while being used as an electron acceptor by denitrifying polyphosphate-accumulating organisms (PAO clade II) for anoxic phosphate uptake. An improved operational strategy to control the PAO-GAO competition in aerobic granular sludge reactors and favor the development and growth of PAOs under unfavorable conditions for the P removal process (e.g., high temperatures) is also described in the work conducted by Bassin et al. (2012a).

Previous studies focusing on granular sludge reactors, without being concerned with the intracellular storage of polymers, are also published in the literature. One example is an investigation by QIN and LIU (2006), in which microbial granules with excellent settling properties were cultivated in four sequencing batch reactors

(SBR_1–SBR_4) operated under alternating aerobic-anaerobic conditions and submitted to different nitrogen loads. The cycle time was 6 h (5 min of filling, 230 min of aerobic reaction, 119 min of anaerobic reaction, 2 min of sedimentation, and 4 min of effluent discharge) and the HRT was 12 h. The dissolved oxygen (DO) concentration was greater than 50% air saturation during the aerobic phase. In the anaerobic stage, nitrogen gas was used. The solids retention time was approximately 20 days after the granule formation.

The reactors were inoculated with activated sludge originating from a municipal wastewater treatment plant and were fed, from the bottom, with synthetic medium containing ethanol as the only carbon source, ammonium chloride, sodium bicarbonate, and other required nutrients. The influent COD was maintained constant at 500 mg/L, while the incoming ammonium concentration of the four reactors was gradually increased from 37.5 to 112.5 mgN/L. These values led to a gradual increase in the nitrogen load from 0.15 (in SBR_1) to 0.45 kg/(m^3 day) (in SBR_4) (QIN and LIU 2006).

Granules were observed in the reactors after 40 days of operation under alternating aerobic-anaerobic conditions. These had a spherical shape and a compact structure. The biomass concentration in the reactors was significant, varying from 3.0 gVSS/L in SBR_1 to 5.5 gVSS/L in SBR_4, indicating that the granulation promoted better biomass retention in the reactors. The SVI decreased from 230 mL/g (activated sludge inoculated into the reactor) to 30–12 mL/g (granular sludge submitted to increasing nitrogen loads from SBR_1 to SBR_4). These SVI values are still lower than those obtained during the granular sludge reactor operation under exclusively aerobic conditions, where the typical values varied between 50 and 100 mL/g (DANGCONG et al. 1999; TAY et al. 2001b). The settling velocity of the granules cultivated in the four reactors was greater than 60 m/h, which is much higher than that shown by conventional activated sludge (2–5 m/h) (*apud* QIN and LIU 2006).

In an experiment without the addition of an external carbon source during the anaerobic phase, QIN and LIU (2006) observed that 95% of the influent COD was removed during the first hour of the aerobic phase. During this period the ammonium concentration slowly decreased. After the COD had been completely removed, the ammonium concentration began to decrease rapidly due to nitrification. During the aerobic phase, ammonium was completely converted into nitrite and nitrate. In the subsequent anaerobic phase, the denitrification was observed at very reduced levels (partial denitrification). Approximately 13–27 mg/L N-NO_x was denitrified when the external carbon source was not available, which represent less than 10% of the N-NO_x formed by nitrification.

Based on the results obtained without the addition of an external carbon source and seeking to achieve complete denitrification, ethanol was used as the external carbon source in a mass ratio of 2:1 ethanol/N-NO_x at the beginning of the anaerobic (anoxic) phase. At this condition, almost all of the nitrite and nitrate produced in the aerobic phase by nitrification were rapidly denitrified into N_2 in the anaerobic phase (QIN and LIU 2006).

In contrast to the abovementioned studies, in which different processes occurred in different regions (aerobic and anoxic/anaerobic) within the granules due to the presence of an oxygen concentration gradient and the availability of intracellular polymers as a carbon source for denitrification, in the study by QIN and LIU (2006), the situation was similar to that at most wastewater treatment plants. Thus, in this case, the advantages of granular sludge technology (e.g., saving of organic matter for the denitrification by the use of stored substrate in the form of intracellular polymers) did not apply (QIN and LIU 2006).

The respirometric activities of ammonium-oxidizing and nitrite-oxidizing bacteria were inferred by the specific oxygen uptake rates (SOURs) associated with the specific ammonium ($SOUR_{NH4}$) and nitrite ($SOUR_{NO2}$) oxidation rates, respectively. The activity of the heterotrophic bacteria was determined by SOUR associated with carbon oxidation ($SOUR_H$). It was observed that both in the absence and presence of an external carbon source in the anaerobic phase, the activity of the ammonium-oxidizing bacteria increased with an increase in the applied nitrogen load, while the nitrite-oxidizing bacteria showed no significant variation under these conditions. In the absence of an external carbon source, it was also observed that the activity of the heterotrophic population decreased when the microbial consortium was enriched with nitrifiers, making heterotrophs less dominant. The supply of external carbon stimulated the activity of heterotrophs and denitrifiers (which were comprised mostly of facultative heterotrophic bacteria), especially when high organic loads were applied (QIN and LIU 2006).

The activity of the denitrifying bacteria was determined through the specific NO_x reduction rate, which varied from 12.0 to 25.3 mgN/(gVSS h) when the reactors were fed with an external carbon source at the beginning of the anaerobic phase. These values are an order of magnitude higher than those obtained during the operation of the reactors without the addition of an external carbon source, which varied from 1.9–3.5 mgN/(gVSS h). The results showed that the activity of denitrifying bacteria was essentially determined by the availability of the external carbon source in the anaerobic phase (QIN and LIU 2006).

Due to their large surface area, high degree of porosity, and good settling capacity, the aerobic granules play an important role in the treatment of toxic chemical compounds and heavy metals. GAI et al. (2008) studied the mechanism of Cu^{2+} adsorption by granular sludge. These authors observed that 70% of the copper was adsorbed through ion exchange with Na^+, Ca^{2+}, and Mg^{2+}, which were released by the granules. XU and LIU (2008) proposed that ion exchange, binding with EPS, and chemical precipitation are the main mechanisms involved in the biosorption of Cd^{2+}, Cu^{2+}, and Ni^{2+} by aerobic granules. These authors also observed that alcohol, carboxylate, and ether groups actively participate in the capture of the metals. NANCHARAIAH et al. (2008) cultivated aerobic granules in wastewaters containing nitrilotriacetic acid. The chelating agent significantly accelerated the granulation process, while the granules formed removed this compound almost completely. ZHU et al. (2008) cultivated granules in an SBR with a high height/diameter ratio, aiming at the degradation of 4-chloroaniline. Mature granules were obtained in 90 days, and the specific degradation rate of 4-chloroaniline varied from 0.14 to 0.27 g/(gVSS day).

Most of the reactors employed to cultivate aerobic granules were fed with synthetic wastewater. As previously mentioned in this section, studies have focused mainly on issues related to COD and nutrient (N and P) removal under different operating conditions. There have been few research studies in which aerobic granulation was applied to the treatment of industrial wastewaters.

As mentioned in Sect. 4.4.5, ARROJO et al. (2004) evaluated the formation of aerobic granules in two sequencing batch reactors (SBR_1 and SBR_2) fed with industrial wastewater originating from a laboratory where the analysis of dairy products is carried out. In SBR_1 an anoxic phase was included at the beginning of each cycle, while SBR_2 was submitted only to aerobic conditions. The same authors also evaluated the nitrogen removal process. At the beginning of the experiment in SBR_1 (submitted to organic and nitrogen loads of 1 gCOD/(L day) and 0.1 gN-NH_4^+/(L day), respectively), the nitrogen removal was related to the assimilation of ammonium for biomass growth, since nitrification did not occur during this period. Nitrification activity was only observed after 1 month of operation. When the ammonium concentration was doubled due to changes in the composition of the wastewater (after day 150 of operation), denitrification began to occur in both systems. Thus, nitrification and denitrification were the main mechanisms involved in the nitrogen removal after this period. The nitrogen removal efficiency reached 80% on day 188 of operation in both reactors. The organic and nitrogen loads applied in both systems were between 1 and 7 COD/(L day) and 0.1 and 0.7 gN/(L day), respectively.

The nitrogen removal efficiency was 70% in both units even though SBR_2 was only operated under aerobic conditions. The influent COD was between 500 and 1800 mg/L, and the ammonium concentration varied from 30 to 180 mgN-NH_4^+/L throughout the operational period. The COD removal efficiencies were between 85 and 95%. The ammonium and nitrate concentrations in the effluent varied from 0 to 20 mgN-NH_4^+/L and 0 to 40 mgN-NO_3^-/L, respectively. Similar values were obtained by GARRIDO et al. (2001), who treated industrial wastewater in a 28 m^3 SBR containing flocculent biomass. However, it should be taken into consideration that the organic and nitrogen loads applied were greater in the SBR with granular sludge (around five times greater considering the maximum values applied), and even so the effluent characteristics, in terms of nitrogen concentration and COD, were similar. Such results highlight the advantages of aerobic granulation when compared with conventional activated sludge technology (ARROJO et al. 2004).

The concentrations of the compounds were measured during the operating cycles (particularly on day 260 of operation when the organic and nitrogen loads were 5 gCOD/(L day) and 0.4 gN-NH_4^+/(L day), respectively). The results revealed that almost all of the biodegradable organic matter and the nitrate (remaining from the previous cycle) were completely removed in both systems during the first 10 min of the cycle. Nitrate was consumed via denitrification, while the biodegradable organic compounds were partially oxidized aerobically, partially used as electron donors for denitrification, and partly accumulated in the form of biomass. Ammonium was oxidized to nitrate during the aerobic period, particularly after the anoxic period in SBR_1 and throughout the cycle in SBR_2, immediately after the biodegradable COD had been completed consumed. Even when organic loads of 5–7 gCOD/(L day) were applied, the occurrence of nitrification was observed (ARROJO et al. 2004).

The specific denitrification activity was 0.28 and 0.30 gN/(gVSS day) in SBR_1 and SBR_2, respectively, while the specific nitrification activity was approximately 0.033 gN/(gVSS day) in SBR_2 and 0.023 gN/(gVSS day) in SBR_1. In the case of SBR_2, the fact that the granules presented denitrifying activity under aerobic conditions may be explained by the fact that during the first few minutes of the cycle, a large quantity of biodegradable organic matter was fed to the system. Thus, during this period, the diffusion of this organic matter fraction to the interior of the granules was greater in comparison to the diffusion of oxygen. Also, the dissolved oxygen present during the filling period of SBR_2, in a concentration of 3 mg/L, was probably consumed by the external layer of the granules. Thus, the internal layers, maintained under anoxic conditions, received sufficient carbon source and nitrate to allow the denitrification process to be carried out (ARROJO et al. 2004). These findings have also been reported in BEUN et al. (2001), as mentioned earlier in this section.

An important parameter to be considered in this study was the total suspended solids (TSS) concentration in the industrial wastewater fed to the reactors (SBR_1 and SBR_2), which varied from 200 to 900 mg/L and contained a large fraction of VSS (in the range of 75–97%). The suspended solids concentration in the influent was partially removed in both SBRs. The TSS content of the effluent was lower than that of the influent when the reactors were fed with industrial wastewater. TSS concentrations were in the range of 50 to 800 mg/L (ARROJO et al. 2004).

The presence of TSS in the effluent of the SBRs was related to the TSS in the influent, to the detachment of small pieces of granules and to the occurrence of small flocs which were washed out of the reactor. Effluent TSS content was strongly influenced by both the duration of the discharge period and the relation between the particulate COD and the biomass (CODp/VSS) in the systems. The TSS concentration in the effluent increased when the CODp/VSS ratio was greater than 0.12 gCODp/gVSS. In order to decrease the solids content of the effluent, the authors applied a strategy in which the discharge period was reduced. A significant reduction in the TSS concentration of the effluent occurred, that is, from 450 to 200 mgTSS/L and then to 150 mgTSS/L, when the discharge was gradually reduced from 3 to 1 min and then to 0.5 min, respectively. This was related to the higher washout of the small suspended biomass aggregates when the duration of this phase was reduced, decreasing the growth of small flocs in the reactors. Thus, the strategy employed was shown to have a positive effect on the effluent quality in terms of solids concentration (ARROJO et al. 2004).

On day 296 of operation, both reactors began to be fed with the same synthetic wastewater, free of solids, which was used during the start-up of SBR_1. The TSS concentration of the effluent was around 200 mgTSS/L, which is lower than the value observed when the reactors were fed with industrial wastewater and when the discharge time was 3 min (450 mgTSS/L). This observation indicates that the solids content of the effluent is not dependent only on the biomass washout from the reactor but also on the solids content of the influent wastewater (ARROJO et al. 2004).

The results obtained by these authors should certainly be taken into consideration on applying the granular sludge technology. Biomass washout, which in some cases can occur, may be the consequence of the high solids concentration in the influent. ARROJO et al. (2004) noted that the presence of suspended solids in the effluent may be one of the main obstacles in the development of SBRs with granules at full scale. The TSS removal efficiency is a crucial aspect which must be taken into consideration when applying granular sludge technology, since pollutants are often found in the form of colloids or particles. SCHWARZENBECK et al. (2004) used aerobic granules for the treatment of wastewater containing malt and a high content of particulate organic matter (0.9 gTSS/L) and observed that the particles with average diameters of less than 25–50 μm were removed with 80% efficiency. The removal efficiency of particles smaller than 50 μm was only 40%. The authors observed that the capacity of the granular sludge to remove particulate organic matter was due to its incorporation by the biofilm matrix and the metabolic activity of the protozoan population which covered the surface of the granules. However, INIZAN et al. (2005) observed that the aerobic granules were not able to remove suspended solids present in wastewaters originating from the pharmaceutical industry.

In this context, the advantages of granular SBRs, such as a high conversion capacity and good granule sedimentation properties, can frequently be counteracted by the presence of suspended solids in the effluent. This factor can represent a paradox since granular SBRs are generally described in the literature as systems which can improve the bioreactor performance in relation to the retention of solids. In some situations, a posttreatment process (e.g., filtration membranes, surface filters, secondary settling tanks) can be added to remove the TSS content in order to comply with environmental legislation, although this procedure will reduce the economic attractiveness of aerobic granulation technology.

CASSIDY and BELIA (2005) operated a granular sludge reactor fed with slaughterhouse wastewater containing total COD of 7685 mg/L, soluble COD of 5163 mg/L, total Kjeldahl nitrogen of 1057 mg/L, and volatile suspended solids content (VSS) of 1520 mg/L. The authors obtained a COD and phosphorus removal efficiency of 98%, while both the removal of nitrogen and VSS reached 97%. To achieve this performance, CASSIDY and BELIA (2005) submitted the reactor to DO of 40% air saturation, a condition considered to be ideal for nitrogen removal, according to BEUN et al. (2001). Also, in order to maintain the granule stability under conditions of limited DO concentration, the authors followed the recommendations of DE KREUK et al. (2005b) and included an anaerobic feeding period. TSUNEDA et al. (2006) observed a nitrogen removal rate of 1.0 kgN/(m^3 day) and a nitrogen removal efficiency of 95% in a system containing autotrophic granules, applied in the treatment of wastewater originating from a metal refinery containing 1.0–1.5 gN-NH_4^+/L and up to 22 g/L of sodium sulfate.

Aerobic granulation technology has been applied not only in laboratory scale. A research study in pilot scale was carried out in the Netherlands in order to demonstrate the applicability of aerobic granulation technology to municipal wastewater treatment. The hydraulic capacity of the plant was 5 m^3/h, the main installation being comprised of two parallel granular sludge sequencing batch reactors ($GSBR_1$ and $GSBR_2$), with a diameter and a height of 6 and 0.6 m, respectively. The pretreatment of the pilot plant consisted of a primary settling tank, optionally followed by a pressurized sand filter (DE BRUIN et al. 2005).

The influent had an average COD of 560 mg/L, suspended solids of 225 mg/L, Kjeldahl nitrogen of 58.4 mg/L and total phosphorus of 10 mg/L. $GSBR_1$ was always operated as a bubble column reactor, while $GSBR_2$ was initially operated as an airlift reactor and later transformed into the bubble column configuration. The cycle time of the reactors varied from 2.5 to 4 h. The feeding period, under anaerobic conditions, was 50–60 min, and the settling time was 20–30 min. The remainder of the cycle referred to the aerobic phase (DE BRUIN et al. 2005).

The pilot plant was inoculated with activated sludge originating from a conventional treatment plant. Initially, the granulation process showed good progress, which resulted in granules with good settling properties. However, the granulation did not continue due to the low operating temperatures (around 5 °C) and the inadequate control of the aeration, which led to high concentrations of nitrate in the effluent and low phosphate removal efficiencies. Thus, aeration control was implemented. Moreover, the reactor walls were insulated and the influent flow preheated in order to control the process temperature (DE BRUIN et al. 2005).

The reactor start-up was carried out with the new modifications aiming to obtain high phosphate removal and nitrification efficiencies. The average removal of phosphate was 65%, and nitrification was incomplete. Since the ammonium concentrations in the effluent were high and the settling characteristics of the sludge deteriorated, the organic load, in terms of COD, was reduced by 50%. Under this condition, complete nitrification was achieved and greater concentrations of nitrate in the effluent were observed. At the same time, phosphate removal reduced considerably, and the granulation process did not improve (DE BRUIN et al. 2005).

Based on the results obtained, some operational conditions were modified. It was observed that during the start-up period of the reactors, it was not possible to maintain complete phosphate removal and complete nitrification at the same time. Bearing in mind that the selection of organisms with a low growth rate, in this case polyphosphate-accumulating organisms (PAOs), was considered to be a prerequisite for granulation, the COD was increased once again, and the beginning of aerobic granulation was observed. In approximately 3 months, around 30% of the sludge in $GSBR_1$ consisted of granules. However, due to an operational failure, 75% of the sludge was lost from both reactors (DE BRUIN et al. 2005).

The reactors were again inoculated with the same activated sludge as used before, and this time rapid granulation was observed. This enhancement of the granulation process was attributable to the sample collection method. With the previously used method, many of the granules were crushed/broken during their collection. One month after start-up, the granulation in $GSBR_1$ was complete, and the SVI_{30} was only 55 mL/g. The granulation in $GSBR_2$ proceeded more slowly and corresponded to 50% after 1 month of the reactor start-up. The SVI_{30} of the sludge present in this reactor was around 70 mL/g. As the granulation process proceeded, the sludge washout decreased which resulted in the accumulation of biomass in the reactors (DE BRUIN et al. 2005).

TAY et al. (2005) studied the development of aerobic granules in a pilot-scale sequencing batch reactor, inoculated with pre-cultivated granular sludge in a small reactor in the format of a column. The authors observed that the granules inoculated disintegrated within the first days of operation, although granules were subsequently formed. On the other hand, granules which were maintained in the laboratory-scale reactor did not disintegrate. The results obtained were related to the different hydrodynamic conditions of the reactors in pilot and laboratory scale. Factors such as the reactor diameter and the wall effect, as well as the size and location of the air diffusers, may have influenced the hydrodynamic conditions in the reactors which, in turn, determined the granule properties (TAY et al. 2005).

In general, the results obtained show the potential for the application of granules in larger scales, since it was demonstrated that granules can be formed with municipal wastewater and significant levels of nutrient removal can be achieved (DE BRUIN et al. 2005). In addition, the research in pilot scale highlighted the difficulty in obtaining successful granulation when scaling up from laboratory to larger scales. It was clear that greater care and better control of the operational conditions are required in order for the formation of aerobic granules to be carried out satisfactorily, which can lead to good reactor performance.

A study on the applicability showed that the granular sludge process is a very promising technology from the economic point of view (DE BRUIN et al. 2004). Based on the total annual costs, granular sludge reactors operating in sequencing batch mode (GSBR) with pretreatment and posttreatment can be more attractive than the conventional activated sludge process, with potential savings varying within 6–16%. One study on the sensitivity showed that the GSBRs are less sensitive to the price of land and more sensitive to the flow of water originating from rain. Due to the potential application of high volumetric loads, the area occupied by GSBR systems represents only 25% of the area used by conventional flocculent sludge systems. However, granular sludge systems involving only primary treatment may not reach the established effluent discharge standards, mainly due to the presence of solids in the effluent resulting from the washout of biomass with unsatisfactory settling properties. With stricter standards being imposed with regard to effluent discharge, the activated sludge plants must incorporate a posttreatment stage, such as sand filtration, or be transformed into bioreactors with membranes. In this case, GSBR systems with primary and posttreatment may represent an attractive alternative (DE BRUIN et al. 2004).

Data on the application of aerobic granular sludge (AGS) in full-scale installations are still scarcely reported in the literature. Few works have provided detailed information on this issue so far. Pronk et al. (2015) reported the implementation of the technology for the treatment of domestic sewage in the Netherlands. This work provided interesting data related to the operation of the plant. The AGS plant was operated in parallel with the existing activated sludge reactor. The AGS reactor treats 28,600 m^3/day of sewage, which corresponds to 41% of the total influent flow. Average influent COD and BOD_5 during the field tests were reported to be 506 and 224 mg/L, respectively.

The AGS reactor was operated in sequencing batch mode with simultaneous feeding and effluent withdrawal (fill-and-draw operation), reaction and final sludge settling, sludge withdrawal, and idle periods. Nitrogen removal was mainly obtained by simultaneous nitrification and denitrification (SND). However, non-mixed periods under anoxic conditions obtained by a recycle from top to bottom of the reactor were used as a strategy to improve nitrogen removal performance. During aeration, dissolved air concentration was kept between 1.8 and 2.5 mg/L. The cycle time was set at 6.5 h at dry weather conditions, and it is reduced to 3 h at rainy weather conditions by shortening the aeration period. The AGS reactor was inoculated with surplus sludge (1 g/L) coming from other full-scale AGS plant located in Epe, the Netherlands. No granules were present in the seed biomass, which presented a SVI_{30} and SVI_5 of 140 and 90 mL/g, respectively (Pronk et al. 2015).

The authors reported the development of a stable and robust granule bed with more than 8 g/L of biomass after 5 months of start-up period. SVI_5 and SVI_{30} were found to be 45 mg/L and 35 mL/g, respectively. More than 80% of the granular biomass was larger than 0.2 mm and more than 60% larger than 1 mm. Effluent nitrogen and phosphorus concentrations lower than 7 mgN/L and 1 mgP/L were achieved during both summer and winter, complying with the environmental regulations. Maximum nitrogen and phosphorus volumetric conversion rates were observed to be 0.17 and 0.24 kg/(m^3 day). The energy usage was around 60% lower than in conventional activated sludge reactors, while the volume required for the AGS plant was 33% smaller than the existing activated sludge plant (Pronk et al. 2015).

Li et al. (2014) also reported successful granulation in a full-scale SBR treating 50,000 m^3/day of domestic sewage in Yancang wastewater treatment plant, located in Haining, China. The wastewater has the following characteristics: COD of 200–600 mg/L, BOD of 50–105 mg/L, ammonium of 28–40 mg/L, and total phosphorus of 2–4 mg/L. The AGS reactor was inoculated with sludge from the secondary settler of an oxidation ditch. The full-scale AGS plant was divided in four tanks, operated in fill-and-draw mode. The wastewater was fed from the top of the reactors. The volume exchange ratio varied from 50 to 70%.

After inoculation, the SBR cycle consisted of 40 min filling, around 240 min aeration, 60 min settling, and 30 min effluent withdrawal. Subsequently the settling time was reduced to 40–50 min. After around 330 days of operation, the SVI_{30} of the granular biomass reached only 47.1 mL/g, while the diameter of the granules was 0.5 mm. X-ray fluorescence analysis has shown that metal (e.g., iron) and other inorganic ions (calcium, silicon, phosphorus) present in the raw wastewater precipitated in the sludge and acted as a core, enhancing granulation. The authors also evaluated the polysaccharide concentration in the biomass. The amount of polysaccharide in the full-scale SBR biomass was higher than in the continuous flow activated sludge-based reactors (i.e., anaerobic-oxic reactor and oxidation ditch). They reported that these EPS components could have mediated cohesion and adhesion of cells, contributing to maintain a stable structure of aerobic granules. Moreover, the cyclic operation of the SBR consisting of a periodic feast-famine regime, short settling time, and no return sludge was favorable for the development of fast-settling particles (Li et al. 2014).

Fig. 4.11 Full-scale demonstrations of aerobic granular sludge technology. In (**a**), the granular reactors are displayed in the bottom of the picture (Provided by Royal Haskoning DHV). (**a**) Nereda® WWTP Garmerwolde, The Netherlands. Operation since 2013. (**b**) Nereda® WWTP Frielas, Portugal. Operation since 2014. (**c**) Nereda® WWTP Deodoro, Brazil. Operation since 2016

Many full-scale aerobic granular sludge treatment plants have been implemented in Europe, Africa, Australia and South America. Figure 4.11 shows three examples of wastewater treatment plants designed by the company Royal HaskoningDHV, which is current market leader in applying aerobic granular sludge process for industrial and municipal application. Since 2005 they have full-scale applications with their Nereda® aerobic granular sludge technology. Current (2016) project portfolio amounts to approx. 30 ranging from smaller industrial applications up to 2.4 M PE municipal wastewater treatment plants and the amount of references is increasing rapidly.

4.8 Final Considerations and Future Perspectives

The aerobic granulation technology, when compared with conventional processes which make use of flocculent biomass, offers many advantages from the technical-operational point of view. The settling characteristics of granular sludge are by far superior to those of conventional activated sludge, which leads to a shorter time required for the settling of particles, lower SVI values, and a greater biomass concentration in the reactor. The latter characteristic favors the stability of the biological system, making it less vulnerable to shock loads of organic or toxic compounds, and also increases the treatment capacity of the bioreactor.

The cultivation of granular sludge is mostly carried out in sequencing batch reactors (SBRs). The mode of operation of these reactors is easier when combined with the use of biomass in the form of granules. The filling and treated effluent discharge phases are to be carried out simultaneously, particularly when a plug-flow regime is adopted, with an upward flow through the settled sludge bed. Considering that the settling of granules is rapid, a longer period can be allocated to the reaction phase, that is, to the actual biological treatment. Another positive characteristic of the use of aerobic granules in an SBR is the reduction in the area required for the installation of the biological treatment system since, in most cases, a secondary settling tank is not needed. The greater retention of biomass in the system, which allows an increase in the volumetric conversion rates, also contributes to reducing the volume of the granular sludge reactor, reducing even further the space requirement.

With the use of aerobic granules, the removal of organic compounds and nutrients (nitrogen and phosphorus) can be carried out simultaneously. The presence of different regions within the granules, both aerobic and anoxic/anaerobic, allows both heterotrophic and autotrophic nitrifiers to coexist in their structure, favoring a high microbial diversity. The stratification of the granule structure means that aerobic heterotrophic and autotrophic microorganisms are present in the external layer, below which there is a layer with a reduced level of oxygen where facultative anaerobic bacteria are located. Finally, in the central region of the granules, anaerobic/anoxic microorganisms predominate due to the absence of oxygen. In fact, the distribution of aerobic and anaerobic/anoxic regions is strongly influenced by the dissolved oxygen concentration and the diameter of the granules. However, the control of the latter remains a challenge.

While most of the heterotrophic microorganisms are responsible for the degradation of organic matter, some of them, such as polyphosphate-accumulating organisms (PAOs), may also be present. These are responsible for phosphate removal and may also play an important role in the denitrification process (in this case being known as denitrifying PAOs, DPAOs) taking place in the anoxic zone of the granules. The DPAOs use compounds stored intracellularly (polyhydroxyalkanoates, PHA) as a carbon source. Some operational conditions of SBRs must be ensured for the selection of PAOs, such as the application of a feeding period under anaerobic conditions. This procedure also induces the formation of stable granules (important for preventing their washout from the reactor) even at low dissolved oxygen concentrations, since it favors the development of bacteria with a lower growth rate. Moreover, it contributes to reduce the investment and running costs and facilitate operation in larger scales.

Studies on granular sludge systems for the treatment of industrial wastewater have indicated that the use of this technology is suitable for obtaining good removal efficiencies for organic matter and nutrients (N and P). However, in some cases, the use of a pre- or posttreatment is recommended to ensure that the wastewater complies with the discharge limits imposed by the environmental regulators, particularly considering that the effluent may contain high concentrations of suspended solids.

The most important aspect of the aerobic granulation process is the start-up period, in which the formation of granules takes place. When granules originating from other systems in operation are used as the inoculum, the time required for aerobic granulation to occur can be minimized. Another drawback associated with aerobic granulation relates to the uncertainty regarding criteria such as the stability and robustness of the granules, characteristics which can often be affected due to operation under certain specific conditions.

Although several studies have been carried out with respect to aerobic granulation technology, some issues have not been fully explored in this research area. A better understanding of the mechanisms involved in the formation of the granules is one point to be considered. Greater attention should be given to the oxygen concentration gradients present in the granular sludge, which are determined by the distribution and activity of different microbial populations inside the granules. These directly influence the conversion processes which occur within the granules. Since the oxygen concentration profiles are the result of different factors which influence one another, such as the diffusion coefficients, conversion rates, granule size, pore size and distribution, and permeability and density, it is difficult to perform experimental studies to investigate the effect of each factor individually. In this regard, the development of mathematical models certainly represent a valuable tool to provide the understanding of important factors which affect the reactor performance and the distribution of microbial populations responsible for the biodegradation processes. Furthermore, such models are also useful for optimizing the process and for the scaling-up, design, and control of reactions in pilot/industrial scale.

Increasing the scale of granular sludge systems alters the hydrodynamics, which play an important role in the formation and stability of aerobic granules. Thus, additional studies in pilot/industrial scales should be conducted. Another promising issue to be investigated is a detailed research at the microbiological level in order to elucidate the relation between certain microorganisms and the stability of the granular sludge. Furthermore, the exopolymers produced by granular biomass is also of particular interest for better understanding the formation of these aggregates. The influence of the granule structure on the conversion processes and the mass transport involved in these processes also need to be studied in greater depth.

Finally, laboratory and pilot-scale research on aerobic granular sludge for the treatment of several types of wastewaters should be encouraged as it consists of an important step for full-scale implementation and expansion of the technology worldwide.

References

ADAV, S. S., CHEN, M. Y., LEE, D. J., REN, N. Q. Degradation of phenol by aerobic granules and isolated yeast *Candida tropicalis*. *Biotechnology and Bioengineering*, v. 96, p. 844-852, 2007a.

ADAV, S. S., LEE, D. J. Physiological characterization and interactions of isolates in phenol degrading aerobic granules. *Applied Microbiology and Biotechnology*, v. 78, p. 899-905, 2008.

ADAV, S. S.; LEE, D. J., LAI, J. Y. Effects of aeration intensity on formation of phenol-fed aerobic granules and extracellular polymeric substances. *Applied Microbiology and Biotechnology*, v. 77, p. 175-182, 2007b.

ADAV, S. S., LEE, D.-J., SHOW, K.-Y., TAY, J.-H. Aerobic granular sludge: recent advances. *Biotechnology Advances*, v. 26, n. 5, p. 411-423, 2008a.

ADAV, S. S., LEE, D.-J., TAY, J.H. Extracellular polymeric substances and structural stability of aerobic granules. *Water Research*, v. 42, p. 1644-1650, 2008b.

ARROJO, B., MOSQUERA-CORRAL, A., GARRIDO, J. M., MÉNDEZ, R. Aerobic granulation with industrial wastewater in sequencing batch reactors. *Water Research*, v. 38, n. 14-15, p. 3389-3399, 2004.

BASSIN, J.P., PRONK, M., KRAAN, R., KLEEREBEZEM, R., VAN LOOSDRECHT, M.C. Ammonium adsorption in aerobic granular sludge, activated sludge and anammox granules. *Water Research*, v. 15, p. 5257-5265, 2011.

BASSIN, J.P., KLEEREBEZEM, R., DEZOTTI, M., VAN LOOSDRECHT, M.C.M. Simultaneous nitrogen and phosphate removal in aerobic granular sludge reactors operated at different temperatures. *Water Research*, v. 46, n. 12, 3805-3816, 2012a.

BASSIN, J.P., KLEEREBEZEM, R., DEZOTTI, M., VAN LOOSDRECHT, M.C.M. Measuring biomass specific ammonium, nitrite and phosphate uptake rates in aerobic granular sludge. *Chemosphere*, v. 89, n. 10, p. 1161-1168, 2012b.

BEUN, J. J., HENDRIKS, A., VAN LOOSDRECHT, M. C. M., MORGENROTH, E., WILDERER, P. A., HEIJNEN, J. J. Aerobic granulation in a sequencing batch reactor. *Water Research*, v. 33, p. 2283-2290, 1999.

BEUN, J. J., VAN LOOSDRECHT, M. C. M., HEIJNEN, J. J. Aerobic granulation. *Water Science and Technology*, v. 41, n. 4-5, p. 41-48, 2000.

BEUN, J. J., VAN LOOSDRECHT, M. C. M., HEIJNEN, J. J. Aerobic granulation in a sequencing batch airlift reactor. *Water Research*, v. 36, p. 702-712, 2002.

BEUN, J. J., VAN LOOSDRECHT, M. C. M., HEIJNEN, J. J. N-removal in a granular sludge sequencing batch airlift reactor. *Biotechnology and Bioengineering*, v. 75, n. 1, p. 82-92, 2001.

BRDJANOVIC, D., LOGEMANN, S., VAN LOOSDRECHT, M. C. M., HOOIJMANS, C. M., ALAERTS, G. J., HEIJNEN, J. J. Influence of temperature on biological phosphorus removal: process and molecular ecological studies. *Water Research*, v. 32, n. 4, p. 1035-1048, 1998.

CAMMAROTA, M. C., SANT'ANNA, G. L. Metabolic blocking of exopolysaccharides synthesis: effects on microbial adhesion and biofilm accumulation *Biotechnology Letters*, n. 20, p. 2283-2290, 1998.

CASSIDY, D. P., BELIA, E. Nitrogen and phosphorus removal from an abattoir wastewater in a SBR with aerobic granular sludge. *Water Research*, v. 39, n. 19, p. 4817-4823, 2005.

CHEN, G., STREVETT, K. A. Bacterial deposition in aqueous media: a surface thermodynamic investigation. *Environmental Engineering Science*, v. 20, p. 237-248, 2003.

CROCETTI, G. R., BANFIELDA, J. F., KELLER, J., BOND, P. L., BLACKALL, L. L. Glycogen-accumulating organisms in laboratory-scale and full-scale wastewater treatment processes. *Microbiology*, v. 148, p. 3353-3364, 2002.

CROCETTI, G. R., HUGENHOLTZ, P., BOND, P. L., SCHULER, A., KELLER, J., JENKINS, D., BLACKALL, L. L. Identification of polyphosphate-accumulating organisms and design of 16S rRNA-directed probes for their detection and quantification. *Applied Environmental Microbiology*, v. 66, n. 3, p. 1175-1182, 2000.

DANGCONG, P., BERNET, N., DELGENES, J.P., MOLETTA R Aerobic granular sludge—a case report. *Water Research*, v. 33, n. 3, p. 890-893, 1999.

DE BRUIN, L. M. M., DE KREUK, M. K., VAN DER ROEST, H. F. R., UIJTERLINDE, C., VAN LOOSDRECHT, M. C. M. Aerobic granular sludge technology: an alternative to activated sludge. *Water Science and Technology*, v. 49, n. 11-12, p. 1-7, 2004.

DE BRUIN, L. M. M., VAN DER ROEST, H. F., DE KREUK, M. K., VAN LOOSDRECHT, M. C. M.. Promising results pilot research aerobic granular sludge technology at WWTP Ede. In: BATHE, S., DE KREUK, M., McSWAIN, B., SCHWARZENBECK, N. Aerobic Granular Sludge. Water and Environmental Management Series, IWA Publishing, London, 135-142, 2005.

DE KREUK, M. K.. Aerobic Granular Sludge. Scaling-up a new technology. PhD Thesis, Technical University of Delft, Delft, The Netherlands, 2006.

DE KREUK, M. K., McSWAIN, B. S., BATHE, S., TAY, S. T. L., SCHWARZENBECK, N., WILDERER, P. A. Discussion Outcomes. In: BATHE, S., DE KREUK, M., McSWAIN, B., SCHWARZENBECK, N. Aerobic Granular Sludge. Water and Environmental Management Series, IWA Publishing, London, 155-169, 2005a.

DE KREUK, M. K., HEIJNEN, J. J., VAN LOOSDRECHT, M. C. M. Simultaneous COD, nitrogen and phosphate removal by aerobic granular sludge. *Biotechnology and Bioengineering*, v. 90, n. 6, p. 761-769, 2005b.

DE KREUK, M. K., PRONK, M., VAN LOOSDRECHT, M. C. M. Formation of aerobic granules and conversion processes in an aerobic granular sludge reactor at moderate and low temperature. *Water Research*, v. 39, n. 18, p. 4476-4484, 2005c.

DE KREUK, M. K., VAN LOOSDRECHT, M. C. M. Selection of slow growing organisms as a means for improving aerobic granular sludge stability. *Water Science and Technology*, v. 49, p. 9-17, 2004.

DURMAZ, B., SANIN, F. D. Effect of carbon to nitrogen ratio on the composition of microbial extracellular polymers in activated sludge. *Water Science and Technology*, v. 44, n. 10, p. 221-229, 2001.

FALKENTOFT, C.M. Simultaneous removal of nitrate and phosphorus in a biofilm reactor; the aspect of diffusion. Doctoral dissertation. Lyngby: Technical University of Denmark, 2000.

FIGUEROA, M., MOSQUERA-CORRAL, A., CAMPOS, J. L., MÉNDEZ, R. Treatment of saline wastewater in SBR aerobic granular reactors. *Water Science and Technology*, v. 58, n. 2, p. 479-485, 2008.

FRIED, J., LEMMER, H. On the dynamics and function of ciliates in sequencing batch biofilm reactors. *Water Science and Technology*, v. 47, p. 189-196, 2003.

GAI, L.-H., WANG, S.-G., GONG, W.-X., LIU, X.-W., GAO, B.-Y., ZHANG, H.-Y. Influence of pH and ionic strength on Cu(II) biosorption by aerobic granular sludge and mechanism. *Journal of Chemical Technology and Biotechnology*, v. 83, n. 6, p. 806-813, 2008.

GARRIDO, J. M., OMIL, F., ARROJO, B., MÉNDEZ, R., LEMA, J. M. Carbon and nitrogen removal from wastewater of an industrial dairy laboratory with a coupled anaerobic filter-sequencing batch reactor system. *Water Science and Technology*, V. 43, p. 315-321, 2001.

GHIGO, J. M. Are there biofilm-specific physiological pathways beyond a reasonable doubt? *Research in Microbiology*, v. 154, p. 1-8, 2003.

GROTENHUIS, J. T. C., VAN LIER, J. B., PLUGGE, C. M., STAMS, A. J. M., ZEHNDER, A. J. B. Effect of ethylene glycol-bis(β-aminoethyl ether)-N, *N*-tetraacetic acid (EGTA) on stability and activity of methanogenic granular sludge. *Applied Microbiology and Biotechnology*, v. 36, p. 109-114, 1991.

HARTMANN, C., OZMUTLU, O., PETERMEIER, H., FRIED, J., DELGADO, A. Analysis of the flow field induce by the sessile peritrichous ciliate *Opercularia asymmetrica*. *Journal of Biomechanics*, n. 40, p. 137-148, 2007.

HEIJNEN, J. J., VAN LOOSDRECHT, M. C. M. Method for acquiring grain-shaped growth of a microorganism in a reactor. Patent Cooperation Treaty (PCT). US, European patent, Technische Universiteit Delft: 13, 1998.

HU, L., WANG, J., WEN, X., QIAN, Y. The formation and characteristics of aerobic granules in sequencing batch reactor (SBR) by seeding anaerobic granules. *Process Biochemistry*, v. 40, p. 5-11, 2005.

HWANG, K. J., YOU, S. F., DON, T. M. Disruption Kinetics of Bacterial Cells during Purification of Poly-beta-hydroxyalkanoate Using Ultrasonication. *Journal of the Chinese Institute of Chemical Engineers*, v. 37, p. 209, 2006.

INIZAN, M., FREVAL, A., CIGANA, J., MEINHOLD, J. Aerobic granulation in a sequencing batch reactor (SBR) for industrial wastewater treatment. *Water Science and Technology*, v. 52, n. 10-11, p. 335-343, 2005.

JIANG, H. L., TAY, J. H., LIU, Y., TAY, S. T. L. Ca^{2+} augmentation for enhancement of aerobically grown microbial granules in sludge blanket reactors. *Biotechnology Letters*, v. 25, n. 2, p. 95-99, 2003.

JIANG, H. L., TAY, J. H., TAY, S. T. L. Aggregation of immobilized activated sludge cells into aerobically grown microbial granules for the aerobic biodegradation of phenol. *Letters in Applied Microbiology*, v. 35, p. 439-445, 2002.

KERRN-JESPERSEN, J.P., HENZE, M. Biological phosphorus uptake under anoxic and aerobic conditions. *Water Research*, v. 27, p. 617-624, 1993.

KIM, D. J., SEO, D. Selective enrichment and granulation of ammonia oxidizers in a sequencing batch airlift reactor. *Process Biochemistry*, v. 41, n. 5, p. 1055-1062, 2006.

KIM, S. M., KIM, S. H., CHOI, H. C., KIM, I. S. Enhanced aerobic floc-like granulation and nitrogen removal in a sequencing batch reactor by selection of settling velocity. *Water Science and Technology*, v. 50, n. 6, p. 157-162, 2004.

KJELLEBERG, S., HERMANSSON, M. Starvation-induced effects on bacterial surface characteristics. *Applied and Environmental Microbiology*, v. 48, p. 497-503, 1984.

KUBA, T., SMOLDERS, G., VAN LOOSDRECHT, M.C.M., HEIJNEN, J.J. Biological phosphorus removal from wastewater by anaerobic– anoxic sequencing batch reactor. *Water Science and Technology*, v. 27, p. 241–252, 1993.

LETTINGA, G., VAN VELSEN, F. M., HOBMA, S. W., DE ZEEUW, W., KLAPWIJK, A. Use of upflow sludge blanket reactor concept for biological wastewater treatment, especially for anaerobic treatment. *Biotechnology and Bioengineering*, v. 22., p. 699-734, 1980.

LOPEZ-VAZQUEZ, C. M., BRDJANOVIC, D., HOOIJMANS, C. M., GIJZEN, H. J., VAN LOOSDRECHT, M. C. M. Factors affecting the occurrence of phosphorus accumulating organisms (PAO) and glycogen accumulating organisms (GAO) at full-scale enhanced biological phosphorus removal (EPBR) wastewater treatment plants. *Water Research*, v. 42, n. 10-11, p. 2349-2360, 2008.

LOPEZ-VAZQUEZ, C. M., OEHMEN, A., HOOIJMANS, C. M., BRDJANOVIC, D., GIJZEN, H. J., YUAN, Z., VAN LOOSDRECHT, M. C. M. Modelling the PAO-GAO competition: effects of carbon source, pH and temperature. *Water Research*, v. 43, n. 2, p. 450-462, 2009.

LI, J., DING, L.-B., CAI, A., HUANG, G.-X., HORN, H. Aerobic sludge granulation in a full-scale sequencing batch reactor. *BioMed Research International*, Article ID 268789, 12 pages 2014.

LIN, Y. M., LIU, Y., TAY, J. H. Development and characteristics of phosphorous-accumulating granules in sequencing batch reactor. *Applied Microbiology and Biotechnology*, v. 62, p. 430-435, 2003.

LIU, Y., LIU, Q.-S. Causes and control of filamentous growth in aerobic granular sludge sequencing batch reactors. *Biotechnology Advances*, v. 24, p. 115-127, 2006.

LIU, Y., QIN, L., TAY, J. H. Effect of settling on aerobic granulation in sequencing batch reactor. *Biotechnology and Bioengineering*, v. 21, p. 47-52, 2004b.

LIU, Y. Q., TAY, J. H. Influence of cycle time on kinetic behaviors of steady-state aerobic granules in sequencing batch reactors. *Enzyme and Microbial Technology*, v. 41, p. 516-522, 2007.

LIU, Y., TAY, J. H. State of the art of biogranulation technology for wastewater treatment. *Biotechnology Advances*, v. 22, p. 533-563, 2004.

LIU, Y., TAY, J.-H. The essential role of hydrodynamic shear force in the formation of biofilm and granular sludge. *Water Research*, v. 36, p. 1653-1665, 2002.

LIU, Y.-Q., LIU, Y., TAY, J.-H. The effects of extracellular polymeric substances on the formation and stability of biogranules. *Applied Microbiology and Biotechnology*, v. 65, p. 143-148, 2004a.

LIU, Y.-Q., TAY, J. H. Influence of starvation time on formation and stability of aerobic granules in sequencing batch reactors. *Bioresource Technology*, v. 99, n.5, p. 980-985, 2008.
LIU, Y., YANG, F., WANG, F., ZHANG, H., ZHANG, X. Study on the stability of aerobic granules in a SBAR-Effect of superficial upflow air velocity and carbon source. In: BATHE, S., DE KREUK, M., McSWAIN, B., SCHWARZENBECK, N. Aerobic Granular Sludge. Water and Environmental Management Series, IWA Publishing, London, 35-43, 2005.
McSWAIN, B. S., IRVINE, R. L., WILDERER, P. A. The effect of intermittent feeding on aerobic granule structure. *Water Science and Technology*, v. 49, n. 11-12, p. 19-25, 2004.
MOSQUERA-CORRAL, A., DE KREUK, M. K., HEIJNEN, J. J., VAN LOOSDRECHT, M. C. M. Effects of oxygen concentration on N-removal in an aerobic granular sludge reactor. *Water Research*, v. 39, n. 12, p. 2676-2686, 2005.
MOY, B. Y. P., TAY, J. H., TOH, S. K., LIU, Y., TAY, S. T. L. High organic loading influences the physical characteristics of aerobic sludge granules. *Letters in Applied Microbiology*, v. 34, p. 407-412, 2002.
NANCHARAIAH, Y. V., JOSHI, H. M., MOHAN, T. V. K., VENUGOPALAN, V. P., NARASIMBAN, S. V. Formation of aerobic granules in the presence of a synthetic chelating agent. *Environ Pollut.*, v. 153, n. 1, p. 37-43, 2008.
NIELSEN, P. H., JAHN, A. Extraction of EPS. In: WINGERDEN, J., NEU, T. R., FLEMMING, H. C. (Eds). Microbial Extracellular Polymeric Substances. Springer, Berlin, p. 21-47, 1999.
OEHMEN, A., SAUNDERS, A. M., VIVES, M. T., YUAN, Z., KELLER, J. Competition between polyphosphate and glycogen accumulating organisms in enhanced biological phosphorus removal systems with acetate and propionate as carbon sources. *Journal of Biotechnology*, v. 123, n. 1, p. 22-32, 2006.
PICIOREANU, C., VAN LOOSDRECHT, M. C. M., HEIJNEN, J. J. Mathematical modeling of biofilm structure with a hybrid differential-discrete cellular automaton approach. *Biotechnology and Bioengineering*, v. 58, n. 1, p. 101-116, 1998.
POCHANA, K., KELLER, J. Study of factors affecting simultaneous nitrification and denitrification (SND). *Water Science and Technology*, v. 39, n. 6, p. 61-68, 1999.
PRONK, M., DE KREUK, M. K., DE BRUIN, B., KAMMINGA, P., KLEEREBEZEM, R., VAN LOOSDRECHT, M. C. M. Full scale performance of the aerobic granular sludge process for sewage treatment. *Water Research*, v. 84, p. 207-217, 2015.
QIN, L., LIU, Y. Aerobic granulation for organic carbon and nitrogen removal in alternating aerobic-anaerobic sequencing batch reactor. *Chemosphere*, v. 63, p. 926-933, 2006.
QIN, L., TAY, J.-H, LIU, Y. Selection pressure is a driving force of aerobic granulation in sequencing batch reactors. *Process Biochemistry*, v. 39, p. 579-584, 2004.
RUIJSSENAARS, H. J., STINGELE, F., HARTMANS, S. Biodegradability of food-associated extracellular polysaccharides. *Current Microbiology*, v. 40, p. 194-199, 2000.
SCHMIDT, J. E., AHRING, B. K. Extracellular polymers in granular sludge from different upflow anaerobic sludge blanket (UASB) reactors. *Applied Microbiology Biotechnology*, v. 42, p. 457-462, 1994.
SCHWARZENBECK, N., ERLEY, R., WILDERER, P. A. Aerobic granular sludge in an SBR-system treating wastewater rich in particulate matter. *Water Science and Technology*, v. 49, n. 11-12, p. 41-46, 2004.
SHEN, C. F., KOSARIC, N., BLASZCZYK, R. The effect of selected heavy metals (Ni, Co and Fe) on anaerobic granules and their extracellular polymeric substance (EPS). *Water Research*, v. 27, p. 25-33, 1993.
SMOLDERS, G.J.F., VAN DER MEIJ, J., VAN LOOSDRECHT, M.C.M., HEIJNEN, J.J. Stoichiometry and pH influence of the anaerobic metabolism. *Biotechnology and Bioengineering*, v. 43, p. 461-470, 1994.
SUTHERLAND, I. W. Biofilm exopolysaccharides. In: WINGENDER, J., NEU, T. R., FLEMMING, H.-C. (eds). Microbial extracellular polymeric substances. Characterization, structure and function. Springer, Berlin, p. 73-92, 1999.

TAY, J. H., LIU, Q. S., LIU, Y. Microscopic observation of aerobic granulation in sequential aerobic sludge blanket reactor. *Journal of Applied Microbiology*, v. 91, p. 168-175, 2001a.

TAY, J. H., LIU, Q. S., LIU, Y. The effects of shear force on the formation, structure and metabolism of aerobic granules. *Applied Microbiology Biotechnology*, v. 57, p. 227-233, 2001b.

TAY, J. H., LIU, Q. S., LIU, Y. The role of cellular polysaccharides in the formation and stability of aerobic granules. *Letters in Applied Microbiology*, n. 33, p. 222-226, 2001c.

TAY, J. H., LIU, Q. S., LIU, Y., SHOW, K.-Y., IVANOV, V., TAY, S. T.-L. A comparative study of aerobic granulation in pilot- and laboratory scale SBRs. In: In: BATHE, S., DE KREUK, M., McSWAIN, B., SCHWARZENBECK, N. Aerobic Granular Sludge. Water and Environmental Management Series, IWA Publishing, London, 125-133, 2005.

TAY, J. H., YANG, S. F., LIU, Y. Hydraulic selection pressure-induced nitrifying granulation in sequencing batch reactors. *Applied Microbiology and Biotechnology*, v. 59, p. 332-337, 2002b.

TAY, S. T. L., IVANOV, V., YI, S., ZHUANG, W. Q., TAY, J. H. Presence of anaerobic *Bacteroides* in aerobically grown microbial granules. *Microbial Eco*logy, v. 44, p. 278-285, 2002a.

THIRD, K. A., BURNETT, N., CORD-RUWISCH, R. Simultaneous nitrification and denitrification using stored substrate (PHB) as the electron donor in an SBR. *Biotechnology and Bioengineering*, v. 83, n. 6, p. 706-720, 2003.

TIJHUIS, L., VAN BENTHUM, W. A. J., VAN LOOSDRECHT, M. C. M., HEIJNEN, J. J. Solids retention time in spherical biofilms in a biofilm airlift suspension reactor. *Biotechnology and Bioengineering*, v. 44, n. 8, p. 867-879, 1994.

TSUNEDA, S., NAGANO, T., HOSHINO, T., EJIRI, T., NODA, N., HIRATA, A. Characterization of nitrifying granules produced in an aerobic upflow fluidized bed reactor. *Water Research*, v. 37, n. 20, p. 4965-4973, 2003.

TSUNEDA, S., OGIWARA, M., EJIRI, Y., HIRATA, A. High-rate nitrification using aerobic granular sludge. *Water Science and Technology*, v. 53, n. 3, p. 147-154, 2006.

VAN LOOSDRECHT, M. C. M., EIKELBOOM, D., GJALTEMA, A., MULDER, A., TIJHUIS, L., HEIJNEN, J. J. Biofilm structures. *Water Science and Technology*, v. 32, p. 235-243, 1995.

VARON, M., CHODER, M. Organization and cell-cell interaction in starved *Saccharomyces cerevisiae* colonies. The *Journal of Bacteriology*, v. 182, p. 3877-3880, 2000.

VILLASEÑOR, J. C., VAN LOOSDRECHT, M. C. M., PICIOREANU, C., HEIJNEN, J. J. Influence of different substrates on the formation of biofilms in a biofilm airlift suspension reactor. *Water Science and Technology*, v. 41, n. 4-5, p. 323-330, 2000.

WANG, X. H., ZHANG, H. M., YANG, F. L., XIA, L. P., GAO, M. M. Improved stability and performance of aerobic granules under stepwise increased selection pressure. *Enzyme and Microbial Technology*, v. 41, n. 3, p. 205-211, 2007.

WANG, Z. W., LIU, Y., TAY, J. H. Distribution of EPS and cell surface hydrophobicity in aerobic granules. *Applied Microbiology and Biotechnology*, v. 69, p. 469-473, 2005.

WANG, Z. W., LIU, Y., TAY, J. H. The influence of short-term starvation on aerobic granules. *Process Biochemistry*, v. 41, p. 2373-2378, 2006.

WEBER, S. D., LUDWIG, W., SCHLEIFER, K.-H., FRIED, J. Microbial composition and Structure of Aerobic Granular Sewage Biofilms. *Applied and Environmental Microbiology*, v. 73, n.19, p. 6233-6240, 2007.

WILEN, B. M., ONUKI, M., HERMANSSON, M., LUMLEY, D., MINO, T. Microbial community structure in activated sludge floc analysed by fluorescence in situ hybridization and its relation to floc stability. *Water Research*, v. 42, n. 8-9, p. 2300-2308, 2007.

WINKLER, M-K.H., BASSIN, J.P., KLEEREBEZEM, R., DE BRUIN, L.M.M., VAN DEN BRAND, T.P.H., VAN LOOSDRECHT, M.C.M. Selective sludge removal in segregated aerobic granular biomass system as a strategy to control PAO-GAO competition at high temperatures. *Water Research*, v. 45, n. 11, p. 3291-3299, 2011.

XU, H.; LIU, Y. Mechanisms of Cd^{2+}, Cu^{2+} and Ni^{2+} biosorption by aerobic granules. *Sep Purif Technology*, v. 58, n.3, p. 400-411, 2008.

YANG, S. F., LI, X. Y., YU, H. Q. Formation and characterisation of fungal and bacterial granules under different feeding alkalinity and pH conditions. *Process Biochemistry*, v. 43, p. 8-14, 2008.

ZHANG, X., BISHOP, P. L. Biodegradability of biofilm extracellular polymeric substances, *Chemosphere*, v. 50, p. 63-69, 2003.

ZHENG, Y. M., YU, H. Q., SHENG, G. P. Physical and chemical characteristics of granular activated sludge from a sequencing batch airlift reactor. *Process Biochemistry*, v. 40, p. 645-650, 2005.

ZHU, L., XU, X. Y., LUO, W. G., TIAN, Z. J., LIN, H. Z., ZHANG, N. N. A comparative study on the formation and characterization of aerobic 4-chloroaniline-degrading granules in SBR and SABR. *Applied Microbiology and Biotechnology*, v. 79, n.5, p. 867-874, 2008.

ZITA, A., HERMANSSON, M. Determination of bacterial cell surface hydrophobicity of single cells in cultures and in wastewater *in situ*. *FEMS Microbiology Letters*, v. 152, p. 299-306, 1997.

Chapter 5
New Processes for Biological Nitrogen Removal

João Paulo Bassin

5.1 Introduction

The approach toward the biological treatment of wastewaters has changed several times over the years. In the 1960s and 1970s, the main concern revolved around the removal of organic material from wastewaters given its high polluting capacity in the environment. In this context, several aerobic and anaerobic systems were developed aiming at the degradation of organic compounds.

In the 1980s, it was observed that even in the absence of organic material in wastewaters, the polluting effect on the environment remained intense, notably when nitrogen compounds were present. The increase in the global population and the consequent increase in the quantity of human waste as well as the use of synthetic nitrogen fertilizers produced from atmospheric N_2 by the Haber-Bosch process, intensified in the past 40 years, are two factors which have contributed significantly to enhance the pollution levels caused by nitrogen compounds.

Nitrogen removal is of crucial importance. When it is available, together with phosphorus, other important nutrient, enhanced primary production occurs, which is exemplified by the excessive growth of phytoplankton species (algal bloom). This phenomenon is referred to as eutrophication. During this process, the microbial population responsible for the degradation of organic matter originating from phytoplankton species grows exponentially, and as a consequence, the oxygen demand is substantially increased. The death of fish and other aquatic organisms by asphyxiation, an increase in the cost of treating eutrophicated waters, and even the inappropriateness of these waters for many uses are some of the consequences of eutrophication.

J.P. Bassin (✉)
Chemical Engineering Program, COPPE, Federal University of Rio de Janeiro,
Rio de Janeiro, RJ, Brazil
e-mail: jbassin@peq.coppe.ufrj.br

M. Dezotti et al., *Advanced Biological Processes for Wastewater Treatment*,
DOI 10.1007/978-3-319-58835-3_5

In this context, it became clear that the application of processes aimed at the removal of nitrogen and phosphorus was essential in order to preserve the quality of the receiving water bodies. Thus, the conventional processes for biological nutrient removal (involving nitrification and denitrification stages) began to be used extensively.

In the following decade (1990s), the increase in the installation costs of traditional technologies for wastewater treatment, the increased strictness of the limits imposed on effluent discharges (the treatment plants requiring significant modifications in order to comply with environmental regulations), the limitations of the conventional processes regarding the treatment of high-strength nitrogen wastewaters, and the appearance of innovative ideas motivated the development of new technologies for biological nitrogen removal.

During this period, novel bacteria involved in the nitrogen cycle were identified and isolated from various environments (natural or engineered bioreactors), which culminated in the development of new nitrogen removal processes. In fact, these new processes still have rather limited application compared with conventional process, particularly for operation in larger scales. However, as will be discussed in Sect. 5.2, most of them have enormous potential for use in wastewater treatment, as they were developed as an attempt to overcome the limitations of the conventional nitrification-denitrification processes encountered under certain conditions.

In comparison with physicochemical processes, such as ammonia stripping and precipitation with magnesium ammonium phosphate, the recently developed biological processes for nitrogen removal offer a considerable economic advantage. The physicochemical processes require significant quantities of chemical products and thus lead to the production of a greater amount of chemical sludge. Despite the argument that physicochemical processes can allow the recovery of ammonium, only a small quantity of the ammonium is recovered in comparison with its general use, for instance, as a fertilizer. Additionally, these techniques, in general, require a greater amount of energy than biological nitrification-denitrification processes. In view of all of these factors, biological nitrogen removal is the option most commonly recommended and used (VAN LOOSDRECHT 2008).

The new nitrogen removal processes are the subject of this chapter. Before they are presented, the new aspects related to the microbial transformation of nitrogen in the context of wastewater treatment are first discussed.

5.2 New Processes for Biological Nitrogen Removal

5.2.1 Introduction and Contextualization

Complex interactions occur between the different nitrogen species (such as ammonium, nitrite, and nitrate) and in the different transformation mechanisms. Organic nitrogen is made up of various compounds including amino acids, urea, uric acid, and nitrogen bases. By means way of hydrolysis and mineralization, organic nitrogen is converted into ammonium nitrogen. Ammonium is one of the most important nitrogen compounds in surface waters and other ecosystems for several reasons: (1) it is the preferred nutrient of various species of plants and autotrophic bacteria; (2) it is chemically reduced and thus can be easily oxidized in natural aquatic environments resulting in the consumption of dissolved oxygen; and (3) non-ionized ammonia (NH_3) is toxic to several forms of aquatic life even at low concentrations (<0.2 mg/L) (KADLEC and KNIGHT 1996).

Under aerobic conditions, ammonium is oxidized to nitrite, which is further oxidized to nitrate. This process is called nitrification. Two main groups of bacteria are involved in nitrification: ammonium-oxidizing bacteria (AOB) and nitrite-oxidizing bacteria (NOB). Nitrification (Eq. 5.1) is the first step in the conventional nitrogen removal process. The complementary step is denitrification, in which the reduction of nitrate to nitrogen gas occurs (Eq. 5.2). Equation 5.3 represents the combined process involving the two previous steps. The conventional nitrification-denitrification process is the commonly used approach for nitrogen removal in modern wastewater treatment plants. It is very efficient and, when properly operated, is stable and reliable. The operation cost is moderate and the process can be controlled relatively easily (METCALF and EDDY 2003).

$$NH_4^+ + 2O_2 + 2HCO_3^- \rightarrow NO_3^- + 3H_2O + 2CO_2 \tag{5.1}$$

$$5C + 4NO_3^- + 2H_2O \rightarrow CO_2 + 4HCO_3^- + 2N_2 \tag{5.2}$$

$$4NH_4^+ + 8O_2 + 5C + 4HCO_3^- \rightarrow 2N_2 + 10H_2O + 9CO_2 \tag{5.3}$$

Despite its extensive application, the conventional nitrogen removal process occurs very slowly, a characteristic associated with the low microbial activity and the reduced yield of this process. It is used for the treatment of wastewaters containing low concentrations of ammonium, generally less than 100 mgN/L. For the treatment of high-strength nitrogen wastewaters, such as those from anaerobic digesters, pig waste, landfill leachates, and certain industrial effluents, conventional nitrogen removal processes become limited, mainly due to sizing and operation issues. The space limitation and economic restriction are some problems which hinder the achievement of the desired performance in the larger existing treatment plants, particularly when the nitrogen load to be treated is high. Table 5.1 shows some examples of wastewaters containing high concentrations of ammonium.

The water originating from the dewatering of digested sludge is generally returned to the beginning of the treatment and mixed with the influent wastewater. On carrying out a nitrogen mass balance at the Dokhaven sewage treatment plant (Rotterdam, the Netherlands), it was observed that although the reject water contributes very little to the total flow, it corresponded to 15% of the nitrogen load (VAN DONGEN et al. 2001a; MULDER et al. 2001). A separate (sidestream) treatment of this stream rich in ammonium nitrogen, as indicated in Fig. 5.1, could reduce the

Table 5.1 Wastewaters containing high concentrations of ammonium

Wastewater	COD[a] (mg/L)	BOD_5[b] (mg/L)	Total nitrogen (mg/L)	Reference
Reject water[c]	232–12,587	81–750	260–958	Gil and Choi (2004)
	390–2720	n.m.	943–1513	Jenicek et al. (2004)
	610	140	910	Wyffels et al. (2003)
Fine fraction of pig waste	n.m.	2912	707	Chen et al. (2004)
	3969	1730	1700	Obaja et al. (2003)
	9000–13,000	n.m.	3100–4300	Poo et al. (2004)
	6456	n.m.	695	Tilche et al. (1999)
Landfill leachate	2000–5000	1500–4000	500–1000	Chung et al. (2003)
	n.m.	45	310	Ilies and Mavinic (2001)
	1300–1600	n.m.	160–270	Jokela et al. (2002)
	9660–20,560	n.m.	780–1080	Kalyuzhnyi and Gladchenko (2004)
Tannery	300–1400	n.m.	50–200	Abeliovich (1992)
	1940–2700	n.m.	123–185	Murat et al. (2003)
Slaughterhouse	1400–2400	n.m.	170–200	Keller et al. (1997)[d]
Starch production	3000	990	1060	Kalyuzhnyi and Gladchenko (2004)[e]
	5000–10,000	2000–5000	800–1100	Abeling and Seyfried (1993)[e]
Pectin production	15,000–22,000	n.m.	1280–2990	Austermann-Haun et al. (1999)
	8100	n.m.	1600	Deng Petersen et al. (2003)

n.m not mentioned
[a]Chemical oxygen demand
[b]Biochemical oxygen demand obtained in 5 days
[c]Water originating from dewatering of sludge digested in sewage treatment plants
[d]After treatment in anaerobic pond
[e]After treatment in anaerobic digester

nitrogen load originating from the sludge digesters and could contribute significantly to reducing the nitrogen concentration of the treatment plant effluent and allow the discharge limits to be reached.

Thus, the application of sidestream processes is particularly important when the treatment plant requires upgrading due to stricter effluent discharge standards or due to an increase in the nitrogen load. With the addition of a relatively small reactor volume, the nitrogen concentration of the effluent can be reduced. An additional advantage is that this allows for the construction of a reactor which is independent from the main treatment process, which is clearly a much simpler approach compared with modifying and expanding the existing treatment plants. Moreover, if the ammonium is not totally converted in the main treatment process, each kg of ammo-

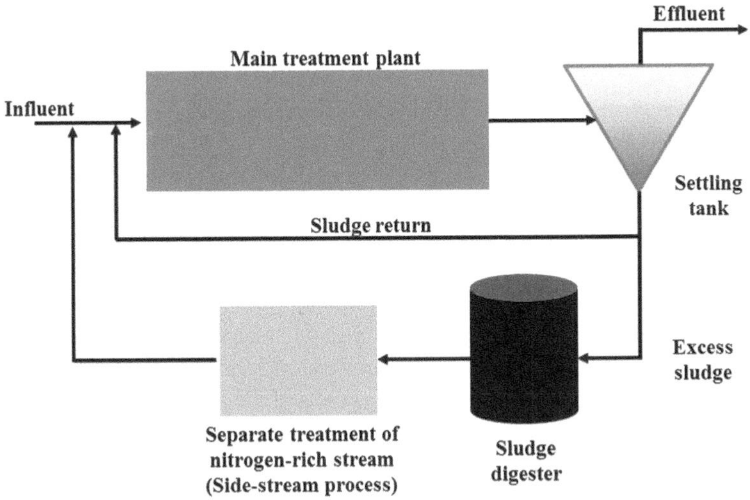

Fig. 5.1 Schematic representation of a separate (sidestream) treatment for a stream rich in ammonium nitrogen originating from a sludge digester

nium removed in the sidestream process will result in 1 kg less in the treatment plant effluent.

Landfill leachates are highly polluting and must be captured and treated. In general, the leachate is returned to the upper layer of the landfill, which leads to a reduction in the concentration of organic material. Conversely, the nitrogen concentration gradually increases, since the landfill acts as an anaerobic bioreactor (CLABAUGH 2001). Similarly, to reject water originating from dewatered sludge, some landfill leachates are characterized by a high concentration of ammonium and a low organic content (ILIES and MAVINIC 2001).

Pig wastes can be separated into two fractions: coarse and fine. The coarse fraction can be used as manure for soils while the fine fraction is treated. The composition of the fine fraction can vary depending on the separation method and the composition of the animal feed. In many cases, besides nitrogen and phosphorus, high concentrations of organic matter can be present, which is not favorable for the application of autotrophic nitrogen removal processes.

Wastewaters coming from industrial processes may contain high concentrations of nitrogen, especially when they are firstly treated in anaerobic digesters. Examples of these are the wastewaters generated in pharmaceutical plants (CARRERA et al. 2003); tanneries (MURAT et al. 2003); slaughterhouses (KELLER et al. 1997); potato, alcohol, and starch processing industries (ABELING and SEYFRIED 1993); and formaldehyde production (CAMPOS et al. 2003).

In recent years, the paradigm that the only way to biologically convert ammonium into nitrogen gas was the complete oxidation of ammonium to nitrate (nitrification) followed by the reduction of nitrate to nitrogen gas (denitrification) became obsolete. With the discovery of new metabolic pathways, more sustainable processes for nitrogen removal were developed and these have undergone continual improvement as research in this area has advanced. Among the most relevant new processes for removal of nitrogen, the following can be mentioned: partial nitrification and denitrification (SHARON, single-reactor high-activity ammonia removal over nitrite), CANON (completely autotrophic nitrogen removal over nitrite), OLAND (oxygen-limited autotrophic nitrification-denitrification), and aerobic/anoxic deammonification (DEMON). The last three are based on partial nitritation and on the relatively recently discovered anammox (anaerobic ammonium oxidation) process. The SHARON technology may also be coupled to anammox, giving rise to the so-called SHARON-anammox process.

In general, the specific application of alternatives available for the removal of nitrogen needs to be evaluated in relation to the aspects involving costs, chemical and energy requirements, operation experience, and the reliability and environmental impact of the process. However, the selection of the best alternative is usually based on the cost criteria. The new processes meet the objective of reducing the operating cost. In most new processes, the aim is to remove nitrogen via nitrite (which is used as electron acceptor) and not via nitrate as in the conventional process.

The nitritation-denitritation process (Eqs. 5.4 and 5.5) is one example of an alternative process for nitrogen removal via nitrite. It consumes less oxygen in the nitrification (partial nitrification up to nitrite) and requires less organic carbon for the denitrification (nitrite and not nitrate should be reduced to nitrogen gas) (SCHMIDT et al. 2003). Another advantage is a lower production of sludge. The application of the combined process of partial nitritation-anammox (described later in this chapter) brings even more advantages. In this combined process, organic matter is not required, since the nitrogen removal is carried out by autotrophic bacteria. This process is especially recommended for the treatment of wastewaters with a low organic carbon-to-nitrogen (C/N) ratio (RUIZ et al. 2003) and which contain high ammonium concentrations (between 100 and 5000 mgN/L) (MULDER 2003). Within this range, the autotrophic removal of nitrogen is more advantageous in relation to the conventional nitrification and denitrification process, requiring a lower amount of energy and chemical products.

$$NH_4^+ + 1.5O_2 + 2HCO_3^- \rightarrow NO_2^- + 3H_2O + 2CO_2 \tag{5.4}$$

$$3C + 4NO_2^- + 2H_2O + CO_2 \rightarrow 4HCO_3^- + 2N_2 \tag{5.5}$$

In comparison with the traditional nitrogen removal process, the oxygen consumption in the nitritation-denitritation and partial nitritation-anammox processes are 25 and 60% lower, respectively. When the denitritation is applied after the nitritation process, a 40% saving can be obtained in terms of organic carbon. With the application of the anammox process downstream of the nitritation process, the saving is even greater, since this process does not require an organic carbon source (AHN 2006). HAO et al. (2001) and NIELSEN et al. (2005) note that for the treatment of highly concentrated streams, the relatively high installation cost of the combined partial nitritation-anammox processes is compensated for by the lower operating costs and by the good nitrogen removal performance.

The three processes (conventional, nitritation-denitritation, and partial nitritation-anammox), despite being very distinct in terms of oxygen and organic matter requirements, require similar levels of alkalinity, bearing in mind that 1 mol H^+ is produced per mol of nitrogen converted. Thus, these processes are not associated with high costs for the pH control when the wastewater to be treated has a good buffering capacity (1 mol HCO_3^- per mol of NH_4).

The main disadvantage of the autotrophic removal of nitrogen is the reduced growth rate of ammonium-oxidizing bacteria and anammox microorganisms. The performance of reactors in which slow-growing bacteria predominate can be improved through the application of high cell retention times. One alternative in this regard is the use of support media for the development of biofilms or the creation of conditions for the self-aggregation of the cells in the form of granules (VÁZQUEZ-PADÍN et al. 2009). Figure 5.2a, b show, in an illustrative and simplified manner, two different perspectives of the nitrogen cycle updated with the autotrophic removal of nitrogen. As can be observed, the detection of new microorganisms (such as anammox bacteria) increases substantially the complexity of the nitrogen cycle. The anammox process will be described in the next section. Further details about the new processes developed for nitrogen removal, many of them based on the anammox process, will be addressed further on in this chapter.

5.2.2 Anammox Process

5.2.2.1 Brief History

In 1977, Engelbert Broda predicted, based on thermodynamic and evolutionary theories, that the oxidation of ammonium under anoxic conditions with nitrate or nitrite as the electron acceptor was possible (BRODA 1977). The oxidation of ammonium using an electron acceptor which is not oxygen had been previously predicted by researchers of marine environments based on mass balance studies (RICHARDS 1965) and later in combination with thermodynamic explanations (CLINE and RICHARDS 1972). These predictions were taken into consideration to some degree, although at the time few researchers were convinced that biological oxidation of ammonium could occur under anoxic conditions.

In 1985, the removal of ammonium was observed for the first time under anoxic conditions in a pilot-scale denitrifying reactor at a yeast powder production plant called Gist-Brocades, now part of the company DSM, in Delft, the Netherlands (HEIJNEN 1988; MULDER 1989; VAN DE GRAAF et al. 1990; MULDER et al. 1995). In this reactor, nitrate was added in order to obtain a combined process of sulfide oxidation and nitrate reduction. As an unexpected result, ammonium (considered not to be reactive under anoxic conditions) was also removed (MULDER et al. 1995). This new biological process was called the anammox process, an

(a)

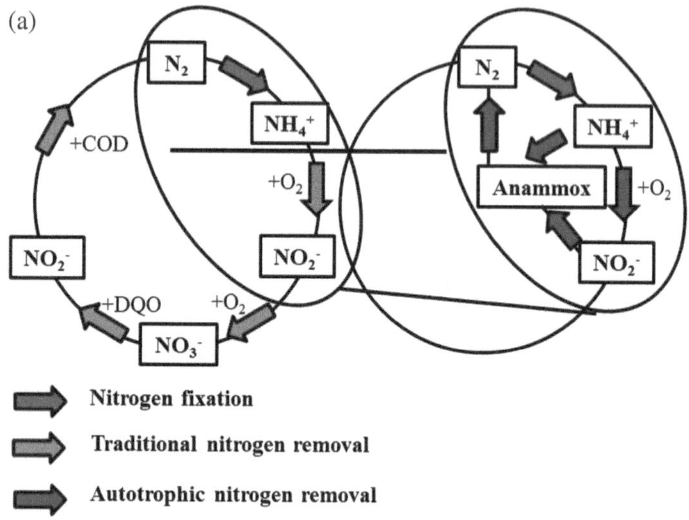

(b)

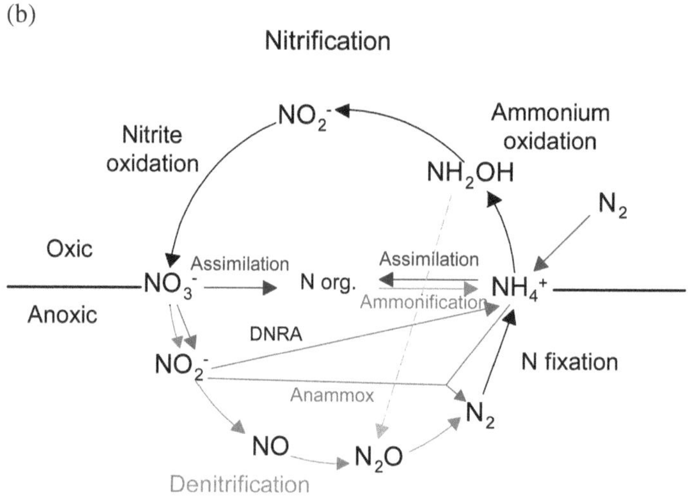

Fig. 5.2 Schematic representation of the nitrogen cycle: (**a**) simplified comparison between the conventional nitrogen removal process and the anammox process; (**b**) main oxic and anoxic reactions involved in the nitrogen cycle. The different reactions are indicated in different colors: nitrogen fixation (*dark gray*), ammonification (*orange*), assimilation (*purple*), nitrification (*black*), denitrification (*green*), dissimilatory reduction of nitrate to ammonium (DNRA) (*blue*), and anammox (*red*). The *dotted gray line* indicates the formation of N_2O by ammonium-oxidizing bacteria (adapted from Madigan et al. 2000)

abbreviation for anoxic ammonium oxidation (MULDER 1989; VAN DE GRAAF et al. 1990) or anaerobic ammonium oxidation (MULDER 1989; VAN DE GRAAF et al. 1995), the name by which this process is most commonly known. A bench-scale reactor based on the pilot-scale denitrification reactor was run in the laboratory, and the enrichment of the organism responsible for the anammox process was achieved only after nitrite (and not nitrate) had been identified as the actual electron acceptor (VAN DE GRAAF et al. 1996).

Ten years after the first observations of the ammonium oxidation reaction under anoxic conditions in Delft (the Netherlands), in Germany (HIPPEN et al. 1996, 1997), and in Switzerland (BINSWANGER et al. 1997; SIEGRIST et al. 1998), the production of nitrogen gas instead of nitrate (leading to nitrogen loss from the liquid phase) was reported in rotating biological contactors (RBCs) treating ammonium-rich wastewaters originating from landfill leachate. Since the only electron donor present in significant quantities in these wastewaters was ammonium, it was expected that all of the nitrogen in the form of ammonium would be converted to nitrate, instead of being lost to the atmosphere in the form of N_2. The conversion of soluble nitrogen compounds into N_2 in these reactors was initially related to the denitrification carried out by nitrifying organisms, previously described by POTH and FOCHT (1985). Later on, it was attributed to a combination of nitrifying organisms and the bacteria responsible for the anammox process.

Anammox bacteria have been found not only in wastewater treatment plants but also in natural environments, such as in different marine ecosystems (e.g., the Black Sea) and rivers (KUYPERS et al. 2003), where they substantially influence the nitrogen cycle. Depending on the organic load, up to 70% of the N_2 production in marine sediments can be attributed to the anammox process (DALSGAARD and THAMDRUP 2002).

Enriched cultures (purity of 50–90%) obtained by some researchers (SCHMID et al. 2000; EGLI et al. 2001; STROUS et al. 2006; TSUSHIMA et al. 2007a; LÓPEZ et al. 2008) were the only sources of information on the anammox process, since the numerous attempts to isolate the organisms responsible for this process had failed (STROUS et al. 1999a). In order to observe that the enriched organisms were in fact responsible for the anammox process, cells were physically purified up to 99.6% through density gradient centrifugation (Percoll method) (STROUS et al. 1999a). The fact that the purified cells were able to carry out the characteristic anammox conversion (ammonium to nitrite) at a high rate confirmed that the enriched microorganism was indeed responsible for the ammonium oxidation under anoxic conditions. The concentrated cell solution allowed the sequencing of the 16S rRNA gene, which provided evidence that the organism in question belonged to the *Planctomycetes* phylum (STROUS et al. 1999a) and was named *Candidatus Brocadia anammoxidans*.

5.2.2.2 Conversions Involved in the Anammox Process and Characteristics of the Organisms Responsible for this Process

The main substrates of the anammox process are ammonium, nitrite, and bicarbonate (HCO_3^-) (VAN DE GRAAF et al. 1996). The coupling of the nitrogen atom of ammonium (electron donor) and the nitrogen atom of nitrite (electron acceptor) for the formation of nitrogen gas comprises the catabolic reaction (Eq. 5.6).

$$NH_4^+ + NO_2^- \rightarrow N_2 + 2H_2O \quad (5.6)$$

As an autotrophic biological process, HCO_3^- is the carbon source for the production of biomass in the anabolic reactions (Eq. 5.7). The oxidation of nitrite to nitrate generates the electrons required for the HCO_3^- reduction process (VAN DE GRAAF et al. 1996). The catabolic reaction is carried out 15 times for the fixation of one molecule of CO_2 with nitrite acting as the electron donor, leading to the anaerobic production of nitrate in the anabolism. In autotrophic processes, the anoxic generation of nitrate can serve as a measure of the growth of anammox biomass and is a good indicator of the anammox activity (VAN LOOSDRECHT 2008).

$$HCO_3^- + 2.1NO_2^- + 0.2NH_4^+ + 0.8H^+ \rightarrow CH_{1.8}O_{0.5}N_{0.2} + 2.1NO_3^- + 0.4H_2O \quad (5.7)$$

The combination of catabolic (Eq. 5.6) and anabolic (Eq. 5.7) reactions using the biomass yield determined experimentally (0.066 mol C/mol NH_4^+, STROUS et al. 1998) results in the following overall reaction (Eq. 5.8):

$$\begin{aligned} &NH_4^+ + 1.13NO_2^- + 0.066HCO_3^- + 0.053H^+ \\ &\rightarrow 0.99N_2 + 0.066CH_{1.8}O_{0.5}N_{0.2} + 0.14NO_3^- + 2H_2O \end{aligned} \quad (5.8)$$

The stoichiometry of Eq. (5.8) is very close to that obtained experimentally (STROUS et al. 1998), represented in Eq. (5.9). As can be observed, ammonium and nitrite are consumed in almost equimolar proportions (1:1.3). The excess of nitrite (0.3 mol of nitrite per mol of ammonium) is oxidized anaerobically to nitrate. The electrons derived from this oxidation are probably used for the fixation of CO_2 (VAN DE GRAAF et al. 1996). The main product of the anammox process is N_2, although a small part of the nitrogen fed to the system is converted into nitrate. For the reduction of the nitrate produced through nitrification, organic carbon is needed, which does not present a problem since most real wastewaters contain at least a small quantity of biodegradable organic matter which can be used for this purpose.

$$\begin{aligned} &NH_4^+ + 1.32NO_2^- + 0.066HCO_3^- + 0.13H^+ \\ &\rightarrow 1.02\,N_2 + 0.066CH_{1.8}O_{0.5}N_{0.2} + 0.26NO_3^- + 2.03H_2O \end{aligned} \quad (5.9)$$

The organisms responsible for the anammox process grow very slowly, as evidenced through the stoichiometry. The duplication time is several days under ideal operating conditions (STROUS et al. 1998; TSUSHIMA et al. 2007b). According to SCHMID et al. (2003), the duplication time is 11 days. Other authors, such as VAN DER STAR et al. (2008), have reported duplication times of between 5.5 and 7.5 days, values calculated based on the maximum conversion capacity. The same authors indicated the possibility of obtaining even shorter duplication times of around 3 days. Other researchers have reported the possibility of obtaining a duplication time of only 1.8 days under ideal operating conditions (ISAKA et al. 2006). One possible explanation for these divergent results is the method used for the determination of the growth rate. ISAKA et al. (2006) determined the growth rate based on the direct counting of the anammox bacteria, while in the other studies it was based on the biomass yield and nitrogen removal rate.

Autotrophic growth is a very costly process in terms of energy, and thus it is always associated with low growth rates when compared with heterotrophic growth. Consequently, the start-up period of the anammox process is very long and it takes a considerable time to achieve appreciable reaction rates. The use of reactors with efficient biomass retention is crucial to obtaining the enrichment of the anammox culture (JETTEN et al. 2001). The low growth rate of the anammox bacteria and the difficulty associated with obtaining enriched cultures of these microorganisms may hinder research involving the anammox process (STROUS et al. 1998). However, the low growth rate does not represent a limitation to the nitrogen removal capacity, which can reach values of 5–10 mgN/(m^3 day) due to the fact that the anammox microorganisms form compact biofilms or granules, enabling high concentrations of biomass to be reached in the bioreactor.

It should be noted that the extremely slow growth of anammox bacteria cannot be explained simply by autotrophy. The energy obtained from the catabolism (calculated per mol of electrons) is comparable with that of the autotrophic nitrification process, although the growth rate is much lower. Other plausible explanations for the slow growth may be related to the fact that the anammox bacteria have a low intrinsic conversion rate (of ammonium to nitrite) or that the enrichment of the culture is carried out under non-ideal growth conditions.

There are many uncertainties regarding the reaction intermediates in the catabolism of anammox bacteria. However, there is a general consensus that hydrazine (N_2H_4) is an intermediate. The production of hydrazine from hydroxylamine can be used as a method to detect active anammox biomass. The oxidation of this compound to N_2 is an energy-generating stage. Nitrite is not converted directly to hydrazine, but via hydroxylamine and/or nitric oxide (VAN DE GRAAF et al. 1997; STROUS et al. 2006). A schematic representation of the three possible metabolisms is illustrated in Fig. 5.3. Important enzymes involved in the process are hydroxylamine oxidoreductase (HAO), purified by SCHALK et al. (2000); hydrazine oxidase (HZO), purified by SHIMAMURA et al. (2007); and the nitrite reductases ccNir, partially purified by SCHALK (2000), and cd1Nir, found in the genome of *Kuenenia* (STROUS et al., 2006). Since all of these enzymes are able to carry out several reactions involving the conversion of nitrogen, it is still not clear which enzyme is responsible for a certain reaction.

According to VAN DONGEN et al. (2001a), the enzyme hydrazinase converts hydroxylamine into hydrazine. The hydrazine formed is oxidized by hydroxylamine oxidoreductase (HAO) to nitrogen gas, a stage in which four protons and four electrons are released. When nitrite is present in the system, the four electrons released allow the conversion of nitrite into hydroxylamine by the enzyme nitrite reductase. When nitrite is not present in the system and the anammox process is operated

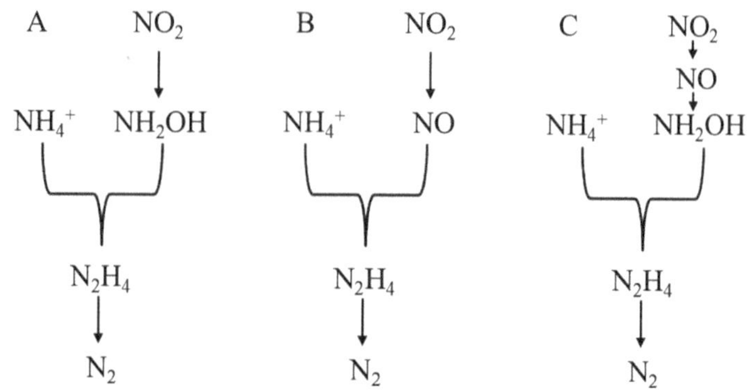

Fig. 5.3 Catabolic reactions of the anammox process with hydrazine acting as the main intermediate. Other potential intermediates are hydroxylamine (**a**), nitric oxide (**b**), or hydroxylamine and nitric oxide (**c**) (adapted from VAN DER STAR 2008)

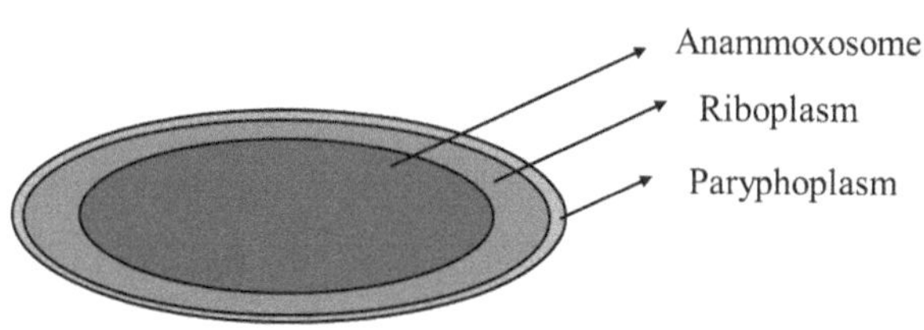

Fig. 5.4 Schematic representation of the different compartments of anammox bacteria (adapted from LINDSAY et al. (2001) and FUERST (2005)). The cytoplasm is divided into the paryphoplasm (external compartment), the riboplasm (where ribosomes and chromosomes are found), and the anammoxosome (where most or all of the cytochrome *c* is present and catabolism probably occurs)

under limited nitrite conditions, the electrons leave the system in a different way. This process generally occurs through the reaction of hydrazine disproportionation to ammonium and nitrogen gas according to reaction 5.10.

$$3N_2H_4 \rightarrow 4NH_3 + N_2 \qquad (5.10)$$

The disintegration of hydrazine occurs more slowly than the formation of hydroxylamine. As a consequence, hydrazine accumulates in the system. Since the disintegration of hydrazine into ammonium and nitrogen gas occurs, ammonium is expected to accumulate in the system.

It should be noted that while N_2O is usually the intermediate compound associated with denitrifying bacteria, this compound is not part of the physiology of anammox bacteria. This means that this powerful greenhouse gas is not produced by the anammox organisms.

The main compartment of anammox bacteria is anammoxosome. Anammoxosome is surrounded by riboplasm (where the ribosomes and chromosomes are located), which in turn are surround by the paryphoplasm (LINDSAY et al. 2001; VAN

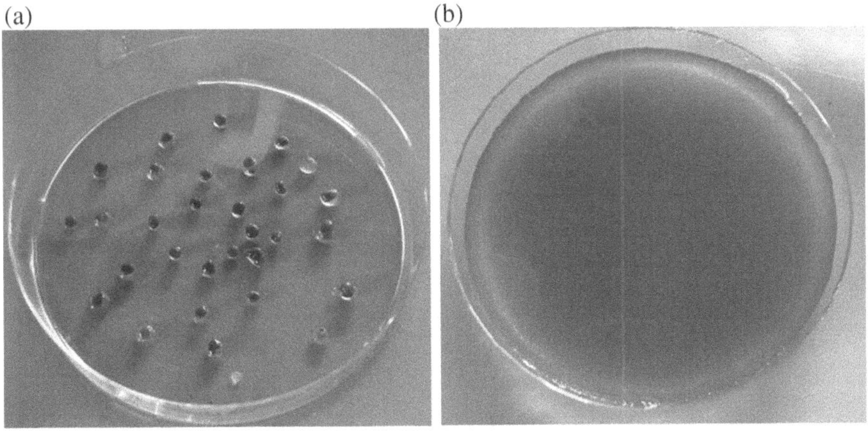

Fig. 5.5 Anammox biomass in the form of granules (a) and in suspension—free cells (b)

NIFTRIK et al. 2008), as shown in Fig. 5.4. Considering that there is no consensus regarding the characteristics of the membrane between the paryphoplasm and the riboplasm, the classification of the paryphoplasm (whether as a true internal compartment or as a region which is similar to the periplasm in Gram-negative bacteria) is still a matter under debate.

Microscopic observations suggest that the anammox bacteria, as in the case of other *Planctomycetes*, do not have peptidoglycans, although they exhibit a protein cell membrane. The lipids of the anammox bacteria contain a combination of fatty acids bound to esters (a typical characteristic of bacteria and eukaryotic cells) and ethers (typically found in Archaea). Lipid membranes are essential for the establishment of gradients of ions and metabolites. The anammox bacteria contain a variety of membrane lipids which are quite special and unique in nature (SINNINGHE DAMSTÉ et al. 2002, 2005; KUYPERS et al. 2003). The anammox cells are spherical shaped (coccus) and have a diameter of less than 1 μm. The anammox biomass presents a reddish brown color (Fig. 5.5), which is probably due to the high content of cytochromes (JETTEN et al. 1999).

The anammoxosome has been considered as the locus of catabolism, with the function of generating energy, in a way analogous to the function of mitochondria in eukaryotic cells (LINDSAY et al. 2001; VAN NIFTRIK et al. 2004). This hypothesis implies that the proton-motive force is created through the anammoxosome membrane for the coupling of energy generation and anabolism. The presence of important enzymes (hydrazine/hydroxylamine oxidoreductase) in the anammoxosome indicates that the anammox catabolism occurs in this compartment.

A biochemical model (Fig. 5.6) has been proposed in which the anaerobic oxidation of ammonium is catalyzed by various *c*-type cytochromes and proteins (STROUS et al. 2006). In this model, nitrite is firstly reduced to nitric oxide by a *c*-type cytochrome and *d1*-type cytochrome containing the enzyme nitrite reductase (NirS). It is assumed that the nitric oxide and ammonium are combined, forming

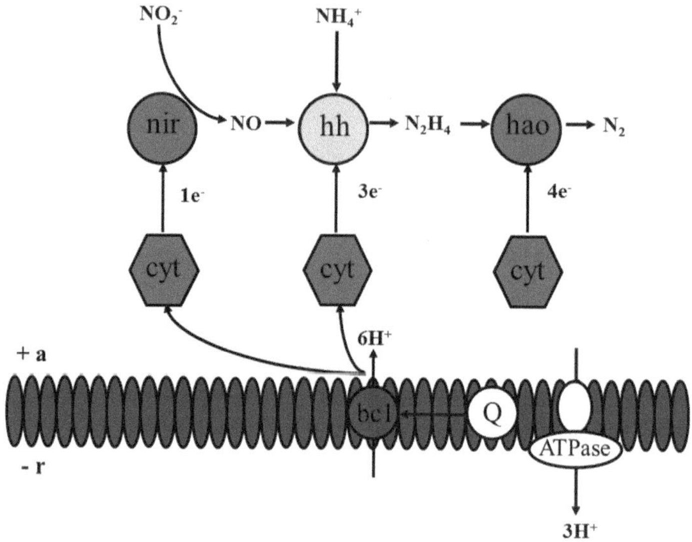

Fig. 5.6 Biochemical model representing anaerobic oxidation of ammonium coupled with the anammoxosome membrane in anammox bacteria, resulting from the proton-motive force and subsequent synthesis of ATP via ATPases bound to the membrane. *bc1* cytochrome *bc1* complex, *cyt* cytochrome, *hao* hydrazine/hydroxylamine oxidoreductase, *Q* coenzyme Q, *a* anammoxosome compartment, *r* riboplasm compartment (adapted from STROUS et al. 2006)

hydrazine through the action of hydrazine hydrolase, and this compound is finally oxidized to nitrogen gas by a *c*-type cytochrome protein called hydrazine/hydroxylamine oxidoreductase (SCHALK et al. 2000; SHIMAMURA et al. 2007). The four electrons derived from this oxidation are transferred to the cytochrome *c* electron carriers (CIRPUS et al. 2005; HUSTON et al. 2007), to ubiquinone, to cytochrome *bc1* complex, to cytochrome *c* electron carriers, and finally to nitrite reductase and hydrazine hydrolase.

In this model, the anammox reaction establishes a proton gradient through the translocation of protons from the riboplasm to the anammoxosome, which results in an electrochemical proton gradient directly from the anammoxosome to the riboplasm. This gradient contains chemical potential energy (chemical proton gradient in the form of a difference in the pH in which the riboplasm is more alkaline compared with the anammoxosome) and electrical potential energy (electrical proton gradient in the form of a difference in the charge considering that the riboplasm is negatively charged in relation to the anammoxosome). The differences in both the pH and charge cause a proton displacement from outside to inside the anammoxosome, providing the proton-motive force. This mechanism can be used for the synthesis of ATP catalyzed by adenosine triphosphatases (ATPases) located in the anammoxosome membrane, as shown in Fig. 5.6. The protons passively move back to the riboplasm (due to the electrochemical proton gradient) through the pores formed by the ATPases. The globular and hydrophobic domain of the ATPases where ATP is synthesized will be located in the riboplasm, and its hydrophobic domain where protons are translocated will be located in the anammoxosome membrane. The ATP synthesized will be released into the riboplasm.

The anammox bacteria are dependent on the electrochemical ion gradient through the membrane for ATP synthesis. Since the anammox catabolism is slow, only a few protons are translocated within a certain time, while the dissipation of the electrochemical gradient resulting from the passive diffusion is independent of the growth rate and proceeds at the normal rate. Thus, the passive diffusion of protons through the membrane is relatively important and leads to a greater energy expenditure in the case of anammox bacteria. To give an idea, the expenditure due to the passive diffusion of protons in mitochondria corresponds to 10% (HAINES 2001). Thus, it is clear that the presence of a special, less permeable membrane is essential for the metabolism of anammox cells. Additionally, the intermediates of the anammox reaction, such as hydrazine, easily diffuse through the membrane.

From a bioenergy perspective, the loss of energy associated with the loss of one hydrazine molecule from the anammox cell is equivalent to 15 catabolic cycles. The explanation for this is as follows: when one molecule of hydrazine is lost, the hydrazine reserve needs to be reestablished. Presumably, hydrazine is formed via the reduction of nitrite to nitric oxide and the subsequent combination of nitric oxide with ammonium. The equivalents required (four electrons) must come from the oxidation of the reserve material, such as glycogen, which is derived from CO_2. Considering that only one molecule of CO_2 is fixed for every 15 mol of ammonium oxidized (15 catabolic cycles), a 10% loss of hydrazine would cause a complete loss of cell viability. Thus, the limitation of the diffusion of both protons and anammox intermediates is extremely important for the metabolism of anammox bacteria. Since anammox catabolism occurs in the anammoxosome, the lipid membrane of this compartment (described previously) needs to have a rigid structure in order to limit the diffusion of intermediates which are important to the process outside the anammoxosome. This is actually a specific adaptation of anammox bacteria for their unusual metabolism.

Due to the dense structure of the anammoxosome membrane, specific transporters are required to regulate the transport of ammonium and nitrite. The genome of *Candidatus Kuenenia stuttgartiensis* (one of the species of anammox bacteria, as will be described below) encodes four ammonium transporters (Amt), four formate/nitrite transporters (FocA), and two nitrate/nitrite transporters (NarK) of unknown location (STROUS et al. 2006).

While the anammoxosome membrane must be rigid, the cytoplasmic membrane must be flexible and permeable for the maintenance of homeostasis, controlling the intracellular ion concentrations and transport processes. Thus, with a rigid anammoxosome membrane and a flexible cytoplasmic membrane, the cell can overcome a problem associated with the presence of only one membrane, which would have to be impermeable and permeable at the same time. Moreover, the use of an intracytoplasmic compartment (anammoxosome) for the synthesis of ATP through the proton-motive force results in complete control of this force, allowing efficient energy transduction.

Until now, four genera of anammox bacteria have been described. The similarity of the 16S rRNA gene sequences of the species varies between 87 and 99% (SCHMID et al. 2007). Despite this relatively large phylogenetic distance, all of the anammox organisms belong to the same family, called *Anammoxaceae* (JETTEN et al. 2008), forming a monophyletic group belonging to the phylum *Planctomycetes* (STROUS et al. 1999a). One characteristic of *Planctomycetes* is its unusually high level of cellular organization, each cell consisting of one or more internal compartments bound by membranes with variable and unknown functions (LINDSAY et al. 2001; FUERST 2005).

The different anammox bacteria include those which have been enriched from activated sludge reactors, such as *Candidatus Kuenenia stuttgartiensis* (SCHMID et al. 2000), *Candidatus Brocadia anammoxidans* (STROUS et al. 1999a), *Candidatus Brocadia fulgida* (KARTAL et al. 2004), and *Candidatus Anammoxoglobus propionicus* (KARTAL et al. 2007), as well as those detected in marine environments, particularly in sediments and zones with minimal oxygen content, such as *Scalindua brodae*, *Scalindua wagneri*, and *Scalindua sorokinii* (SCHMID et al. 2003).

Since none of the anammox bacteria have been obtained as a pure culture, all of the species take the name *Candidatus*. *Brocadia* and *Kuenenia* have been generally enriched in laboratory experiments using a synthetic medium similar to that employed by VAN DE GRAAF et al. (1996) regardless of the inoculum used (SCHMID et al. 2000; VAN DONGEN et al. 2001b; CHAMCHOI and NITISORAVUT 2007). The addition of fatty acids was reported to lead to the enrichment of *Anammoxoglobus* (KARTAL et al. 2007) and *Brocadia fulgida* (KARTAL et al. 2004, 2008). Despite the considerable diversity in different enrichment systems, *Scalindua sorokinii* is the dominant species in marine environments.

Some studies have indicated that besides the conversion of ammonium and nitrite, anammox bacteria belonging to the genera *Brocadia*, *Anammoxoglobus*, and *Kuenenia* are also able to metabolize fatty acids such as propionate, acetate, and formate (GÜVEN et al. 2005; KARTAL et al. 2008). The oxidation of these fatty acids to CO_2 is coupled with the reduction of nitrate (via nitrite) to ammonium. Thus, the anammox bacteria are able to produce their own substrate (ammonium and nitrite) for their catabolism (Fig. 5.7).

It is still unknown whether the conversion of nitrate to ammonium represents an additional catabolic reaction, which would constitute an energy source. Surprisingly, studies have indicated that the fatty acids are not incorporated into the biomass, being completely converted into CO_2 (KARTAL et al. 2008). Bearing in mind that the anammox bacteria also carry out the fixation of CO_2 via acetate (the reverse reaction in relation to the previously described acetate oxidation), this characteristic is rather surprising, although already established for anammox bacteria.

In addition to the conversion of fatty acids, *Kuenenia stuttgartiensis* has been shown to be able to oxidize Fe^{2+} to Fe^{3+} using nitrate as the electron acceptor, as well as to reduce Fe^{3+} to Fe^{2+} and Mn^{4+} to Mn^{2+} using formate as an electron donor (STROUS et al. 2006). As in the case of the fatty acids conversion, the metabolism and the growth on these substrates are unknown.

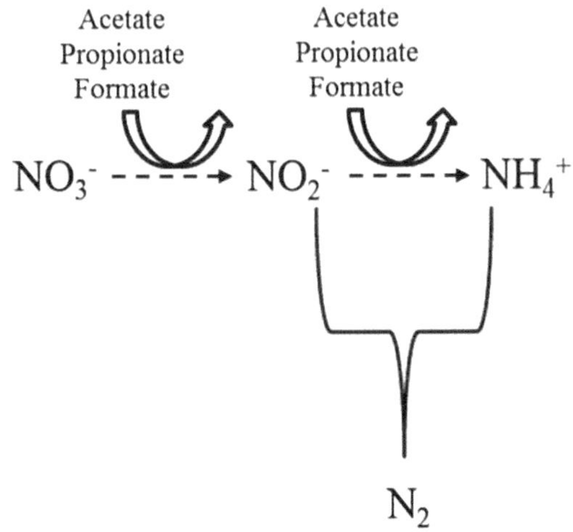

Fig. 5.7 Metabolic versatility of anammox bacteria. Besides the conversion of ammonium and nitrite (*solid lines*), anammox bacteria can use short-chain fatty acids as electron donors for ammonification (*dashed lines*) (adapted from VAN DER STAR 2008)

5.2.2.3 Factors Influencing Anammox Bacteria Activity

Substrates and Products

The concentration of nitrite is an important parameter that needs to be controlled in the anammox process. However, a certain concentration of nitrite is required during the start-up of the anammox system. If the concentration of nitrite is very low, substrate limitation may lead to a low growth rate. On the other hand, a very high concentration can lead to inhibition. Several ranges of nitrite concentration which can cause inhibition have been reported in the literature, and thus there is no consensus observed in the previous studies.

STROUS et al. (1999b) observed that the anammox process (using *Candidatus Brocadia anammoxidans*) was completely inhibited when the nitrite concentration was higher than 100 mgN/L. DAPENA-MORA et al. (2007) observed that 350 mgNO_2-N/L caused 50% inhibition of the anammox process. FUX (2003) maintained the nitrite concentration at around 40 mgN/L for several days and observed an irreversible inactivation of anammox bacteria. To restore the anammox activity due to inhibition by nitrite, trace amounts of intermediates of the anammox process, such as hydroxylamine and hydrazine, were added, even after long periods of exposure to high nitrite concentrations (STROUS et al. 1999b).

In this regard, studies have shown that the nitrite tolerance differs according to the genera of the anammox bacteria. EGLI et al. (2001) carried out experiments with *Candidatus Kuenenia stuttgartiensis* and observed that the anammox process was only inhibited when submitted to nitrite concentrations higher than 182 mgN/L. The experiments carried out by STROUS et al. (1999b) showed that with an increase in the nitrite concentration, the stoichiometry of the process changed. The stoichiometry of the ammonium and nitrite consumption increased from 1.3 g of nitrite per gram of ammonium (when the nitrite concentration was 0.14 gN/L) to almost 4 g of nitrite per gram of ammonium (when the nitrite concentration was 0.7 gN/L).

The significant change in the stoichiometry at high nitrite concentrations suggests that the microorganisms, when submitted to these conditions, do not use only ammonium as the electron donor, but they must also generate an internal electron donor to reduce the nitrite. A change in the stoichiometry has also been observed at high temperatures. DOSTA et al. (2008) reported nitrite-to-ammonium consumption ratios of 1.38:1 at 30 °C and 1.05:1 at 18 °C.

STROUS et al. (1999b) reported that the anammox process is not inhibited by ammonium or by nitrate at concentrations below 1 gN/L. However, DAPENA-MORA et al. (2007) observed a 50% drop in activity at high concentrations of ammonium and nitrate (770 and 630 mgN/L, respectively). Considering that chemolithoautotrophic organisms essentially use inorganic carbon as a carbon source, the influent concentration of bicarbonate is an important factor which can affect the enrichment of the anammox culture.

DEXIANG et al. (2007) observed low anammox activity at low bicarbonate-to-ammonium ratios (2.3:1). The reduction in the activity under these conditions may be due to the limitation of CO_2. On the other hand, a high bicarbonate concentration (bicarbonate-to-ammonium ratio of 4.7:1) can also lead to inhibition. In this case, the inhibition is related to the formation of significant quantities of free ammonia due to the increase in the pH to values above 8.

Oxygen

Anammox bacteria are strictly anaerobic and are inhibited by the presence of dissolved oxygen in the medium. The inhibition caused by low oxygen concentrations has been described in several reports in the literature as being reversible. Based on experiments in which oxygen was provided intermittently, STROUS et al. (1997) concluded that the anammox process was reversibly inhibited by oxygen, which would make it possible to obtain partial nitritation and anammox in a single reactor. EGLI et al. (2001) affirmed that the anammox metabolism is reversibly inhibited at low oxygen levels (0.25–2% of air saturation) although it is probably irreversible at high levels (>18% of air saturation).

Organic Carbon

The treatment of landfill leachates and effluents from sludge digesters, which contain a high concentration of nitrogen, can be carried out through the new nitrogen removal processes, such as the combined partial nitritation-anammox. However, besides the substantial amount of nitrogen, these wastewaters may also contain a high concentration of organic matter. Nevertheless, they are considered to be potentially treated in an anammox system. During anaerobic digestion, organic matter is easily biodegraded and converted into biogas. Thus, only the portion with low biodegradability remains in the digester effluent.

RUSCALLEDA et al. (2008) observed that anammox bacteria and denitrifying organisms can coexist and are important for the treatment of wastewaters with high concentrations of slowly biodegradable organic matter, such as digester effluent and leachates. In these wastewaters, in particular, the growth of denitrifying heterotrophic organisms is limited by the low availability of easily biodegraded organic carbon. As a consequence, these microorganisms cannot become dominant and do not hamper the growth of anammox organisms.

However, in several studies reported in the literature, a negative effect of the presence of organic matter on the growth of anammox bacteria was observed (JETTEN et al. 1999; MOLINUEVO et al. 2009; TANG et al. 2010). In the presence of certain quantities of organic matter, the slow-growing anammox bacteria are no longer able to compete with the denitrifying organisms for nitrite, given the higher growth rate of the latter. Furthermore, the denitrification reaction is thermodynamically more favorable than the anaerobic oxidation of ammonia, since the Gibbs free energy of anammox and denitrification reactions are of the order of −355 kJ/mol (JETTEN et al. 1999) and −427 kJ/mol (RITTMANN and MCCARTY 2001), respectively. Thus, the denitrifying heterotrophs will grow more rapidly when organic carbon is present in combination with ammonium and nitrite, hindering the development of anammox microorganisms.

As in the case of nitrite, there is no consensus in the literature regarding the concentration of organic matter at which denitrifying microorganisms hamper the growth of anammox bacteria. CHAMCHOI et al. (2008) reported that concentrations of organic matter above 300 mg/L (in terms of COD) or COD/N ratios greater than 2 caused the inactivation of anammox organisms in a UASB reactor fed with milk with a high fat content as a source of organic carbon. TANG et al. (2010) reported that the denitrifying microorganisms began to dominate the system when a high COD/NO_2-N ratio (2.9:1) was applied. MOLINUEVO et al. (2009) observed complete inhibition of the anammox process when the COD was 292 mg/L.

Since the anammox process removes only 90% of the nitrogen present in the form of ammonium/nitrite and 10% of the nitrogen remains in the effluent in the form of nitrate, the coexistence of anammox and denitrifying organisms is favorable. Under anoxic conditions, nitrate can be reduced to nitrite by denitrifiers, and nitrite can then be used by anammox bacteria for the anaerobic oxidation of ammonium (KUMAR and LIN 2010).

It should be noted that not all types of organic matter can be used in processes in which anammox and denitrifying bacteria coexist. As reported by GÜVEN et al. (2005), the anammox activity is completely and irreversibly inhibited by methanol and ethanol. This aspect needs to be taken into consideration, given that methanol is commonly used to remove nitrate in post-denitrification systems. Inhibition by methanol may be caused by the formation of formaldehyde by the enzyme hydroxylamine oxidoreductase (PAREDES et al. 2007).

On the other hand, some carbon sources do not have an inhibitory effect on the anammox activity. Therefore, they can be used by anammox bacteria. As mentioned in Sect. 5.2.2.2, where the different types of anammox metabolisms were described, some genera of anammox bacteria are able to oxidize acetate and propionate.

Studies regarding the adaptation of anammox bacteria to wastewaters containing toxic components have also been described in the literature. TOH and ASHBOLT (2002) and TOH et al. (2002) observed the acclimation of anammox organisms to a synthetic medium simulating a coke oven effluent, which contained not only a high concentration of organic compounds (COD of 2000–2500 mg/L) but also some chemical compounds such as phenol (300–800 mg/L), cyanides (10–90 mg/L), and thiocyanates (300–500 mg/L). The initial attempt to enrich the anammox bacteria failed, although the gradual addition of 50–500 mg/L of phenol allowed the adaptation of these organisms.

Temperature and pH

A pH range of 6.7–8.3 is considered to be ideal for anammox bacteria, with an optimum value of 8.0 (STROUS et al. 1999b). The ideal temperature is between 30 and 40 °C (STROUS et al. 1999b; EGLI et al. 2001). Experiments carried out by DOSTA et al. (2008) to evaluate the short-term effect of temperature on the anammox activity showed that the maximum non-adapted anammox biomass activity was obtained at between 35 and 40 °C, while a temperature of 45 °C caused an irreversible decrease in the anammox activity due to cell lysis. Small differences in the optimum temperature were found for *K. stuttgartiensis* (40 °C) and *B. anammoxidans* (37 °C) (STROUS et al. 1999b; EGLI et al. 2001).

Although the optimum temperatures for the anammox process are relatively high, CEMA et al. (2007) and ISAKA et al. (2006) managed to operate the anammox process in a rotating biological contactor (RBC) and an anaerobic biofilter, respectively, at a temperature of 20 °C. The gradual adaptation of the biomass appears to be a key factor in successfully operating the anammox process at temperatures lower than those considered ideal for the process (SZATKOWSKA and PLAZA 2006).

In order to start up the anammox system at low temperatures, one strategy is to produce the desired quantity of biomass in a separate reactor, which should be operated at temperatures close to those considered ideal. Later, the biomass can be gradually adapted to lower temperatures in the same reactor, and finally the adapted biomass can be inoculated into the reactor maintained under low temperature conditions (DOSTA et al. 2008).

Some researchers who have carried out studies on samples of anammox bacteria originating from sediments have reported anammox activity at low temperatures, suggesting that local environmental factors influence the characteristics of these bacteria. RYSGAARD et al. (2004) observed anammox activity in artic sediments at a temperature ranging from −1.3 to 30 °C. The optimum temperature was found to be 12 °C. Similar results were obtained by DALSGAARD and THAMDRUP (2002), who observed an optimum temperature of 15 °C for marine sediments from the Baltic Sea.

It should be noted that, in contrast to anammox bacteria in wastewater treatment systems, anammox bacteria in marine environments are dependent on another process to obtain the nitrite required in the process. In marine environments, nitrate is much more abundant than nitrite, and thus the anammox process requires an additional step for the reduction of nitrate to nitrite. Since the dissolved oxygen concentration decreases progressively through the sediment, in the deeper layers, nitrate reducers can cause the accumulation of nitrite, allowing the occurrence of the anammox process (DALSGAARD et al. 2005).

Although the physical properties of sludge and the bacterial populations can remain constant during the reactor operation at lower temperatures, the nitrogen conversion rate is substantially reduced. This drawback can be minimized by applying a strategy described by ISAKA et al. (2006), who achieved a high nitrogen conversion (8.1 kgN/(m^3 day)) through reducing the hydraulic retention time (HRT) and adding appropriate and non-inhibiting concentrations of nitrite to the influent.

Biomass Concentration

The anammox activity is highly influenced by the biomass concentration. According to STROUS et al. (1999b), anammox bacteria are only active when the cell concentrations are greater than 10^{10}–10^{11} cells/mL, even in highly enriched cultures. It is possible that the presence of contaminant cells, 1 in 200–500, is required to sustain the growth, since these cells could ensure vitamin supplementation and the removal of toxic components (KUENEN and JETTEN 2001).

PYNAERT et al. (2004) described a hypothesis in which the presence of ammonium-oxidizing bacteria is required for the reactivation of anammox organisms after the biological system has undergone some disturbances. Through the production or accumulation of hydroxylamine or hydrazine by the bacteria responsible for the oxidation of ammonium, anammox bacteria can reactive their metabolism. Once the process is reestablished, the tendency is that the ammonium-oxidizing bacteria do not participate in the anammox process. This supposition has also been described by STROUS (2000) based on the fact that the addition of the intermediates hydroxylamide and hydrazine was needed in order to restart the anammox process after its inhibition.

Suspended Solids

Flocculating agents are generally used to remove organic and inorganic colloidal substances from wastewaters prior to the anammox process. The effect of these flocculants on the anammox process was the focus of a study carried out by DAPENA-MORA et al. (2007). Concentrations of up to 1 g/L of a positively charged polymeric compound used as a flocculant did not have a negative effect on the anammox activity.

In a study carried out by YAMAMOTO et al. (2008), a large amount of influent suspended solids present in partially nitrified digested liquid adhered to the material covering the anammox biomass, which was growing on a support material. Consequently, the anammox activity decreased and the performance was significantly adversely affected. The use of a flocculant improved the settleability of the influent suspended solids and reduced its accumulation in the reactor. However, the flocculant was also retained on the surface of the support media, leading to a reduction in the anammox activity.

The precipitation of salts can also lead to the unstable operation of anammox reactors. TRIGO et al. (2006) operated an anammox reactor with membranes, which functioned as a barrier to retain inorganic salts which precipitated and accumulated in the biomass. The precipitation of these salts on the biomass surface led to a reduction in the nitrogen removal from 100 to 10 mg/(L day).

Light and Reactor Mixing Velocity

A study carried out by VAN DE GRAAF et al. (1996) indicated that anammox bacteria are sensitive to visible light. These authors observed a decrease in the anammox activity from 30 to 50%. The results influenced the operating conditions and thus the reactors were covered to avoid the negative effect of light. The effect of shear stress on the anammox process was evaluated by ARROJO et al. (2006), who observed that stirring velocity of up to 180 rpm did not have a negative effect on the performance of the anammox process. However, when the stirring speed was increase to 250 rpm, the anammox activity and the average diameter of the flocs were reduced by 40% and 45%, respectively. In addition, an accumulation of nitrite was observed under these conditions.

5.2.2.4 Application of Anammox Process

The anammox process offers several advantages for the removal of ammonium from wastewaters. In conventional systems, ammonium is removed through nitrification (oxidation of ammonium to nitrate) followed by denitrification (reduction of nitrate to nitrogen gas). The anammox process should always be combined with a partial nitritation process, in which half of the ammonium is oxidized to nitrite (and not to nitrate). The nitrite produced reacts with the remaining ammonium to form nitrogen gas. The main advantages of the combined autotrophic process of partial nitritation + anammox are:

- Lower energy requirement considering the lower aeration costs.
- An organic carbon source (external oxidant) is not required, which leads to a lower production of sludge with a consequent reduction in the sludge disposal costs.
- Absence of CO_2 emissions due to the autotrophic nature of the nitritation and anammox processes (CO_2 is consumed and not produced).

The anammox process is especially appropriate for the treatment of wastewaters containing high concentrations of ammonium but with low organic content. Given

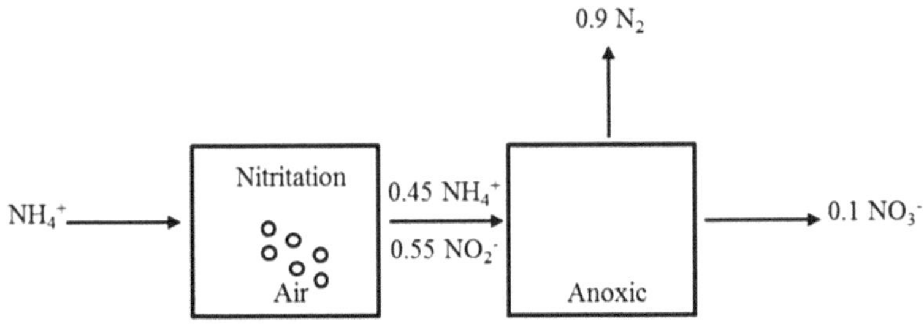

(a) Partial nitritation + anammox process in separate reactors (two stages)

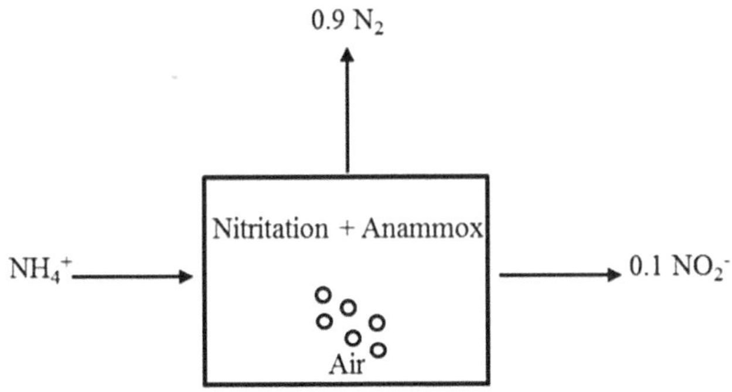

(b) Partial nitritation + anammox process in the same reactor (one stage)

Fig. 5.8 Removal of ammonium through the anammox process carried out with two-reactor configurations: (**a**) two-reactor configuration, nitritation occurs in the first reactor (aerated) and the anammox process in the second reactor (anoxic); (**b**) both processes occur in the same aerated reactor (adapted from VAN DONGEN et al. 2001b)

that the anammox bacteria are characterized by their interaction with other bacteria, since they require nitrite generated by another microbial group, two main reactor configurations are possible for the removal of ammonium through the anammox process:

1. A system comprised of two reactors (Fig. 5.8a) in which part of the ammonium is firstly oxidized to nitrite (partial nitritation) in the reactor maintained under aeration. Subsequently, the ammonium/nitrite mixture is sent to the second reactor, maintained under anoxic conditions, where it is subjected to the anammox process (VAN DONGEN et al. 2001b). The partial nitritation process can be carried out in a SHARON (single-reactor high-activity ammonia removal over nitrite) system, which will be described in detail below.
2. A configuration comprising only one reactor (Fig. 5.8b), in which partial nitritation and the anammox process occur in the same reactor, maintained under aeration. In this configuration the nitrification occurs in the aerobic region of the floc or granule, while the anammox reaction occurs in the deepest zone of the biofilm, maintained under anoxic conditions (HIPPEN et al. 1997; KUAI and VERSTRAETE 1998). The alternation of aerobic (aeration on) and anoxic

(aeration off) periods is another approach to achieving nitritation and anammox reactions in a single reactor.

Currently, the anammox process is employed in both configurations for the removal of ammonium from sludge digester effluents, leachates, and different industrial wastewaters. Besides the application of the anammox process for the removal of ammonium, the combined removal of ammonium and nitrate has also been employed in laboratory scale (KALYUZHNYI et al. 2006; PATHAK et al. 2007). In this process, partial denitrification (reduction of nitrate to nitrite) is coupled with the anammox process. In cases where the electron donors for the denitrification consist of fatty acids, the anammox bacteria are able to carry out the reduction of nitrate (GÜVEN et al. 2005). In the next section, the two systems (two reactors and single reactor) aimed at partial nitritation and anammox will be described, with emphasis on the different types of reactors used and the practical implementation of these systems.

Partial Nitritation and Anammox in Two Separate Reactors (Two Stages)

The combined process of partial nitritation and anammox in two separate reactors makes use of the advantages of the first process to carry out the conversion of half of the ammonium only up to nitrite (and not up to nitrate), providing the substrates required for the anammox process (ammonium and nitrite) in proportions suitable for the generation of nitrogen gas.

One of the challenges during the operation of the first reactor is obtaining an effluent with an ammonium-to-nitrite ratio similar to the stoichiometric ratio of 1:1.32, proposed by STROUS et al. (1998) to represent the anammox reaction. In practice, however, this ratio should be close to 1:1 in order to prevent inhibition by nitrite and provide an excess of ammonium. Given that there is no need for organic compounds or anoxic periods, a partial nitritation reactor can provide the desired ammonium/nitrite mixture, without the need to control the refeeding. One factor which makes this conversion possible is that after 50% of the ammonium has been oxidized the decrease in the pH hinders the oxidation of the residual ammonium.

By limiting the supply of oxygen in a nitrifying reactor with sludge retention, the same results can be obtained, although control of the refeeding may be required (STROUS et al. 1997). It is important that the composition of the anammox reactor influent remains constant, considering the toxicity of nitrite, regardless of the strategy used to obtain adequate proportions of ammonium and nitrite in the first reactor (partial nitritation).

The application of the two-reactor configuration is particularly appropriate when biodegradable organic compounds and toxic compounds are present, since these are degraded in the step prior to partial nitritation and do not reach the anammox reactor (VÁZQUEZ-PADÍN et al. 2009; LACKNER et al. 2008).

As previously mentioned, nitrogen removal based on partial nitritation with the anammox process offers many advantages. Besides the fact that the external addition of carbon is not required, it generates a low quantity of sludge and requires 40% less oxygen than the conventional process, which leads to energy savings (AHN 2006). Furthermore, the operation of a system comprised of two separate reactors (one for partial nitritation and one for the anammox process) is more flexible in comparison with the configuration in which both processes occur in a single reactor. Additionally, since the two processes occur in different units, the process performance is more stable (WYFFELS et al. 2004). Detailed description of the operating conditions required to achieve partial nitritation and anammox conversions is provided below.

Partial Nitritation

In this section, firstly the factors which affect the nitrification process will be presented, some of which, or their combination, represent the basis of the development of partial nitritation technologies. Several studies on this subject will then be described.

In practice, all of the factors involved in achieving partial nitritation are related to the inhibition or limitation of the second stage of nitrification (nitratation or the formation of nitrate). The crucial point in the control of partial nitrification is to obtain a nitrifying reactor with the stable accumulation of nitrite. In order to force biological processes to follow the nitrite route, different strategies have been used (BERNET et al. 2005), which include controlling the temperature, hydraulic retention time, pH, dissolved oxygen, and presence of free ammonia. Table 5.2 details how these factors influence the growth and activity of the microorganisms responsible for nitrification.

One approach to achieving partial nitrification is based on the difference in the activation energies of ammonium (68 kJ/mol) and nitrite (44 kJ/mol) oxidation. The high activation energy of the ammonium oxidation reaction leads to the velocity of this process having a greater degree of dependence on the temperature in comparison with the nitrite oxidation reaction. Only at temperatures above 25 °C is it possible for ammonium-oxidizing bacteria to become dominant to the detriment of nitrite-oxidizing bacteria (VAN DONGEN et al. 2001a; BROUWER et al. 1996). If this condition is combined with a low hydraulic retention time and low cell retention time, the bacteria which oxidize nitrite can be selectively washed out of the system (HELLINGA et al. 1998).

The pH has a strong influence on the system due to the fact that at low values of this parameter the nitrite-oxidizing bacteria grow more rapidly than the ammonium-oxidizing bacteria. Thus, the hydraulic retention times (or dilution rate) required to maintain the ammonium oxidizers and wash out the nitrite oxidizers are more flexible at higher pH values (HELLINGA et al. 1998).

In relation to the pH ranges considered ideal for nitrification, some main effects of this parameter on nitrifying bacteria have been identified: activation/deactivation of nitrifying bacteria, nutritional effects associated with the alkalinity and inorganic

Table 5.2 Effect of some factors (temperature, pH, free ammonia, nitrous acid) on the growth and activity of nitrifying bacteria

Factor	Effect	Reference
Temperature		
$T > 15$ °C	Ammonium-oxidizing bacteria grow more rapidly than nitrite-oxidizing bacteria	van Dongen et al. (2001a), Brouwer et al. (1996)
$T = 25$ °C	Ammonium-oxidizing bacteria become dominant	van Dongen et al. (2001a), Brouwer et al. (1996)
pH		
7.0–8.0	Optimum range for nitrification	Antoniou et al. (1990); Painter and Loveless (1983)
7.9–8.2	Optimum range for ammonium oxidizers	Alleman (1984)
7.2–7.6	Optimum range for nitrite oxidizers	Alleman (1984)
Free ammonia		
150 mg/L	Inhibition of ammonium-oxidizing and nitrite-oxidizing bacteria	Anthonisen et al. (1976)
1–7 mg/L	Inhibition of ammonium-oxidizing bacteria and nitrite accumulation	Anthonisen et al. (1976), Abeling and Seyfried (1992); Kim et al. (2003)
Nitrous acid > 2.8 mg/L	Inhibition of ammonium-oxidizing and nitrite-oxidizing bacteria	Anthonisen et al. (1976)

carbon species, and inhibition by ammonium and nitrous acid (VILLAVERDE et al. 1997). The activation/deactivation of nitrifying bacteria is related to the binding of H^+ or OH^- ions to enzyme groups, blocking the active sites in a reversible manner (QUINLAN 1984). Nutritional effects are mainly associated with the availability of inorganic carbon, which is essential for autotrophic nitrifying microorganisms. At low pH values, the CO_2 species predominate which can be easily removed from water by way of stripping. On the other hand, at high pH values, inorganic carbon is present mainly in the form of carbonate, which is rarely assimilated.

The presence of free ammonia and nitrous acid is strongly associated with the pH value of the medium. The pH affects the substrate concentration in both stages of nitrification, modifying the acid-base equilibrium. With an increase in the pH, for instance, greater concentrations of free ammonia are present in the medium, which can inhibit both the ammonium-oxidizing bacteria and the nitrite-oxidizing bacteria. A reduction in the pH, however, favors the presence of nitrous acid. Both the free ammonia and the nitrous acid can inhibit ammonium-oxidizing and nitrite-oxidizing microorganisms, although the latter are more sensitive than the former, especially in the presence of free ammonia.

Another strategy aimed at avoiding the development of nitrite-oxidizing bacteria and promoting the accumulation of nitrite is to reduce the concentration of dissolved oxygen in the medium. This approach is based on the fact that nitrite-oxidizing bacteria are more sensitive to low dissolved oxygen concentrations in the medium than ammonium-oxidizing bacteria, which have less affinity for oxygen. The oxygen saturation coefficients (Monod kinetics) for the oxidation of ammonium and nitrite are 0.3 and 1.1 mg/L, respectively (WEISMANN 1994).

A possible mechanism for the inhibition of nitrite oxidation caused by low oxygen concentrations is based on the accumulation of hydroxylamine, an intermediate product in ammonium oxidation. In general, ammonium-oxidizing bacteria obtain energy by way of ammonium oxidation to nitrite in a two-step reaction, with hydroxylamine (NH_2OH) as an intermediate. The first step is the oxidation of ammonium, catalyzed by the enzyme ammonia monooxygenase, while the second step involves the oxidation of hydroxylamine, catalyzed by the enzyme hydroxylamine oxidoreductase. At low oxygen concentrations and high ammonium concentrations, the accumulation of hydroxylamine can occur, which can inhibit the nitrite-oxidizing microorganisms starting from concentrations of 250 μM. Concentrations above 2000 μM can also lead to the inhibition of ammonium-oxidizing bacteria. Equations 5.11–5.13 represent the accumulation of hydroxylamine under oxygen-limiting conditions (YANG 1990).

From these equations it can be observed that the four electrons generated from the oxidation of hydroxylamine can be transferred to the oxidation of ammonium when the reduction of the terminal oxygen is interrupted, due to the oxygen deficiency aimed at balancing the number of electrons of this redox reaction (YANG 1990). In a study carried out by HU (1990), the hydroxylamine caused a severe inhibition of *Nitrobacter*, suggesting that nitrite may be accumulated in nitrifying systems. However, hydroxylamine has been practically ignored in terms of nitrification processes, since its concentration is considered to be insignificant.

YANG and ALLEMAN (1992) studied the possibility of hydroxylamine accumulation and its relation to the accumulation of nitrite in a batch system containing a nitrifier-enriched culture. The results obtained by these authors indicated that the quantity of nitrite accumulated increased with an increase in pH, and this was directly related to an increase in the non-ionized hydroxylamine. Hydroxylamine was considered to be the main cause of nitrite accumulation in a nitrifying system maintained under conditions of low dissolved oxygen concentrations and high pH.

$$4H^+ + 2NH_3 + 4e^- + 2e^- \rightarrow 2NH_2OH + 2H_2O \tag{5.11}$$

$$NH_2OH + H_2O \rightarrow HNO_2 + 4e^- + 4H^+ \tag{5.12}$$

$$\begin{aligned}&\text{Global equation:}\\&2NH_3 + 2O_2 \rightarrow NH_2OH + HNO_2 + H_2O\end{aligned} \tag{5.13}$$

Several studies aimed at obtaining nitrification through controlling the dissolved oxygen have been carried out, both in suspended biomass and biofilm systems (Table 5.3). In the case of systems with suspended biomass operated under oxygen-limiting conditions, the complete and stable conversion of ammonium into nitrite was obtained, regardless of the sludge age. However, the sludge age became a critical

Table 5.3 Effect of dissolved oxygen concentration on the nitrification process

System	Dissolved oxygen (mg/L)	Effect	Reference
Activated sludge	<0.5	Accumulation of ammonium and nitrite	Ciudad et al. (2005)
	0.7	Nitrite accumulation around 67% of ammonium fed to system	
	1.0	80% ammonium oxidation, 80% being accumulated as nitrite	
	1.4	99% ammonium oxidation, 70% being accumulated as nitrite	
	2.4	99% ammonium oxidation, 10% being accumulated as nitrite	
Activated sludge	<0.5	Accumulation of ammonium and nitrite	Ruiz et al. (2003)
	1.7	Complete nitrification	
Growth in suspension	0.5	Accumulation of nitrite	Hanaki et al. (1990)
	6	Complete nitrification	
Air-lift reactor	1.0	Stable operation with 100% nitrite accumulation	Kim et al. (2003)
Air-lift reactor	<1.0	Low ammonium conversion and low nitrite accumulation	Garrido et al. (1997)
	1.5	50% conversion of ammonium to nitrite	
	>2.5	Complete nitrification, ammonium oxidation depending on applied load	
Biofilm reactor	0.5	Stable nitrite accumulation (90%) and 100% ammonium removal	Bernet et al. (2001)
	>0.5	Accumulation of nitrate in effluent; decrease in DO causes nitrite accumulation	
Biological aerated filter	2.0–5.0	Accumulation of nitrite up to 60% of the ammonium conversion	Joo et al. (2000)
Moving bed biofilm reactor	3.0	50% conversion of ammonium to nitrite. Complete nitrification after 11 months of operation	Fux et al. (2004)

parameter for obtaining partial nitrification when the operation was not carried out under oxygen-limiting conditions. YANG and ALLEMANN (1992) concluded that a combination of parameters, such as the concentrations of free ammonia, dissolved oxygen, and hydroxylamine, comprise the main factors involved in the accumulation of nitrite in a batch system containing a nitrifier-enriched culture.

As indicated in Table 5.3, the results obtained in biofilm systems are similar to those obtained in reactors with suspended biomass. In general, low oxygen concentrations lead to an accumulation of nitrite. However, in biofilm systems, the nitrite-oxidizing bacteria can be further adversely affected by the actual stratification of the biofilm. In most cases, ammonium-oxidizing bacteria are located in the most external regions of the biofilm, while the nitrite oxidizers are found in a deeper layer (KIM et al. 2003). This spatial distribution means that the nitrite-oxidizing bacteria are more exposed to oxygen-limiting conditions compared with the ammonium oxidizers.

The main reactors used to date to obtain partial nitritation are the continuous stirred-tank reactor (CSTR), membrane bioreactors (MBRs), and sequencing batch reactors (SBRs). In the MBR and SBR, high cell retention times (50–75 days) can be obtained (STROUS et al. 1997). In MBR systems and in other biofilm systems, the cell retention time is difficult to control, in contrast with reactors in which the biomass grows in suspension. Thus, it is difficult to force the washout of nitrite-oxidizing bacteria even under oxygen-limiting conditions (XUE et al. 2009) and the production of nitrite without the accumulation of nitrate may not be achieved (FUX et al. 2004).

In some cases, even applying criteria which favor the selection of ammonium-oxidizing bacteria over nitrite-oxidizing bacteria, such as a high concentration of free ammonia, low oxygen concentration, and a high ammonium load, the suppression of nitrite oxidizers is difficult. Thus, the use of reactor configurations such as CSTR and SBR with suspended biomass is recommended, particularly for operation in real scale. In reactors operated in continuous mode (e.g., CSTR), the criterion for the selection of ammonium-oxidizing bacteria is the hydraulic retention time. In batch reactors (e.g., SBR), the sludge age is the controlling factor regarding the nature of the dominant microbial populations.

The possibility of obtaining an effluent from the partial nitritation process which is ideal for subsequent treatment in an anammox reactor was initially tested by VAN DONGEN et al. (2001a), in a process known as SHARON (single-reactor high-activity ammonia removal over nitrite).

Although this process is not adequate for the treatment of all types of wastewaters, due to its strong dependence on temperature, it is ideal for the removal of nitrogen from high-strength nitrogen wastewater (effluent from sludge digesters, landfill leachates, wastewater from composting processes, and liquid from the sludge drying process), which would consume enormous quantities of dissolved oxygen in the conventional nitrification process (AHN 2006).

In many cases the SHARON process can function as a pretreatment, applied to substantially reduce the ammonium concentration, thus allowing the application of a subsequent conventional system for the final polishing of the wastewater (HELLINGA et al. 1998; MULDER and KEMPEN 1997). In this case, the application of the SHARON system as a sidestream process can be evaluated in terms of load removed and not in terms of effluent quality, since the effluent of the SHARON reactor will later be discharged to the main treatment plant (VAN LOOSDRECHT 2008).

The SHARON process makes use of the different growth rates of ammonium-oxidizing and nitrite-oxidizing bacteria at sufficiently high temperatures (above 26 °C). Thus, it is associated with the selection of ammonium-oxidizing bacteria (from an inoculum originating from a system in which nitrification occurs) in a continuous reactor operating with high specific feed flows. The process conditions are unfavorable for the bacteria responsible for nitrite oxidation and can promote their washout (SCHMIDT et al. 2003; MULDER and VAN KEMPEN 1997).

In the original proposal, operation of a single-stage system using intermittent aeration was envisioned. During the periods in which the reactor is aerated, reduction in the pH with the generation of nitrite (nitritation) is observed. On the other hand, in the periods without aeration, anoxic conditions are established, an external organic carbon source is supplied, and nitrite can be converted to nitrogen gas (denitritation). The latter step would cause an increase in the pH and the production of alkalinity, compensating for the acidifying effect of nitrification. Moreover, the denitritation step is responsible to prevent accumulation of nitrite, which is inhibitory to ammonium oxidizers. Thus, the sequential aerated and anoxic periods were defined as a function of the pH limit values stipulated a priori. Alternatively to the single-stage configuration, partial nitrification and denitritation may be carried out in two separate tanks (two-stage configuration) to decrease the aeration capacity. The nitritation-denitritation (SHARON) process is represented in Eqs. 5.4 and 5.5.

The SHARON process can also be followed by the anammox process, a combination referred to as SHARON-anammox process. In this case, only partial nitritation should be achieved in the SHARON reactor in order to obtain an effluent with NH_4^+/NO_2^- ratio close to 1, which is suitable for the anammox process. In this case, no organic carbon is required while sludge production is low. To reach the desired NH_4^+/NO_2^- ratio of 1, the alkalinity of the wastewater is an important factor to be controlled. Depending on this parameter, the SHARON reaction can convert a fraction or the entire ammonium load to nitrite. Considering that the oxidation of 1 mol of ammonium to nitrite consumes 2 mol of bicarbonate and given that this process is practically interrupted at pH values below 6.5, with an ammonium-to-bicarbonate molar ratio of 1:1, approximately 50% of the ammonium is converted to nitrite, the rest remaining in the form of ammonium.

Conversely, the accurate control of alkalinity does not have great importance in cases where a SHARON system achieves complete ammonium oxidation to nitrite to be followed by heterotrophic denitrification via nitrite (denitritation) (VAN DONGEN et al. 2001a; HELLINGA et al. 1998).

The SHARON process is operated with hydraulic retention times which are higher than the growth rate of the nitrite oxidizers, although not as long as that of the ammonium oxidizers. In general, nitritation is carried out without sludge retention, with a hydraulic retention time (HRT) of 1 day, within a temperature range of 30–40 °C and with pH values of 6.6–7. Under these conditions the nitrification process is stable, with nitrite being the final product (AHN 2006).

Since this process is conducted mainly in continuous systems and consequently without sludge retention (hydraulic retention time = cell retention time), the dilution rate (specific feed flow) must be determined in such a way that the ammonium-oxidizing organisms are able to grow sufficiently to remain in the reactor, while the nitrite oxidizers are washed out.

The SHARON process can also be operated with sludge retention. In this case the aeration time will be the limiting factor for the reactor design due to the greater quantity of oxygen required. The economic balance, taking into consideration the reactor volume and the sludge retention equipment, determines the appropriate choice for the application of the SHARON system with sludge retention. In practice, when the nitrogen concentration is above 0.4–0.5 gN/L, a system without biomass retention is cheaper. Also, a system operated without sludge retention requires less maintenance. The sludge produced in the SHARON reactor will leave with the effluent, which does not represent a problem since the effluent of this process will be sent to the influent of the main treatment plant.

As briefly noted, the SHARON process is not suitable for all types of wastewater since it is dependent on high temperatures. However, for the treatment of sludge digester effluents, the SHARON process is ideal, since the temperature of these effluents varies between 20 and 35 °C, permitting reactor operation at reduced sludge retention times. The absence of sludge retention and the fixed hydraulic residence time, common characteristics of the SHARON process, mean that the volumetric nitrogen load applied is dependent on the ammonium concentration of the influent. Therefore, the process cost is also dependent on the ammonium influent concentration, with the increase in the operation cost being directly proportional to the decrease in this concentration. The composition of the effluent of the SHARON process is also dependent on the rate of bacterial growth involved, which, in turn, varies according to the ammonium concentration of the influent (VAN DONGEN et al. 2001a).

Aeration is required not only to provide oxygen but also for the stripping of CO_2 in the reactor and to control the pH. Nitrite can be reduced to nitrogen gas through denitritation using organic compounds (e.g., methanol) as electron donors, added periodically while the aeration is deactivated (SCHMIDT et al. 2003), or by applying the anammox process.

VAN HULLE et al. (2005) described the start-up of a laboratory-scale SHARON reactor operated at 35 °C without pH control. The effluent of the SHARON reactor was found to be suitable for feeding to the anammox process when the influent of the process consisted of synthetic wastewater containing an ammonium load of 1.5 kgN/m^3 day.

FUX et al. (2002) operated a 2.1 m^3 CSTR reactor in Zurich (Switzerland), which was submitted to an HRT of 1.1 days at 30 °C, without pH control. The reactor was fed with sludge digester effluents from two different treatment plants, and it was possible to obtain an ammonium-to-nitrite ratio of 1:1.32 at pH values of between pH 6.6 and 7.2.

The SHARON technology is currently used successfully in real scale to treat effluents from sludge digesters, removing high amounts of ammonium from those streams. Working large-scale SHARON reactors can be found in many wastewater treatment plants, such as those located in Rotterdam and Utrecht (the Netherlands), both representing the first full-scale demonstrations of the process (MULDER et al. 2001). Many other plants have been installed and are in operation in the Netherlands and other countries. The applied nitrogen load usually varies between 400 and 2500 kgN/day (VAN LOOSDRECHT and SALEM, 2005). Due to the high temperature requirement, spread of the technology in tropical countries is foreseen.

Despite its successful application in real scale, there are some disadvantages associated with the SHARON process. The HRT of sludge digesters is high, which guarantees the stable composition of their effluents to be subsequently fed to a SHARON reactor. The digester effluent is characterized by a high concentration of nitrogen and low content of biodegradable organic compounds. When the HRT of the digesters is shorter than normal or when industrial wastewaters are treated, fluctuations in the composition of the effluent fed to the SHARON process are likely to occur. Thus, some parameters of this process, such as the dissolved oxygen and pH, need to be controlled in the SHARON process in order to obtain the ideal ammonium-to-nitrite ratio for the subsequent anammox process (VOLCKE et al., 2006).

Another disadvantage of the SHARON process is the fact that the maximum volumetric capacity is limited, since the biomass is constantly removed from the system. In order to ensure stable conditions, the minimum HRT of a chemostat is limited to 1.0–1.2 days. In an MBR, SBR, or biofilm systems, the biomass is retained more easily, allowing the HRT to be independent of the cell retention time. In these systems HRT of less than 1 day can be applied, which results in greater volumetric capacities (WYFFELS et al. 2004).

As mentioned above, the temperature is crucial to obtaining good performance in a SHARON reactor. In general, this process is operated at relatively high temperatures (above 26 °C) and low sludge retention time (less than 2 days), a condition under which AOB are at an advantage over nitrite oxidizers and remain inside the reactor system. On the other hand, NOB are potentially washed out. However, when the influent of the SHARON process (generally effluent from sludge digesters) is below 24 °C, the maximum growth rate of the organisms which oxidize ammonium becomes lower than that of the nitrite oxidizers, and thus the nitrite formed is converted to nitrate (FUX et al. 2002).

Taking into consideration that the use of temperature as a selection criterion for the dominant microbial populations is not very favorable in economic terms, and given the difficulty associated with modifying and controlling this parameter in large-scale reactors, other strategies are needed in order to achieve partial nitritation at temperatures of less than 24 °C. One such approach is the inhibition of nitrite-oxidizing bacteria using free ammonia or nitrous acid and another is maintaining the reactor under oxygen-limiting conditions.

Free ammonia and nitrous acid are two potential inhibitors of the nitrifying process. Nitrite-oxidizing bacteria are the most strongly affected since relatively low concentrations are sufficient to promote their inhibition. ABELING and SEYFRIED (1992) reported that free ammonia concentrations can inhibit nitratation without affecting nitritation. The maximum specific rate of nitrite generation and minimum generation of nitrate are obtained with a free ammonia concentration of 5 mg/L, under pH and temperature conditions of 8.5 and 20 °C, respectively.

MAURET et al. (1996) reported that the inhibition of nitrite-oxidizing bacteria can be achieved with free ammonia concentrations in the range of 6.6–8.9 mgN/L. BALMELLE et al. (1992) observed that even at concentrations of only 1 mgN/L free ammonia can inhibit the nitratation stage. GANIGUÉ et al. (2007) showed that it is possible to obtain stable influent for the anammox process during leachate treatment in a sequencing batch reactor. At low pH values, the microbial activity is decreased due to the inhibitory effect of nitrous acid, resulting from the lack of alkalinity (bicarbonate). High pH values indicated a decrease in the oxygen consumption rate due to inhibition by free ammonia. The authors concluded that the pH controlled the partial nitritation process.

On applying partial nitritation followed by the anammox process for the treatment of effluent from an anaerobic digester, YAMAMOTO et al. (2008) observed stable conversion of ammonium to nitrite equivalent to 58%. The authors attributed this result to the inhibition promoted by free ammonia and nitrous acid. It should be noted that the strategy to control the action of bacteria only through the pH and the presence of free ammonia may not provide the desired success (RUIZ et al. 2003; SCHMIDELL and REGINATTO 2005), since the limit of the free ammonia concentration able to promote inhibition of the nitrite oxidation step may increase over time.

Some researchers have reported the possibility for the adaptation of nitrite-oxidizing bacteria to occur even at high free ammonia concentrations. WONG-CHONG and LOEHR (1978) observed that pure cultures of *Nitrobacter* acclimatized to free ammonia were able to tolerate up to 40 mgNH_3-N-L, while non-acclimatized cultures were inhibited at concentrations equivalent to 3.5 mgNH_3-N/L. Other studies on systems with biofilm or biomass in suspension have shown that nitrite-oxidizing organisms can adapt to high concentrations of free ammonia, observing that within a certain period (from 6 to 12 months) the nitrite accumulation decreases and the nitrate concentration increases (FUX et al. 2004; VILLAVERDE et al. 2000).

In order to investigate the accumulation of nitrite using the strategy of maintaining low oxygen levels, WYFFELS et al. (2003), operating a continuous reactor using membranes for total recycling of cells, reached 50% conversion of ammonium to nitrite when the system was submitted to a dissolved oxygen concentration of 0.1 mg/L, at 35 °C and pH 7.9.The complete conversion of ammonium to nitrite was observed when the concentration of oxygen was approximately 0.25 mg/L, a situation in which around 800 mgNO_2^--N/L was reached. TURK and MAVINIC (1987) observed that cells acclimatized under anoxic conditions were able to provide long periods of nitrite accumulation (up to hours), even under aeration conditions. The results obtained by these authors highlighted the possibility of intercalating periods of aeration with periods without aeration, obtaining the desired accumulation of nitrite, without generating considerable quantities of nitrate. This strategy provides greater simplicity in comparison with the use of refined systems to control the oxygen concentration at very low values. In addition, the use of alternating periods with and without aeration to limit the ammonium oxidation only up to nitrite favors the application of simultaneous ammonium and nitrite removal processes, such as the anammox process (MULDER et al. 1995; KUAI and VERSTRAETE 1998).

RUIZ et al. (2003) determined the best conditions for partial nitrification, with the accumulation with nitrite, in a synthetic wastewater with a high ammonium concentration, aimed at decreasing the total amount of oxygen required for the nitrification stage. On operating an activated sludge reactor in laboratory scale, they selected the pH and DO as the operation parameters to evaluate the possibility for nitrite accumulation, without affecting the overall ammonium removal. According to the authors, the pH was not an ideal parameter for promoting nitrite accumulation, considering that when this parameter had values in the range of 6.45–8.95 complete nitrification up to nitrate occurred in the system and at pH values below 6.45 and above 8.85 complete inhibition of the nitrification process occurred. In contrast, when the DO concentration in the reactor was maintained at 0.7 mg/L, it was possible to accumulate over 65% of the ammonium in the form of nitrite with an ammonium conversion of 98%. At DO concentrations below 0.5 mg/L, there was the accumulation of ammonium and at concentrations above 1.7 mg/L, complete nitrification up to nitrate was reached (RUIZ et al. 2003).

WYFFELS et al. (2004) used a membrane bioreactor as the first stage of the autotrophic nitrogen removal process, which was submitted to dissolved oxygen concentrations below 0.1 mg/L. The pH was maintained at 7.9 and the temperature at 35 °C. A reduction in the temperature did not have a significant effect on the ammonium-to-nitrite ratio obtained. In addition, a reduction in the ammonium concentration (which probably contributed to a decrease in the inhibitory effect on the nitrite oxidizers) did not alter the ammonium-to-nitrite ratio obtained. These results indicated that the oxygen limitation was the main operational factor which determined the ammonium-to-nitrite ratio obtained.

In general, the different strategies used in the studies described above to obtain the desired partial nitritation and produce an influent which is ideal for the anammox process can be summarized as follows:

- Operation of the reactor at low dissolved oxygen concentrations (less than 0.5 mg/L)
- Operation of the reactor at high pH values (7.5–8.5), resulting in an increase in the free ammonia concentration and a decrease in the nitrous acid concentration
- Operation of the reactor at high temperature (above 25 °C)
- Operation of the reactor with limited nitrification, which stops the ammonium oxidation before its complete conversion

Anammox

The application of the anammox process is still limited, which is mainly due to the long start-up of the process (up to one year), which is associated with the very low growth rates and low cellular yield of the anammox organisms. The frequent loss of biomass with the effluent increases the start-up period of anammox reactors. Consequently, the use of systems with a good biomass retention capacity is crucial to the success of the process. The cultivation of slow-growing bacteria, such as anammox organisms, is based on the ability of the biomass to form biofilms or aggregates such as flocs or granules (VAN DER STAR 2008).

Several different reactor configurations have been used for the enrichment of anammox bacteria. These include fixed-bed reactors, fluidized-bed reactors, UASB reactors, sequencing batch reactors (SBRs), and air-lift reactors (WYFFELS et al. 2004; STROUS et al. 2002). Of these, the SBRs have been the preferred choice due to their operational simplicity and because of their efficient biomass retention, homogeneous mixture inside the reactor, stability and reliability during long periods of operation, stability under limited-substrate conditions, and high nitrogen conversions (JETTEN et al. 1999; STROUS et al. 1998).

STROUS et al. (1997) initiated the operation of the anammox process in a fixed-bed reactor and in a fluidized-bed reactor with glass and sand particles acting as support material. The authors did not manage to avoid the biomass loss due to sludge flotation caused by gas bubbles. The same situation was observed by DAPENA-MORA et al. (2004) in an air-lift reactor. These authors observed that mechanical mixing in an SBR can be more effective for the elimination of the gas which penetrates the granules compared with the air-lift reactor.

In order to obtain the complete retention of the biomass in anammox systems, membrane bioreactors can be used. In contrast to reactors with granular biomass, the MBR allows the cultivation of slow-growing bacteria with very good biomass retention and does not require biomass with good settling characteristics. Therefore, this represents a good option to obtain anammox cells in suspension.

WANG et al. (2009) used a mixer in an MBR to promote the formation of free anammox cells and obtained a more homogeneous distribution of the substrate and biomass. However, for real-scale applications, reactors with biofilm or granular sludge are preferred in comparison with MBRs since the anammox bacteria easily form granules or biofilms, allowing obtaining a high biomass concentration in a simple and economic manner. Also, the fouling of the membrane is one of the disadvantages of MBR systems. The operating costs associated with the cleaning of the membrane (backwashing or the addition of chemical products) reduce the economic feasibility of the process (TRIGO et al. 2006). Another important point is the fact that wastewaters generally contain a certain quantity of solids, which will also be retained by the membrane. The accumulation of this material also promotes a reduction in the activity of the anammox organisms in the MBR system (YAMAMOTO et al. 2008).

As previously mentioned, the start-up period of anammox systems is one of the disadvantages of the process. Ensuring conditions in which oxygen is absent is essential during the start-up period. In general, this period is characterized by a gradual increase in the nitrite concentration. Although the ideal ammonium-to-nitrite ratio is around 1:1, an excess of ammonium is generally used, leading to a reduced efficiency in terms of the overall nitrogen removal, although ensuring greater process stability.

In order to decrease the start-up time of the anammox process, the system can be inoculated with biomass originating from another anammox reactor. The start-up period of an SBR reactor operated by SLIEKERS et al. (2003) was only 1 day, since the reactor was inoculated with highly active anammox biomass. The continuous addition of pre-enriched anammox biomass was used in the Netherlands to start up

Table 5.4 Different reactor configurations and nitrogen removal rates in two-stage autotrophic nitrogen removal processes

Reactor	Wastewater	Nitrogen load (kgN/m^3 day)	Nitrogen removal (kgN/m^3 day)	Reference
Sequencing batch reactor	Digester effluent	1.0	0.75	Van Dongen et al. (2001b)
Sequencing batch reactor	Digester effluent	2.6	2.4	Fux et al. (2002)
Sequencing batch reactor	Synthetic	1.4	1.1	Dapena-Mora et al. (2004)
Fixed-bed reactor	Synthetic	0.07–0.55	0.35–0.38	Fux et al. (2004)
	Digester effluent	n.a.	3.5	
Fixed-bed reactor	Synthetic	1.3	1.1	Strous et al. (1997)
	Synthetic	0.2–2.0	1.8	
	Digester effluent	2.5	1.5	
Fluidized-bed reactor	Digester effluent	0.48–2.63	2.5	Jetten et al. (1997)
Air-lift reactor	Synthetic	2.3	2.0	Dapena-Mora et al. (2004)
Air-lift reactor	Synthetic	10.7	8.9	Sliekers et al. (2003)
UASB reactor	Synthetic	0.52	0.51	Schmidt et al. (2004)
UASB reactor	Pig waste	0.84–1.02	0.59–0.66	Ahn et al. (2004)

n.a. not available

a real-scale reactor of 70 m^3. The start-up period of the reactor was around 3.5 years, and currently it provides stable operation with a nitrogen removal rate of 9.5 kgN/(m^3 day) (VAN DER STAR et al. 2007).

A variation of the anammox process named DEAMOX (denitrifying ammonium oxidation) has been recently tested in laboratory scale. This process is based on the combination of an anammox reaction and autotrophic denitrification using sulfide as the electron donor for the production of nitrite from nitrate in an anaerobic biofilm (KALYUZHNYI et al. 2006). This process will be detailed further.

Table 5.4 details several studies described in the literature involving the autotrophic removal of nitrogen in two systems, including the anammox process. As can be observed, different reactor configurations were employed. The removal of nitrogen varies between systems, which is also related to the different nitrogen loads applied. The abovementioned systems promote efficient biomass retention aimed at counterbalancing the low cellular yield of anammox bacteria. The highest nitrogen removal rate (8.9 kgN/m^3 day) was reported by SLIEKERS et al. (2002), who operated a laboratory-scale air-lift reactor containing granular sludge.

Partial Nitritation and Anammox in a Single Reactor (One Stage)

In theory, the combination of ammonium oxidation and denitrification can be carried out in biofilm systems submitted to low oxygen concentrations. However, in the conventional denitrification process, the organic carbon source can become a limiting factor since the electron donor for denitrification is more rapidly oxidized than the ammonium.

When ammonium acts as the electron donor (as in the case of the anammox process), this problem does not occur (VAN LOOSDRECHT et al. 2004). The combination of partial nitritation and the anammox reaction (Eqs. 5.4 and 5.8) in a single reactor implies that the aerobic autotrophic microorganisms responsible for the partial nitritation (ammonium-oxidizing bacteria) and the anammox microorganisms act in cooperation throughout the process, allowing sequencing reactions to occur simultaneously (AHN 2006). Although the operational controls needed for one-stage partial-nitritation-anammox systems are similar to those required by two-stage processes, the former require more sensitive controls in terms of dissolved oxygen, nitrogen load, temperature, and biofilm thickness. The oxygen concentration, in particular, is a crucial parameter which needs to be controlled.

The nitrifying microorganisms are responsible for the oxidation of ammonium to nitrite, consuming a large part of the oxygen and creating the anoxic conditions essential for the anammox process to occur, in which ammonium and nitrite (generated in the partial nitritation) are converted to nitrogen gas. So, the cooperation of AOB and anammox organisms is critical for successful operation of single-stage partial nitritation and anammox processes.

In general, the oxygen concentration needs to be low for two main reasons: (1) to avoid the inhibition of anammox bacteria, which are reversibly inhibited by oxygen, and (2) to obtain operating conditions under which it is possible to carry out partial nitritation (STROUS et al. 1997), hindering the growth of nitrite-oxidizing bacteria. NOB have a lower affinity for oxygen than AOB and also a lower affinity for nitrite than anammox bacteria (HANAKI et al. 1990). Therefore, under limiting oxygen conditions, both AOB and anammox are favored over NOB, and partial nitritation and anammox reactions can be achieved in single-stage units.

Specific operating conditions aimed at partial nitritation have been previously presented. The strategy for obtaining washout of the nitrite-oxidizing bacteria through the application of specific feed flows cannot be applied in partial nitritation and anammox systems in a single reactor. A representation of the autotrophic removal of nitrogen in biofilm systems is given in Fig. 5.9.

It should be noted that the dissolved oxygen concentration, a key parameter of the combined processes of partial nitritation and anammox in a single reactor, is related to the biofilm thickness. For a certain surface load of ammonium and under low temperature conditions, a thick biofilm is required. Consequently, the concentration of dissolved oxygen needs to be higher. On the other hand, a thin biofilm requires less dissolved oxygen and, in this case, high oxygen concentrations will lead to complete nitrification and lower nitrogen removal (HAO et al. 2002; KOCH et al. 2000).

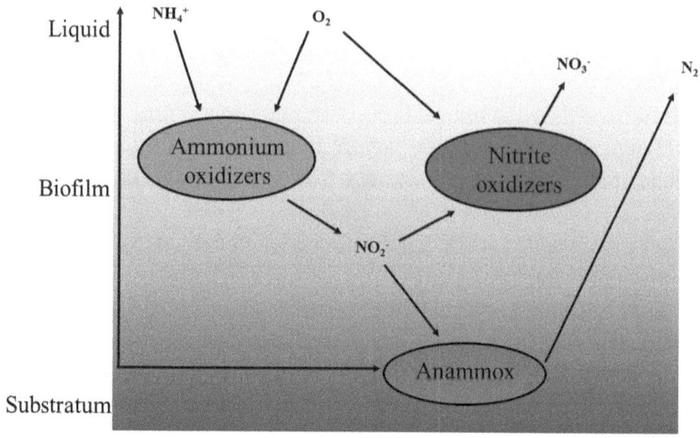

Fig. 5.9 Schematic diagram of the autotrophic nitrogen removal in a biofilm process (adapted from VAN LOOSDRECHT 2008)

Fig. 5.10 Effect of dissolved oxygen concentration on nitrite accumulation in a biofilm system (adapted from VAN LOOSDRECHT 2008)

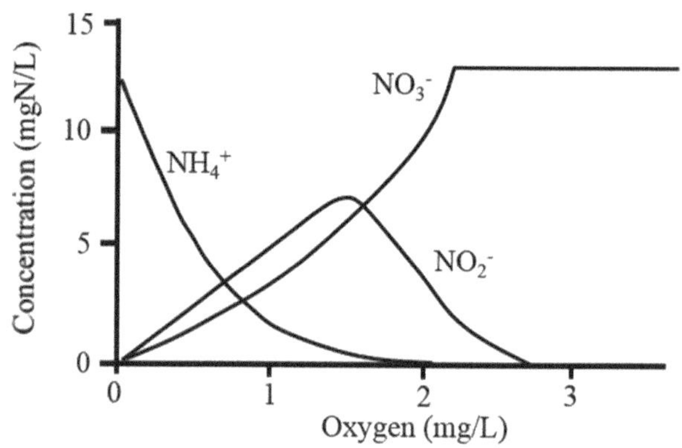

The lower the applied nitrogen load, the lower the dissolved oxygen concentration needed to obtain partial nitrification will be. In this context, it is clear that considering only the dissolved oxygen in order to obtain the accumulation of nitrite is not sufficient, since other factors are involved, such as the biofilm thickness and nitrogen load applied. Figure 5.10 shows the dynamics of the nitrite/nitrate formation in the presence of different dissolved oxygen concentrations in a biofilm system, observed experimentally by GARRIDO et al. (1997) and theoretically explained by PICIOREANU et al. (1997).

Two main strategies can be used to start up a system aimed at the autotrophic removal of nitrogen in a single reactor. The first consists of inoculating a nitrifying biomass into an anammox reactor which is operating in a satisfactory manner and supplying aeration to the reactor in order to maintain micro-aerobic conditions. The second strategy is based on the operation of a reactor aimed at partial nitritation under limited-oxygen conditions in order to obtain an ammonium-to-nitrite ratio of

1:1. Later anammox biomass is inoculated (PYNAERT et al. 2004; GONG et al. 2007).

The high nitrifying activity can protect anammox bacteria from oxygen, besides providing nitrite. Also, on inoculating the partial nitritation reactor with a biomass enriched with anammox bacteria, the start-up process for autotrophic removal of nitrogen in a single reactor is accelerated, and considerable nitrogen removal can be obtained within 1 or 2 months. If the reactor is not inoculated with an anammox biomass, a period of several months or even years may be required in order to achieve significant results for nitrogen removal.

The second strategy based on a single reactor for the autotrophic removal of nitrogen is known as the CANON (completely autotrophic nitrogen removal over nitrite) process, and this appears to be more suitable given the considerable reduction in the anammox activity when the first strategy is applied (SLIEKERS et al. 2002, 2003; LIU et al. 2008). Moreover, only a small amount of anammox biomass is required for the start-up of the CANON process. Further details about the CANON process (e.g., representative equation) can be found below.

The single-stage process is generally associated with a higher volumetric nitrogen removal rate and lower investment cost when compared with the two-stage configuration, since an additional reactor for the partial nitritation is not needed (WYFFELS et al. 2004). Nevertheless, the difficulty related to regulating the dissolved oxygen concentration and the incomplete removal of nitrogen during the treatment of highly concentrated wastewaters are some of the problems encountered when using a single-reactor configuration (HAO et al. 2001; NIELSEN et al. 2005).

Some mathematical models have been developed to understand and predict the behavior of systems under different operating conditions and the effect of the ammonium surface load. The main results obtained have shown that the ammonium load is associated with the biofilm thickness, a thin biofilm having a limited anammox activity, and the stable formation of nitrite is a limiting factor. On the other hand, the anammox process can occur in biofilm systems, although the time rather than the nitrogen load is the key factor in the process. In these systems, it has been predicted that a period of between 5 and 10 years is required to achieve an anammox population which allows the maximum conversion rates to be reached (VAN LOOSDRECHT et al. 2004).

Several types of reactors have been employed to carry out the combined partial nitritation and anammox process in a single reactor. These include sequencing batch reactors, rotating biological contactors, moving bed biofilm reactors, and air-lift reactors. Table 5.5 details the results of some studies with these reactors.

Initially, the best configuration for obtaining an efficient retention time for the slow-growing autotrophic biomass appears to be the sequencing batch reactors, which allow the biomass to be maintained in the reactor through applying alternating reaction/settling phases. However, high nitrogen removal rates were also obtained in other types of reactors, such as rotating biological contactors (PYNAERT et al. 2003, 2004) and air-lift reactors (SLIEKERS et al. 2003).

In biofilm reactors or granular sludge reactors, the microorganisms which oxidize ammonium are active in the external regions of the biofilm (or granule),

Table 5.5 Different reactor configurations and nitrogen removal rates in processes for autotrophic removal of nitrogen in a single reactor (partial nitritation + anammox)

Reactor	Wastewater	Nitrogen load (kgN/m^2 day)	Nitrogen removal (kgN/m^2 day)	Reference
Rotating biological contactor	Leachate	1.4–3.2	0.4–1.2	Siegrist et al. (1998)
Rotating biological contactor	Leachate	1.5	0.9	Hippen et al. (2001)
Rotating biological contactor	Synthetic	2.3	1.55	Pynaert et al. (2002)
Moving bed biofilm reactor	Digester effluent	4.8	2.4	Hippen et al. (2001)
Moving bed biofilm reactor	Digester effluent	4–8	2.0	Scyfricd ct al. (2001)

producing an appropriate quantity of nitrite for the anammox organisms, which are located in inner layers. Thus, the anammox bacteria are protected from the oxygen, which is consumed by AOB in the external biofilm (granule) layers (WYFFELS et al. 2004).

A variation of the conventional biofilm reactors is the membrane reactors (GONG et al. 2007), in which hydrophobic membranes, permeable to gases, are used for the transfer of oxygen. In the region close to the membranes, dissolved oxygen is present, and this is where the ammonium-oxidizing bacteria convert ammonium to nitrite. On the other hand, anammox bacteria are active in the region rich in ammonium, close to the liquid phase.

When systems with biofilm or granules are used in the partial nitritation-anammox process, the resistance to mass transfer is generally the limiting stage. When the ammonium concentration in the external region of the biofilm is much higher than the oxygen or nitrite concentration, the diffusion of ammonium to the interior of the biofilm will not limit the speed of the process. If the nitrite produced in the external region is mainly consumed in the internal region, oxygen will become the main limiting factor of the overall process.

SZATKOWSKA et al. (2007) reported that the oxygen transfer was the limiting factor of the process in a pilot-scale moving bed reactor. The same finding was reported by SLIEKERS et al. (2003) for a laboratory-scale air-lift reactor. The oxygen limitation can be attributed to slow oxygen diffusion to the interior of the biofilm/granule or inefficient transfer at the gas-liquid interface.

As mentioned in Sect. 5.2.2.3, nitrite is a potential inhibitor of the anammox process. If it is consumed in the same proportion in which it is produced, its inhibitory effect is not significant. In addition, in situations in which the nitrite concentration is high, a negative effect on the anammox bacteria is not always observed. This observation was noted by VÁZQUEZ-PADÍN et al. (2009), who recorded nitrite concentrations of the order of 25 mgN/L. The presence of a concentration gradient inside the granules, where the anammox bacteria are located, probably results in lower nitrite concentrations in these regions, hindering their inhibitory effect.

Several names have been used for systems in which nitritation-anammox is carried out in a single reactor: CANON (completely autotrophic nitrogen removal over nitrite) process (THIRD et al. 2001), OLAND (oxygen-limited autotrophic nitrification and denitrification) process (KUAI and VERSTRAETE 1998), aerobic/anoxic DEMON (deammonification) process (HIPPEN et al. 1997; WETT 2007), and SNAP (single-stage nitrogen removal using anammox and partial nitritation) (FURUKAWA et al. 2006).

The different ways of referring to the same process arose from the fact that different research groups primarily attributed the anaerobic oxidation of ammonium to different microorganisms. For instance, when the OLAND process was developed, the organisms considered to be responsible for the anaerobic oxidation of ammonium under micro-aerobic conditions were the nitrifying bacteria (KUAI and VERSTRAETE 1998; HELMER et al. 1999). In the case of the CANON process, the anammox bacteria were assumed to be responsible for this conversion. Studies using the fluorescent in situ hybridization (FISH) technique have confirmed that the anaerobic oxidation of ammonium, in all reactors, was carried out by anammox organisms (PYNAERT et al. 2003; HELMER-MADHOK et al. 2002), although PYNAERT et al. (2003) do not exclude a specific function associated with the ammonium-oxidizing microorganisms.

CANON Process

The CANON process is essentially an integration of the combined SHARON-anammox process into a single reactor and can be represented by Eq. 5.14 (SLIEKERS et al. 2003). The scheme of the process is displayed in Fig. 5.12d.

$$NH_3 + 0.85O_2 \rightarrow 0.11NO_3^- + 0.44N_2 + 0.14H^+ + 1.43H_2O \qquad (5.14)$$

In most of the studies on the CANON process reported in the literature, the system was operated in the temperature range of 30–35 °C, with a maximum nitrogen removal rate in the range of 0.075–1.5 kgN/m^3 day (SLIEKERS et al. 2002, 2003). In this temperature range, the bacteria which oxidize ammonium grow more rapidly than those responsible for the nitrite oxidation. In addition, the growth of the anammox microorganisms is stimulated under these temperature conditions.

However, there are studies reported in the literature in which a significant removal of nitrogen (0.5 kgN/m^3 day) was obtained at temperatures of between 20 and 24 °C, such as that carried out by VÁZQUEZ-PADÍN et al. (2009) in a sequencing batch reactor. The difference is that in this system low activity of the nitrite-oxidizing bacteria was observed. The possibility of obtaining a short start-up period and high nitrogen removal rate in autotrophic nitrogen removal systems at temperatures of around 20 °C has been reported by PYNAERT et al. (2004) in a single-stage system and by DOSTA et al. (2008) and ISAKA et al. (2006) in two-stage systems.

Fig. 5.11 Examples of full-scale implementation of anammox-based system for treatment of nitrogen-rich wastewaters. Information on each plant are given below each picture (provided by Paques). (**a**) Rotterdam, the Netherlands. Treatment of sludge digester effluent. Load: 500 kgN/day. Removal of nitrogen by two-step SHARON-anammox process. Operation since 2002. (**b**) Olburgen, the Netherlands. Treatment of digester reject water combined with wastewater from potato processing plant (pretreated in a UASB reactor). Load: 1200 kN/day and 245 kgP/day. Removal of nitrogen by one-stage anammox and removal of phosphorus by Phospaq and struvite production. Operation since 2006. (**c**) Niederglatt, Switzerland. Treatment of sludge digester effluent. Load: 60 kN/day. Removal of nitrogen by one-stage anammox. Operation since 2008. (**d**) China. Treatment of wastewater from monosodium glutamate plant. Load: 11,000 kgN/day. Removal of nitrogen by one-step anammox process. Operation since 2009. (**e**) Santa Catarina, Brazil. Treatment of wastewater from food industry. Load: 720 kgN/day. Removal of nitrogen by one-stage anammox. Operation since 2014. (**f**) Noord-Brabant, the Netherlands. Removal of nitrogen from the effluent of an UASB reactor treating animal waste in one-stage anammox. Load: 6000 kgN/day. Operation since 2013

SLIEKERS et al. (2002) applied a specific strategy to the start-up of CANON reactors. These authors used a biomass enriched with anammox bacteria as the inoculum (80% of bacterial population), this stage being following by the supply of oxygen for the development of nitrifying microorganisms. The operation was carried out at 30 °C, and pH 7.8, with mixing at 100 rpm and a specific air flow of 0.04 vvm. The reactor was fed with synthetic medium containing NH_4^+ and NO_2^-, with a total nitrogen load of 457 mg/L. Helium gas was used in the anaerobic stage. Ammonium oxidation activity tests were carried out under aerobic conditions during the first 2 weeks of operation and no activity was detected. The results for the FISH analysis indicated a large quantity of anammox bacteria, although the presence of ammonium-oxidizing or nitrite-oxidizing bacteria was not detected.

After 5 weeks of operation, the helium gas was replaced with atmospheric air and the nitrogen load was reduced to 131 mgN/L day. The nitrogen was added only in the form of ammonium. Based on the FISH technique, a substantial increase (to 45%) in the aerobic ammonium-oxidizing bacteria was observed, while the anammox bacteria decreased from 80 to 40%. The nitrogen removal rate of the reactor was found to be slow, both in the anoxic stage (anammox) and in the stage with oxygen limitation (CANON), for which the values were 0.315 and 0.064 kgN/m^3 day, respectively.

The same authors also noted that under oxygen-limiting conditions anammox bacteria consume nitrite, while the nitrite-oxidizing bacteria are not active. This suggests that these latter organisms are present only when oxygen is not limiting. When oxygen is deficient, these bacteria, which are inhibited by free ammonia (present in a high concentration in this study), have to compete with ammonium-oxidizing bacteria for the low amount of oxygen available and also with anammox bacteria for nitrite.

The model developed by HAO et al. (2001) to describe the CANON process indicated that the maximum nitrogen removal rate would be reached only when the dissolved oxygen concentration is proportional to the surface load of ammonium. For variable ammonium loads, the dissolved oxygen should be regulated by way of controlled refeeding. The model developed by the same authors showed that the ideal dissolved oxygen concentration in a CANON reactor was around 1 mg/L, although this ideal value was dependent on the thickness and density of the biofilm, the organic matter concentration of the influent, and the temperature.

OLAND Process

The term OLAND was initially introduced by KUAI and VERSTRAETE (1998). The stoichiometry of the OLAND process considers the removal of ammonium in two stages: in the first ammonium is partially oxidized to nitrite and in the second a reaction between ammonium and nitrite occurs with the formation of nitrogen gas. This stoichiometry is very similar to that involved in the CANON process. The OLAND process can be represented by Eq. 5.15 (VERSTRAETE and PHILIPS 1998).

The key to this process is the supply of oxygen, to ensure that the nitrification occurs only up to nitrite. Subsequently, due to the low dissolved oxygen concentrations (final electron acceptor), nitrite is consumed in the oxidation of ammonium.

Initially, the process was considered to depend only on aerobic AOB. However, further studies revealed that at low dissolved oxygen levels and absence of organic electron donors, AOB and anammox coexist in the OLAND reactor, as occurs in CANON systems. Indeed, the representative reactions of both processes are very similar.

In research carried out by DE CLIPPELEIR et al. (2009), high nitrogen removal rates were observed in granular sludge reactors. On operating an OLAND system, PYNAERT et al. (2003) obtained nitrogen removal at a rate of 1.8 kgN/m^3 day, 100 days after the inoculation of the reactor (rotating biological contactor) with an aerobic granular sludge. In an air-lift reactor, SLIEKERS et al. (2003) observed a nitrogen conversion rate of 1.5 kgN/m^3 day.

$$NH_4^+ + 0.75O_2 \rightarrow 0.5N_2 + H^+ + 1.5H_2O \quad (5.15)$$

Aerobic/Anoxic Deammonification or DEMON

The term aerobic/anoxic deammonification or DEMON was firstly used to describe the significant loss of inorganic nitrogen (up to 90%) in the nitrifying stage of a rotating biological contactor used for leachate treatment, operating with high concentrations of ammonium and submitted to low oxygen concentrations (HIPPEN et al. 1997). In fact, aerobic deammonification is based on the principle of the CANON process (illustrated in Fig. 5.12d), in which there is cooperation between nitrifying and anammox bacteria under oxygen-limiting conditions. Both partial nitritation and anammox reactions occur in a single-reactor system.

The DEMON process is operated similarly to a conventional sequencing batch reactor (SBR). The SBR is subjected to intermittent aeration governed by three control mechanisms: time, pH, and dissolved oxygen control. The operating cycles (of around 8 h, as commonly reported in the literature) comprise a simultaneous filling/reaction phase, settling phase, and effluent withdrawal period. During the reaction phase, both partial nitritation and anaerobic ammonium oxidation occur. These two successive reactions affect the pH: the partial nitritation decreases the pH, while the anammox process increases the pH. So, the duration of the aerated and non-aerated intervals is controlled by the pH value, which characterizes the current state of the conversions (WETT 2007).

When oxygen is provided, partial nitrification is promoted, yielding H^+ which causes a gradual pH drop. At the moment when the minimum pH threshold value (lower set-point) is reached, aeration is turned off. When oxygen becomes depleted, nitrite formed during partial nitritation is used to oxidize ammonium through the anammox reaction. This process recovers part of the alkalinity used in the aerated step, leading to pH increase. Additionally alkalinity is provided by the influent reject water, which is continuously supplied to the SBR until pH reaches the maximum threshold value (upper set point). Then aeration is switched on again. The alternating aerated and non-aerated periods continue until the settling phase begins. Dissolved oxygen levels are kept within a narrow and low range, close to

0.3 mg/L. Such accurate control is made to avoid nitrite buildup and repress nitrite oxidation to nitrate (INNEREBNER et al. 2007; WETT 2007).

Besides the aforementioned control mechanisms, monitoring of redox potential and electrical conductivity along with programmed safeguards may prevent unfavorable operating conditions, such as overaeration, which will potentially lead to a reduction in the process performance (WETT 2007).

Full-scale single sludge deammonification systems located in Strass (Austria) and Zurich (Switzerland) were described in the literature. The plant at Strass has a 500 m^3 sequencing batch reactor for the deammonification of reject water from digested sludge dewatering (INNEREBNER et al. 2007). 2.5 years were necessary for the start-up of the DEMON process and biomass enrichment. Long-term results obtained in the pH-controlled deammonification system revealed that it reached the capacity of removing around 300 kg of nitrogen per day (WETT 2006) with annual ammonium removal above 90% (WETT 2007). Furthermore, the specific air demand decreased from 109 to 29 m^3/kgN (WETT 2006). The sidestream deammonification step requires 1.16 kWh/kg of ammonium nitrogen removal, while the energy required for this conversion is around 6.5 kWh in the mainstream treatment (WETT, 2007). In Zurich a 1400 m^3 reactor was reported to treat 500 gN/m^3 day, with nitrogen conversions of over 90% (JOSS et al. 2009).

The main characteristics of some of aforementioned nitrogen removal processes are detailed in Table 5.6.

Figure 5.11 illustrates some full-scale anammox-based reactors for nitrogen removal from several types of wastewaters.

5.2.3 Denitrifying Ammonium Oxidation (DEAMOX) Process

The DEAMOX process was developed to minimize the difficulties found in achieving partial nitritation of ammonium to nitrite, as required in one- or two-stage partial nitritation-anammox processes (e.g., SHARON + anammox and CANON). According to KALYUZHNYI et al. (2006), obtaining proper nitrite concentrations from partial nitritation requires advanced process control which may hamper large-scale industrial applications. Moreover, problems associated with nitrite inhibition are also overcome.

In this sense, the DEAMOX process does not require a separate production of nitrite, but consists of a combination of the anammox reaction with autotrophic denitrification using sulfide as electron donor, yielding the formation of nitrite from nitrate in an anoxic biofilm reactor. The main process conversions taking place in the different reactors are shown in Eqs. 5.16–5.19, while the scheme of the DEAMOX process is displayed in Fig. 5.12e. Interestingly, the DEAMOX process scheme is quite similar to the autotrophic denitrifying fluidized-bed pilot reactor in which the anammox process was first discovered (Mulder et al. 1995).

The technology involves the combination of three different reactors. First an anaerobic reactor (e.g., UASB) is used as a pretreatment step to generate sulfide and

Table 5.6 Comparison of different nitrogen removal processes (GONZÁLEZ-MARTÍNEZ et al. 2011; PLAZA et al. 2003; MULDER 2003; SCHMIDT et al. 2003; AHN 2006)

	Conventional nitrification and denitrification	Partial nitrification + denitritation (SHARON)	Partial nitritation + anammox (SHARON-anammox)	CANON process	OLAND process	DEMON process
Inoculum	Activated sludge	Activated sludge	Nitrifying and anammox	Anammox (80%) and nitrifying bacteria (20%)	Nitrifying bacteria	Anammox (50%) and nitrifying bacteria (50%)
Number of tanks	Two	One or two	Two	One	One	One
Operating conditions	Aerobic/anoxic	Aerobic/anoxic	Aerobic/anoxic	Oxygen limited[a]	Oxygen limited[a]	Oxygen limited[a]
Requirement of oxygen (gO_2/gN)	4.57 (high)	3.43 (medium)	1.72 (low)	1.90 (low)	1.94 (low)	0.3 (very low)
Requirement of COD (gCOD/gN)	2.86 (without assimilation); 4.0 (with assimilation)	1.72 (without assimilation); 2.4 (with assimilation)	0	0	0	0
Alkalinity ($gCaCO_3/gN$)	7.14	7.14	3.57	3.57	3.57	3.57
Energy requirement	High	Low	Low	Low	Low	Very low
N removal performance (%)	95	90	90	90	85	90
Sludge production	High	Low	Very low	Very low	Very low	Very low

	Conventional nitrification and denitrification	Partial nitrification + denitritation (SHARON)	Partial nitritation + anammox (SHARON-anammox)	CANON process	OLAND process	DEMON process
Forms of bacterial growth	Biofilm/suspension	Suspension/biofilm	Suspension/biofilm/granules	Biofilm/ granules	Biofilm/ granules	Biofilm/granules
Cost of construction	High	Low	Low	Very low	Very low	Very low
Cost of nitrogen removal	High	Medium	Very low	Very low	Very low	Very low
Bacterial types	AOB[b] and NOB[c], heterotrophic denitrifying bacteria	AOB[a], heterotrophic denitrifying bacteria	AOB[b], anammox	AOB[b], anammox	Aerobic nitrifiers (anammox[d])	AOB[b], anammox

[a]Dissolved oxygen concentration usually lower than 0.5 mg/L, mostly within 0.1–0.3 mg/L

[b]Ammonium oxidizing bacteria

[c]Ammonium oxidizing bacteria

[d]Although in the early stage of process development aerobic nitrifiers were considered to perform both aerobic ammonium oxidation to nitrite and anoxic removal of ammonium with nitrite as electron acceptor under low oxygen concentrations, results obtained from FISH revealed that anammox was playing the major role in the latter conversion

ammonium (Eq. 5.16). The effluent of the anaerobic system is split and partially fed to a nitrifying reactor where the main product generated is nitrate (nitrite is a minor product) (Eq. 5.17). The remaining is fed to the DEAMOX biofilm reactor where it is mixed with the nitrified effluent to enable autotrophic partial denitrification with sulfide (Eq. 5.18), generating nitrite. Anammox bacteria present in the DEAMOX reactor can then oxidize ammonium using nitrite as electron acceptor (Eq. 5.9). The distribution of the anaerobic/aerobic flows is chosen based on the wastewater composition, especially by the electron donor (sulfide and ammonium) concentrations (KALYUZHNYI et al. 2006, 2007).

KALYUZHNYI et al. (2006) reported stable process performance in the long term (after 400 days) with nitrogen removal of around 90% for incoming nitrogen loads above 1000 mg/(L day) to the DEAMOX reactor. In order to ensure stable operation, the authors recommend that the NO_X-N/NH_4-N ratio should be higher than 1.2 (taking into account the anammox reaction stoichiometry) and influent H_2S/NO_3^- ratio of 1:4 or H_2S/NO_3-N ratio of 0.57 mgH_2S-S/mgNO_3-N (taking into account the stoichiometry of the sulphide-based denitrification) (KALYUZHNYI et al. 2006, 2007; MASŁOŃ and TOMASZEK 2009).

$$\begin{array}{l}\text{Anaerobic reactor:}\\ \text{Organic nitrogen} + SO_4^{2-} \rightarrow NH_4^+ + HCO_3^- + CH_4 + HS^-\end{array} \tag{5.16}$$

$$\begin{array}{l}\text{Nitrifying reactor:}\\ NH_4^+ + O_2 \rightarrow NO_3^- + NO_2^-\end{array} \tag{5.17}$$

$$\begin{array}{l}\text{DEAMOX:}\\ 4NO_3^- + HS^- \rightarrow 4NO_2^- + SO_4^{2-}\end{array} \tag{5.18}$$

$$NH_4^+ + NO_2^- \rightarrow N_2 + 2H_2O \tag{5.19}$$

Some key characteristics of the DEAMOX process are enumerated as follows (Kalyuzhnyi et al. 2006):

- The production of nitrite does not require complex control of the process.
- The maintenance of the DEAMOX reactor under denitrifying conditions enhances the growth of granules and therefore favors the anammox process.
- High nitrite concentrations (possibly toxic) are avoided and the emission of greenhouse gases is reduced.

However, the inhibition of ammonium-oxidizing bacteria in the nitrifying reactor and anammox organisms in the DEAMOX tank by hydrogen sulfide in the treatment of wastewaters with high sulfate concentrations is one of the disadvantages of the process. Furthermore, the applicability of the DEAMOX concept depends on the presence of sulfate in the influent wastewater and on the sulfate-to-nitrogen ratio given that the process is directly dependent on the amount of sulfide in the DEAMOX reactor (Kalyuzhnyi et al. 2006, 2007; MASŁOŃ and TOMASZEK 2009).

5.2.4 NO_X Processes

The control and stimulation of the denitrifying activity of microorganisms of the genus *Nitrosomonas* through the use of nitrogen oxides allows new possibilities for the treatment of wastewater. In the presence of NO_x, these autotrophic microorganisms are able to carry out nitrification and denitrification simultaneously, even under completely aerobic conditions, with N_2 as the main product.

In this process, only around 40% of the ammonium load is converted to nitrite. The oxygen demand is 50% lower than in the conventional nitrification process. Since nitrite is used as the terminal electron acceptor in denitrification, this step consumes less organic matter (COD) (SCHMIDT et al. 2003).

The NO_x compounds (NO/NO_2), added only in trace quantities (NH_4^+/NO_2 ratio varies from 1000/1 to 500/1), act as a regulatory signal which induces the denitrifying activity of the ammonium-oxidizing bacteria (SCHMIDT et al. 2001). Consequently, around 50% of the reducing equivalents [H] are transferred to nitrite, which acts as the terminal electron acceptor instead of oxygen. Thus, the consumption of oxygen in this process is reduced.

The flow diagrams in Fig. 5.12a–f illustrate some of the main processes employed for autotrophic removal of nitrogen described above. Based on our current understanding, the most important conversions of the nitrogen cycle occur during these processes. The traditional nitrogen removal process (nitrification and denitrification) and the combined partial nitrification and denitrification of nitrite (denitritation) are also represented for comparison with the recently developed technologies.

5.3 Final Considerations

New processes based on autotrophic nitrogen removal represent an alternative to the conventional processes, particularly for the treatment of wastewaters containing high concentrations of nitrogen and with a low organic carbon-to-nitrogen (C/N) ratio, which is unfavorable for the application of the conventional nitrification-denitrification process.

The new technologies based on partial nitrification coupled with the reduction of the nitrite formed, through the addition of a carbon source (denitritation) or even through reduction with ammonium (anaerobic oxidation of ammonium), are economically favorable in terms of nutrition requirements and operating costs.

As illustrated in this chapter, the autotrophic removal processes have been extensively studied by various research groups. Not only laboratory-scale tests but also pilot and industrial-scale studies have been described in the literature.

It is true that initially the autotrophic removal of nitrogen appears to be a difficult process to operate due to the need for strict control of substrate concentrations, pH, temperature, and dissolved oxygen. These are among the decisive factors which will determine the success of the treatment, since they influence the presence or absence

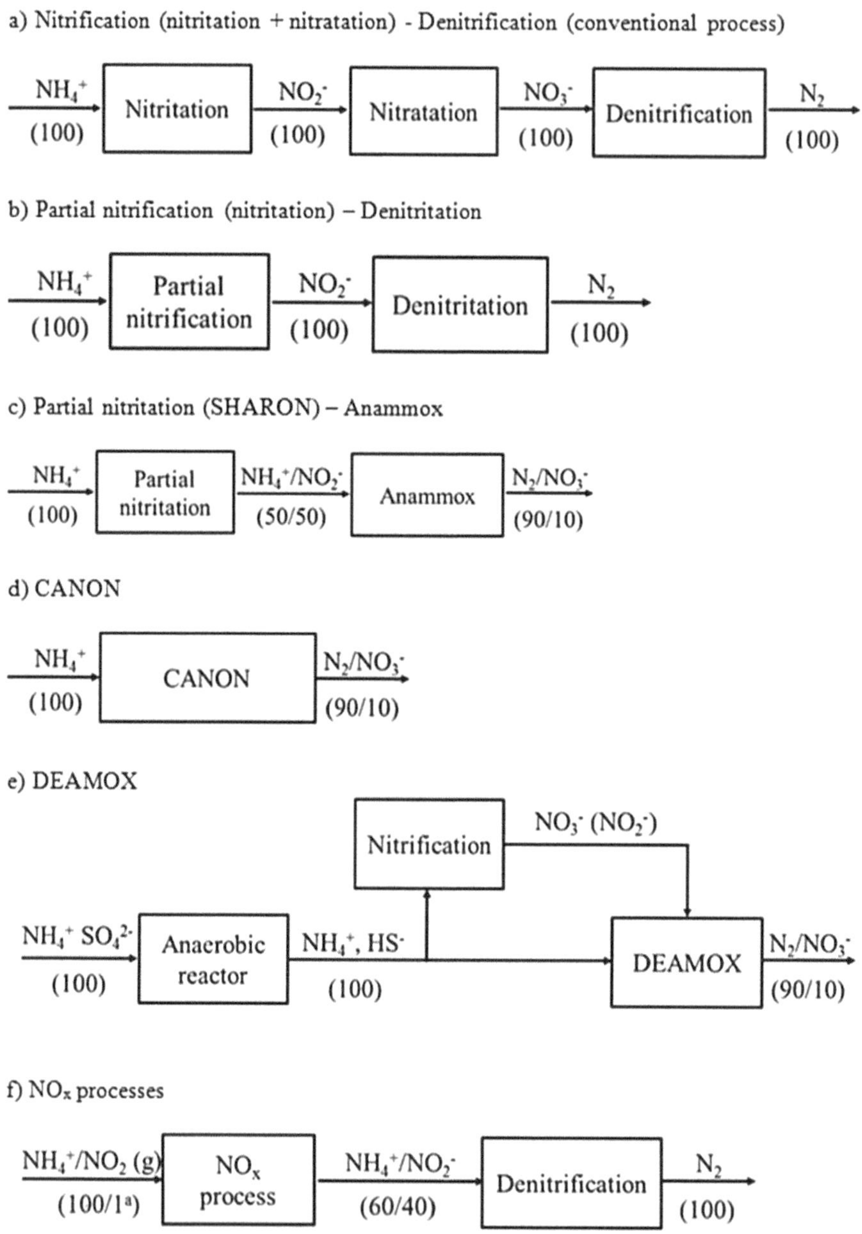

Fig. 5.12 Flow diagrams representing the processes of (**a**) conventional nitrification-denitrification, (**b**) partial nitrification-denitritation, (**c**) partial nitritation-anammox, (**d**) CANON process, (**e**) DEAMOX process, and (**f**) NO_x processes. The numbers in parenthesis represent the quantity of nitrogen in %, which can vary depending on the process; NO_2 (g), gaseous NO_2[a]. In the presence of oxygen, the NO_2 (g) acts as a regulatory signal and not as a substrate, inducing the denitrifying activity of aerobic ammonium-oxidizing bacteria (adapted from SCHMIDT et al. (2003) and KALYUZHNYI et al. (2006))

of inhibitor compounds, which in many cases are used for the selection of specific bacterial populations.

Although the application of systems in which partial nitrification and anammox reaction occur in the same reactor has been carried out with success in many cases, the long-term stability of this process is still difficult to achieve and remains a challenge for future research.

Further research is needed in laboratory scale to gain a better and more detailed understanding of processes such as anammox. This will potentially leads to an increase in the number of real-scale treatment systems in the future.

In order to improve the development and intensify the application of autotrophic nitrogen removal processes, studies should be directed toward finding strategies to minimize long reactor start-up, which represent one of the main drawbacks regarding the application of these processes. In addition, continuous research aimed at understanding the metabolism of anammox bacteria should be carried out, seeking better ways to control the process.

Studies which explore the possibility of making the reactors aimed at the autotrophic removal of nitrogen part of the main treatment process rather than sidestream processes should be stimulated. This research topic referred to as mainstream anammox. In this case the reactors designed for the autotrophic removal of nitrogen (e.g., anammox reactors) need to be adapted to operate at the normal temperature of the treatment plant, which is often much lower than that considered to be optimal for the microorganisms involved.

Regardless of the main focus of the study, the fact that the metabolic pathways of the nitrogen transformation process are complex needs to be taken into account. The way in which different environmental factors affect these processes still needs to be better understood. Knowledge regarding the basic microbial processes involved in these systems is fundamental to increasing the nitrogen removal rate, obtaining satisfactory results for treatment in full scale and expanding the application of reliable treatment strategies.

References

ABELING, U., SEYFRIED, C.F. Anaerobic-aerobic treatment of high strength ammonium wastewater nitrogen removal via nitrite. *Water Science and Technology*, v. 26, p. 1007–1015, 1992.

ABELING, U., SEYFRIED, C.F. Anaerobic-aerobic treatment of potato-starch wastewater. *Water Science and Technology*, 28, p. 165–176, 1993.

ABELIOVICH, A. Transformations of ammonia and the environmental impact of nitrifying bacteria. *Biodegradation*, v. 3, p. 255–264, 1992.

AHN, Y.H. Sustainable nitrogen elimination biotechnologies: a review. *Process Biochemistry*, v. 41, p. 1709–1721, 2006.

AHN, Y.H., HWANG, I.S., MIN, K.S. Anammox and partial denitrification in anaerobic nitrogen removal from piggery waste. *Water Science and Technology*, v. 49, p. 145–153, 2004.

ALLEMAN, J.E. Elevated nitrite occurrence in biological wastewater treatment systems. *Water Science and Technology*, v. 17, p. 409–419, 1984.

ANTHONISEN, A.C., LOEHR, R.C., PRAKASAM, T.B.S., SRINATH, E.G. Inhibition of nitrification by ammonia and nitrous-acid. *Journal Water Pollution Control Federation*, v. 48, p. 835–852, 1976.

ANTONIOU, P., HAMILTON, J., KOOPMAN, B., JAIN, R., HOLLOWAY, B., LYBERATOS, G., SVORONOS, S.A. Effect of temperature and pH on the effective maximum specific growth rate of nitrifying bacteria. *Water Research*, v. 24, p. 97–101, 1990.

ARROJO, B., MOSQUERA-CORRAL, A., CAMPOS, J.L., MÉNDEZ, R. Effects of mechanical stress on Anammox granules in a sequencing batch reactor (SBR). *Journal of Biotechnology*, p. 453–463, 2006.

AUSTERMANN-HAUN, U., Meyer, H., Seyfried, C., Rosenwinkel, K.-H. Full scale experiences with anaerobic/aerobic treatment plants in the food and beverage industry. *Water Science and Technology*, v. 40, p. 305–312, 1999.

BALMELLE, B., NGUYEN, K.M., CAPDEVILLE, B., CORNIER, J.C., DEGUIN, A. Study of factors controlling nitrite build-up in biological processes for water nitrification. *Water Science and Technology*, v. 26, p. 1017–1025, 1992.

BERNET, N., DANGCONG, P., DELGENE'S, J.-P., MOLETTA, R. Nitrification at low oxygen concentration in biofilm reactor. *Journal of Environmental Engineering*, v. 127, p. 266–271, 2001.

BERNET, N., SANCHEZ, O., CESBRON, D., STEYER, J.P., DELGENES, J.P. Modeling and control of nitrite accumulation in a nitrifying biofilm reactor. *Biochemical Engineering Journal*., v. 24, p. 173–183, 2005.

BINSWANGER, S., SIEGRIST, H., LAIS, P. Simultane Nitrifikation/Denitrifikation von stark ammoniumbelasteten Abwassern ohne organische Koblenstoffquellen. *Korresp. Abwasser*. v. 44, p. 1573–1580, 1997.

BRODA, E. Two kinds of lithotrophs missing in nature. *Z Allg Mikrobiol*., v. 17, p. 491–493, 1977.

BROUWER, M., VAN LOOSDRECHT, M.C.M., HEIJNEN, J.J. One reactor system for ammonium removal via nitrite. STOWA report 96-01. STOWA, Utrecht, 1996.

CAMPOS, J.L., SÁNCHEZ, M., MOSQUERA, A., MÉNDEZ, R., LEMA, J.M. Coupled BAS and anoxic USB system to remove urea and formaldehyde from wastewater. *Water Research*, v. 37, p. 3445–3451, 2003.

CARRERA, J., BAEZA, J.A., VICENT, T., LAFUENTE, J. Biological nitrogen removal of high-strength ammonium industrial wastewater with two-sludge system. *Water Research*, v. 37, p. 4211–4221, 2003.

CEMA, G., WISZNIOWSKI, J., ZABCZYNSKI, S., ZABLOCKA-GODLEWSKA, E., RASZKA, A., SURMACZ-GÓRSKA, J. Biological nitrogen removal from landfill leachate by deammonification assisted by heterotrophic denitrification in a rotating biological contactor (RBC). *Water Science and Technology*, v. 55, p. 35–41, 2007.

CHAMCHOI, N., NITISORAVUT, S. Anammox enrichment from different conventional sludges. *Chemosphere*., v. 66, p. 2225–2232, 2007.

CHAMCHOI, N., NITISORAVUT, S., SCHMIDT, J.E. Inactivation of Anammox communities under concurrent operational of anaerobic ammonium oxidation (Anammox) and denitrification. *Bioresource Technology*, v. 99, p. 3331–3336, 2008.

CHEN, M., KIM, J.H., KISHIDA, N., NISHIMURA, O., SUDO, R. Enhanced nitrogen removal using C/N load adjustment and real-time control strategy in sequencing batch reactors for swine wastewater treatment. *Water Science and Technology*, v. 49, p. 309–314, 2004.

CHUNG, J., BAE, W., LEE, Y., KO, G., LEE, S., PARK, S. Investigation of the effect of free ammonia concentration upon leachate treatment by shortcut biological nitrogen removal process. In: *IWA Speciality Symposium on Strong Nitrogenous and Agro-Wastewater*, v. 1, Seoul, Korea, p. 93–104, 2003.

CIRPUS, I.E.Y., DE BEEN, M., OP DEN CAMP, H.J.M., STRUS, M., LE PASLIER, D., KUENEN, G.J., JETTEN, M.S.M. A new soluble 10 kDa monoheme cytochrome c-552 from the anammox bacterium *Candidatus* "Kuenenia stuttgartiensis". *FEMS Microbiology Letters*, v. 252, p. 273–278, 2005.

CIUDAD, G., RUBILAR, O., MUÑOZ, P., RUIZ, G., CHAMY, R., VERGARA, C., JEISON, D. Partial nitrification of high ammonia concentration wastewater as a part of a shortcut biological nitrogen removal process. *Process Biochemistry*, v. 40, p. 1715–1719, 2005.

CLABAUGH, M.M. Nitrification of landfill leachate by biofilm columns. MSc thesis, Virginia Polytechnic Institute and State University, USA, 43p, 2001.

CLINE, J.D., RICHARDS, F.A. Oxygen deficient conditions and nitrate reduction in the eastern tropical North Pacific Ocean. *Limnology and Oceanography*, v. 17, p. 885–900, 1972.

DALSGAARD, T., THAMDRUP, B. Factors controlling anaerobic ammonium oxidation with nitrite in marine sediments. *Applied and Environmental Microbiology*, v. 68, p. 3802–3808, 2002.

DALSGAARD, T., THAMDRUP, B., CANFIELD, D.E. Anaerobic ammonium oxidation (anammox) in the marine environment. *Research in Microbiology*, v. 156, p. 457–464, 2005.

DAPENA-MORA, A., CAMPOS, J.L., MOSQUERA-CORRAL, A., JETTEN, M.S.M., MÉNDEZ, R. Stability of the Anammox process in a gas-lift reactor and a SBR. *Journal of Biotechnology*, v. 110, p. 159–170, 2004.

DAPENA-MORA, A., FERNANDEZ, I., CAMPOS, J.L., MOSQUERA-CORRAL, A., MENDEZ, R., JETTEN, M.S.M Evaluation of activity and inhibition effects on Anammox process by batch tests on the nitrogen gas production. *Enzyme and Microbial Technology*, v. 40, p. 859–865, 2007.

DE CLIPPELEIR, H., VLAEMINCK, S.E., CARBALLA, M., VERSTRAETE, W. A low volumetric exchange ratio allows high autotrophic nitrogen removal in a sequencing batch reactor. *Bioresource Technology*, p. 5010–5015, 2009.

DENG PETERSEN, P., JENSEN, K., LYNGSIE, P., HENDRIK JOHANSEN, N. Nitrogen removal in industrial wastewater by nitration and denitration - 3 years of experience. *Water Science and Technology*, v. 47, 181–188, 2003.

DEXIANG, L., XIAOMING, L., QI, Y., GUANGMING, Z., LIANG, G., XIU, Y. Effect of inorganic carbon on anaerobic ammonium oxidation enriched in sequencing batch reactor. *Journal of Environmental Sciences*, v. 20, p. 940–944, 2007.

DOSTA, J., FERNÁNDEZ, I., VÁZQUEZ-PADÍN, J.R., MOSQUERA-CORRAL, A., CAMPOS, J.L., MATA-ÁLVAREZ, J., MÉNDEZ, R. Short- and long-term effects of temperature on the Anammox process. *Journal of Hazardous Materials*, v. 154, p. 688–693, 2008.

EGLI, K., FANGER, U., ALVAREZ, P.J.J., SIEGRIST, H., VAN DER MEER, J.R., ZEHNDER, A.J.B. Enrichment and characterization of an anammox bacterium from a rotating biological contact treating ammonium-rich leachate. *Archives of Microbiology*, v. 175, p. 198–207, 2001.

FUERST, J.A. Intracellular compartmentation in Planctomycetes. *Annual Review of Microbiology*, v. 59, p. 299–328, 2005.

FURUKAWA, K., LIEU, P.K., TOKITOH, H., FUJII, T. Development of single-stage nitrogen removal using anammox and partial nitritation (SNAP) and its treatment performances. *Water Science and Technology*, v. 53, p. 83–90, 2006.

FUX, C. Biological nitrogen elimination of ammonium-rich sludge digester liquids. PhD thesis, ETH-Zürich, Switzerland, 2003.

FUX, C., BOEHLER, M., HUBER, P., BRUNNER, I., SIEGRIST, H. Biological treatment of ammonium-rich wastewater by partial nitritation and subsequent anaerobic ammonium oxidation (Anammox) in a pilot-plant. *Journal of Biotechnology*, v. 99, p. 295–306, 2002.

FUX, C., HUANG, D., MONTI, A., SIEGRIST, H. Difficulties in maintaining long-term partial nitritation of ammonium-rich sludge digester liquids in a moving-bed biofilm reactor (MBBR). *Water Science and Technology*, v. 49, p. 53–60, 2004.

GANIGUÉ, R., LÓPEZ, H., BALAGUER, M.D., COLPRIM, J. Partial ammonium oxidation to nitrite of high ammonium content urban landfill leachates. *Water Research*, v. 41, p. 3317–3326, 2007.

GARRIDO, J.M., VAN BENTHUM, W.A.J., VAN LOOSDRECHT, M.C.M., HEIJNEN, J.J. Influence of dissolved oxygen concentration on nitrite accumulation in a biofilm airlift suspension reactor. *Biotechnology and Bioengineering*, v. 53, p. 168–178, 1997.

GIL, K.-I., CHOI, E. Nitrogen removal by recycle water nitritation as an attractive alternative for retrofit technologies in municipal wastewater treatment plants. *Water Science and Technology*, v. 49, p. 39–46, 2004.
GONG, Z., YANG, F.L., LIU, S.T., BAO, H., HU, S.W, FURUKAWA, K.J. Feasibility of a membrane-aerated biofilm reactor to achieve single-stage autotrophic nitrogen removal based on Anammox. *Chemosphere*, v. 69, p. 776–784, 2007.
GONZÁLEZ-MARTÍNEZ, A., POYATOS, J.M., HONTORIA, E., GONZALEZ-LOPEZ, J., OSORIO, J. Treatment of Effluents Polluted by Nitrogen With New Biological Technologies Based on Autotrophic Nitrification-Denitrification Processes Recent Pat Biotechnol 5(2), 74–84. 2011.
GÜVEN, D., DAPENA, A., KARTAL, B., SCHMID, M.C., MAAS, B., VAN DE PAS-SCHOONEN, K., SÖZEN, S., MENDEZ, R., OP DEN CAMP, H.J.M., JETTEN, M.S.M., STROUS, M., SCHMIDT, I. Propionate oxidation and methanol inhibition of anaerobic ammonium-oxidizing bacteria. *Applied and Environmental Microbiology*, v. 71, p. 1066–1071, 2005.
HAINES, T.H. Do sterols reduce proton and sodium leaks through lipid bilayers. *Progress in Lipid Research*, v. 40, p. 299–324, 2001.
HANAKI, K., WANTAWIN, C., OHGAKI, S. Effects of the activity of heterotrophs on nitrification in a suspended-growth reactor. *Water Research*, v. 24, p. 289–296, 1990.
HAO, X., HEIJNEN, J.J., VAN LOOSDRECHT, M.C.M. Model-based evaluation of temperatura and inflow variations on a partial nitrification-Anammox biofilm process. *Water Research*, v. 36, p. 4839–4849, 2002.
HAO, X., HEIJNEN, J.J., VAN LOOSDRECHT, M.C.M. Sensitivity analysis of a biofilm model describing a one-stage completely autotrophic nitrogen removal (CANON) process. *Biotechnology and Bioengineering*, v. 77, p. 266–277, 2001.
HEIJNEN, J.J. Biologishe anaëroob-aërobe afvalwaterzuivering bij Gist-Brocades: eindrapport 1977–1986. 's-Gravenhage. N.L., Staatsuitgeverij/DOP, 1988.
HELLINGA, C., SCHELLEN, A.A.J.C., MULDER, J.W., VAN LOOSDRECHT, M.C.M., HEIJNEN, J.J. The Sharon process: an innovative method for nitrogen removal from ammonium-rich wastewater. *Water Science and Technology*, v. 37, p. 135–142, 1998.
HELMER, C., KUNST, S., JURETSCHKO, S., SCHMID, M.C., SCHLEIFER, K.-H., WAGNER, M. Nitrogen loss in a nitrifying biofilm system. *Water Science and Technology*, v. 39, p. 13–21, 1999.
HELMER-MADHOK, C., SCHMID, M., FILIPOV, E., GAUL, T., HIPPEN, A., ROSENWINKEL, K.H., SEYFRIED, C.F., WAGNER, M., KUNST, S. Deammonification in biofilm systems: population structure and function. *Water Science and Technology*, v. 46, p. 223–231, 2002.
HIPPEN, A., HELMER, C., KUNST, S., ROSENWINKEL, H., SEYFRIED, C.F. Six years' practical experience with aerobic/anoxic deammonification in biofilm systems. *Water Science and Technology*, v. 44, p. 39–48, 2001.
HIPPEN, A., ROSENWINKEL, K.-H., BAUMGARTEN, G., SEYFRIED, C.F. Aerobic deammonification: a new experience in the treatment of wastewaters. *Medelingen – Faculteit Landbouwkundige en Toegepaste Biologische Wetenschappen* (Universiteit Gent), v. 61, p. 1967–1974, 1996.
HIPPEN, A., ROSENWINKEL, K.-H., BAUMGARTEN, G., SEYFRIED, C.F. Aerobic deammonification: a new experience in the treatment of wastewaters. *Water Science and Technology*, v. 35, p. 111–120, 1997.
HU, S.S. Acute substrate-intermediate-product related inhibition of nitrifiers. M.S. Thesis, School of Civil Engineering, Purdue University, West Lafalyette, Indiana, 1990.
HUSTON, W.M., HARHANGI, H.R., LEECH, A.P., BUTLER, C.S., JETTEN, M.S.M., OP DEN CAMP, H.J.M., MOIR, J.W.B. Expression and characterization of a major c-type cytochrome encoded by gene kustc0563 from *Kuenenia stuttgartiensis* as a recombinant protein in *Escherichia coli*. *Protein Expression and Purification*, v. 51, p. 28–33, 2007.
ILIES, P., MAVINIC, D. S. The effect of decreased ambient temperature on the biological nitrification and denitrification of a high ammonia landfill leachate. *Water Research*, v. 35, p. 2065–2072, 2001.

INNEREBNER, G., INSAM, H., FRANKE-WHITTLE, I., WETT, B. Identification of anammox bacteria in a full-scale deammonification plant making use of anaerobic ammonia oxidation. *Systematic and Applied Microbiology*, v. 30, p. 408–412, 2007.

ISAKA, K., DATE, Y., SUMINO, T., YOSHIE, S., TSUNEDA, S. Growth characteristic of anaerobic ammonium-oxidizing bacteria in an anaerobic biological filtrated reactor. *Applied Microbiology and Biotechnology*, v. 70, p. 47–52, 2006.

JENICEK, P., SVEHLA, P., ZABRANSKA, J., DOHANYOS, M. Factors affecting nitrogen removal by nitritation/denitritation. *Water Science and Technology*, v. 49, p. 73–79, 2004.

JETTEN, M.S.M., HOM, S.J., VAN LOOSDRECHT, M.C.M. Towards a more sustainable wastewater treatment system. *Water Science and Technology*, v. 35, p. 171–179, 1997.

JETTEN, M.S.M., OP DEN CAMP, J.M., KUENEN, J.G., STROUS, M. Taxonomic description of the family Anammoxaceae. IN: Hedlund, B., Krieg, N.R., Paster, B.J., K.H., Schleifer, Staley, J.T., Ward, N., Whitman, W.B. (eds). Bergey's manual of systematic bacteriology (2 ed.), v. 4. New York: Springer, 2008.

JETTEN, M.S.M., STROUS, M., VAN DE PAS-SCHOONEN, K.T., SCHALK, J., VAN DONGEN, L.G.J.M., VAN DE GRAAF, A.A., LOGEMANN, S., MUYZER, G., VAN LOOSDRECHT, M.C.M., KUENEN, J.G. The anaerobic oxidation of ammonium. *FEMS Microbiology Reviews*, v. 22, p. 421–437, 1999.

JETTEN, M.S.M., WAGNER, M., FUERST, J., VAN LOOSDRECHT, M.C.M., KUENEN, G., STROUS, M. Microbiology and application of the anaerobic ammonium oxidation ('anammox') process. *Current Opinion in Biotechnology*, v. 12, p. 283–288, 2001.

JOKELA, J., KETTUNEN, R., SORMUNEN, K., RINTAL, J. Biological nitrogen removal from municipal landfill leachate: low-cost nitrification in biofilters and laboratory scale in-situ denitrification. *Water Research*, v. 36, p. 4079–4087, 2002.

JOO, S.-H., KIM, D.-J., YOO, I.-K., PARK, K., CHA, G.-C. Partial nitrification in an upflow biological aerated filter by O_2 limitation. *Biotechnol Lett.*, v. 22, p. 937–940, 2000.

JOSS, A., SALZGEBER, D., EUGSTER, J., KONIG, R., ROTTERMANN, K., BURGER, S., FABIJAN, P., LEUMANN, S., MOHN, J., SIEGRIST, H. Full-scale nitrogen removal from digester liquid with partial nitritation and anammox in one SBR. *Environmental Science and Technology*, v. 43, p. 5301–5306, 2009.

KADLEC, R.H., KNIGHT, R.L. Treatment Wetlands, Lewis Publishers, Boca Raton, 893pp, 1996.

KALYUZHNYI, S., GLADCHENKO, M. Sequenced anaerobic-aerobic treatment of high strength, strong nitrogenous landfill leachates. *Water Science and Technology*, v. 49, p. 301–312, 2004.

KALYUZHNYI, S., GLADCHENKO, M., MULDER, A., VERSPRILLE, B. DEAMOX – new biological nitrogen removal process based on anaerobic ammonia oxidation coupled to sulphide-driven conversion of nitrate into nitrite. *Water Research*, v. 40, p. 3637–3645, 2006.

KALYUZHNYI, S., GLADCHENKO, M., MULDER, A., VERSPRILLE, B. Comparison of quasisteady-state performance of the DEAMOX process under intermittent and continuous feeding and different nitrogen loading rates. Biotechnology Journal, v. 2, n. 7, p. 894–900, 2007.

KARTAL, B., RATTRAY, J., VAN NIFTRIK, L., VAN DE VOSSENBERG, J.L., SCHMID, M.C., FUERST, J.A., SINNINGHE DAMSTÉ, J., JETTEN, M.S.M., STROUS, M. *Candidatus* "Anammoxoglobus propionicus" gen. nov., sp. nov., a new propionate oxidizing species of anaerobic ammonium oxidizing bacteria. *Systematic and Applied Microbiology*, v. 30, p. 39–49, 2007.

KARTAL, B., VAN NIFTRIK, L.A., RATTRAY, J., VAN DE VOSSENBERG, J.L.C.M., SCHMID, M.C., SINNINGHE DAMSTÉ, J., JETTEN, M.S.M., STROUS, M. *Candidatus* 'Brocadia fulgida': an autofluorescent anaerobic ammonium oxidizing bacterium. *FEMS Microbiology Ecology*, v. 63, p. 46–55, 2008.

KARTAL, B., VAN NIFTRIK, L.A., SLIEKERS, A.O., SCHMID, M., SCHMIDT, I., VAN DE PAS-SCHOONEN, K., CIRPUS, I., VAN DER STAR, W.R.L., VAN LOOSDRECHT, M.C.M., ABMA, W., KUENEN, J., MULDER, J.W., JETTEN, M.S.M., OP DEN CAMP, H., STROUS, M., VAN DE VOSSENBERG, J. Application, eco-physiology and biodiversity of anaerobic ammonium-oxidizing bacteria. *Reviews in Environmental Science and Biotechnology*, v. 3, p. 255–264, 2004.

KELLER, J., SUBRAMANIAM, K., GÖSSWEIN, J., GREENFIELD, P. Nutrient removal from industrial wastewater using single tank sequencing batch reactors. *Water Science and Technology*, v. 35, p. 137–144, 1997.

KIM, D.J., CHANG, J.S., LEE, D.I., HAN, D.W., YOO, I.K., CHA, G.C. Nitrification of high strength ammonia wastewater and nitrite accumulation characteristics. *Water Science and Technology*, v. 47, p. 45–51, 2003.

KOCH, G., EGLI, K., VAN DER MEER, J.R., SIEGRIST, H. Mathematical modeling of autotrophic denitrification in a nitrifying biofilm of a rotating biological contactor. *Water Science and Technology*, v. 41, p. 191–198, 2000.

KUAI, L., VERSTRAETE, W. Ammonium removal by the oxygen-limited autotrophic nitrification-denitrification system. *Applied and Environmental Microbiology*, v. 64, p. 4500–4506, 1998.

KUENEN, J.G., JETTEN, M.S.M. Extraordinary anaerobic ammonia-oxidizing bacteria. ASM News, p. 456–463, 2001.

KUMAR, M., LIN, J. Co-existence of anammox and denitrification for simultaneous nitrogen and carbon removal – strategies and issues. *Journal of Hazardous Materials*, v. 178, p. 1–9, 2010.

KUYPERS, M.M.M., SLIEKERS, A.O., LAVIK, G., SCHMID, M., JORGENSEN, B.B., KUENEN, J.G., SINNINGHE DAMSTÉ, J.S., STROUS, M., JETTEN, M.S.M. Anaerobic ammonium oxidation by anammox bacteria in the Black Sea. *Nature*, v. 422, p. 608–611, 2003.

LACKNER, S., TERADA, A., SMETS, B.F. Heterotrophic activity compromises autotrophic nitrogen removal in membrane-aerated biofilms: results of a modelling study. *Water Research*, v. 42, p. 1102–1112, 2008.

LINDSAY, M.R., WEBB, R.I., STROUS, M., JETTEN, M.S.M., BUTLER, M.K., FORDE, R.J., FUERST, J.A. Cell compartmentalization in Planctomycetes: novel types of structural organisation for the bacterial cell. *Archives in Microbiology*, 175, p. 413–429, 2001.

LIU, S., YANG, F., GONG, Z., SU, Z. Assessment of the positive effect of salinity on the nitrogen removal performance and microbial composition during the start-up of CANON process. *Applied Microbiology and Biotechnology*, v. 80, p. 339–348, 2008.

LÓPEZ, H., PUIG, S., GANIGUÉ, R., RUSCALLEDA, M., BALAGUER, M.D., COLPRIM, J. Start-up and enrichment of a granular anammox SBR to treat high nitrogen load wastewaters. *Journal of Chemical Technology & Biotechnology*, v. 83, p. 233–241, 2008.

MADIGAN, M.T., MARTINO, J.M., PARKER, J. Brock Biology of Microorganisms. New-Jersey: Prentice-Hall, p 1175. 2000.

MASŁOŃ, A., TOMASZEK, J.A. Anaerobic ammonium nitrogen oxidation in DEAMOX process. *Environment Protection Engineering*, p. 123–130, 2009

MAURET, M., PAUL, E., PUECH-COSTES, E., MAURETTE, M.T., BAPTISTE, P. Application of experimental research methodology to the study of nitrification in mixed culture. *Water Science and Technology.*, v. 34, p. 245–252, 1996.

METCALF and EDDY Inc. Wastewater engineering: treatment and reuse. New York: McGraw Hill (4 Ed.), 2003.

MOLINUEVO, B., GARCIA, M.C., KARAKASHEV, D., ANGELIDAKI, I. Anammox for ammonia removal from pig manure effluents: effect of organic matter content on process performance. *Bioresource Technology*, v. 100, p. 2171–2175, 2009.

MULDER, A. Anoxic ammonia oxidation of wastewater. European Patent Ep327184. Assignee: Gist-Brocades NV, NL, 1989.

MULDER, A. The quest for sustainable nitrogen removal technologies. *Water Science and Technology*, v. 48, p. 67–75, 2003.

MULDER, A., VAN DE GRAAF, A.A., ROBERTSON, L.A., KUENEN, J.G. Anaerobic ammonium oxidation discovered in a denitrifying fluidized bed reactor. *FEMS Microbiology Ecology*, v. 16, p. 177–184, 1995.

MULDER, J.W., VAN KEMPEN, R. N-removal by SHARON. *Water Quality International*, IAWQ (Edit.), p. 30–31, 1997.

MULDER, J.W., Van Loosdrecht, M.C.M., Hellinga, C. Van Kempen, R. Full-scale application of the SHARON process for treatment of rejection water of digested sludge dewatering. Water Science and Technology, 43, p. 127–134, 2001.

MURAT, S., INSEL, G., ARTAN, N., ORHON, D. Performance evaluation of SBR treatment for nitrogen removal from tannery wastewater. In: IWA Speciality Symposium on Strong Nitrogenous and Agro-Wastewater, v. 2. Seoul, Korea, p. 598–605, 2003.

NIELSEN M., BOLLMANN A., SLIEKERS O., JETTEN M., SCHMID M., STROUS M., SCHMIDT I., LARSEN L.H., NIELSEN L.P., REVSBECH N.P. Kinetics, diffusional limitation and microscale distribution of chemistry and organisms in a CANON reactor. *FEMS Microbiology Ecology. v. 51*, 247–256, 2005.

OBAJA, D., MACÉ, S., COSTA, J., SANS, C., MATA-ALVAREZ, J. Nitrification, denitrification and biological phosphorus removal in piggery wastewater using a sequencing batch reactor. *Bioresource Technology*, v. 87, p. 103–111, 2003.

PAINTER, H.A., LOVELESS, J.E. Effect of temperature and pH value on the growth-rate constants of nitrifying bacteria in the activated-sludge process. *Water Research*, v. 17, p. 237–248, 1983.

PAREDES, D., KUSCHK, P., MBWETTE, T.S.A., STANGE, F., MÜLLER, R.A., KÖSER, H. New aspects of microbial nitrogen transformation in the context of wastewater treatment – a review. *Engineering Life Sciences*, v. 7, p. 13–25, 2007.

PATHAK, B.K., KAZAMA, F., SAIKI, Y., SUMINO, T. Presence and activity of anammox and denitrification process in low ammonium-fed bioreactors. *Bioresource Technology*, v. 98, p. 2201–2206, 2007.

PICIOREANU, C., VAN LOOSDRECHT, M.C.M., HEIJNEN, J.J. Modelling the effect of oxygen concentration on nitrite accumulation in a biofilm airlift suspension reactor. *Water Science and Technology*, v. 36, p. 147–156, 1997.

PLAZA E., TRELA J., GUT L., LÖWÉN M., SZATKOWSKA B. Deammonification process for treatment of ammonium rich wastewater. In: *Integration and optimisation of urban sanitation systems, Joint Polish-Swedish Reports, No 10*. Royal Institute of Technology, Stockholm, p. 77–87, 2003.

POO, K., JUN, B., LEE, S., WOO, H., KIM, C. Treatment of strong nitrogen swine wastewater at full-scale sequencing-batch reactor. *Water Science and Technology*, v. 49, p. 315–323, 2004.

POTH, M., FOCHT, D.D. Nitrogen-15 kinetic analysis of nitrous oxide production by *Nitrosomonas europaea*: an examination of nitrifier denitrification. *Applied and Environmental Microbiology*, v. 49, p. 1134–1141, 1985.

PYNAERT, K., SMETS, B.F., BEHEYDT, D., VERSTRAETE, W. Start-up of autotrophic nitrogen removal reactors via sequential biocatalyst addition. *Environmental Science and Technology*, v. 38, p. 1228–1235, 2004.

PYNAERT, K., SMETS, B.F., WYFFELS, S., BEHEYDT, D., SICILIANO, S.D., VERSTRAETE, W. Characterization of an autotrophic nitrogen-removing biofilm from a highly loaded lab-scale rotating biological contactor. *Applied and Environmental Microbiology*, v. 69, p. 3626–3635, 2003.

PYNAERT, K., WYFFLES, S., SPRENGERS, R., BOECKX, P., VAN CLEEMPUT, O, VERSTRAETE, W Oxygen limited nitrogen removal in a lab scale rotating biological contactor treating an ammonium rich wastewater. *Water Science and Technology*, v. 45, p. 357–363, 2002.

QUINLAN, A.V. Prediction of the optimum pH for ammonia-N oxidation by *Nitrosomonas europaea* in well-aerated natural and domestic wastewater. *Water Research*, v. 18, p. 561, 586, 1984.

RICHARDS, F. Anoxic basins and fjords. In: Chemical Oceanography. Riley, J.P., Skirrown G. (eds). New York: Academic Press, v. 1, p. 611–645, 1965.

RITTMANN, B.E., MCCARTY, P.I. Environmental Biotechnology: Principles and Applications. McGraw-Hill, 2001.

RUIZ, G., JEISON, D., CHAMY, R. Nitrification with high nitrite accumulation for the treatment of wastewater with high ammonia concentration. *Water Research*, v. 37, p. 1371–1377, 2003.

RUSCALLEDA, M., LÓPEZ, H., GANIGUÉ, R., OUIG, S., BALAGUER, M., COLPRIM, J. Heterotrophic denitrification on granular anammox SBR treating urban landfill leachate. *Water Science and Technology*, v. 58, p. 1749–1755, 2008.

RYSGAARD, S., GLUD, R.N., RISGAARD-PETERSEN, N., DALSGAARD, D. Denitrification and anammox activity in arctic marine sediments. *Limnology and Oceanography*, p. 1493–1502, 2004.

SCHALK, J. A study of the metabolic pathway of anaerobic ammonium oxidation. PhD thesis. Delft University of Technology, Delft, 107 p., 2000.

SCHALK, J., DE VRIES, S., KUENEN, J.G., JETTEN, M.S.M. Involvement of a novel hydroxylamine oxidoreductase in anaerobic ammonium oxidation. *Biochemistry*, v. 39, p. 5405–5412, 2000.

SCHMID, M., TWACHTMANN, U., KLEIN, M., STROUS, M., JURETSCHKO, S., JETTEN, M., METZGER, J.W., SCHLEIFER, K.-H., WAGNER, M. Molecular evidence for genus level diversity of bacteria capable of catalyzing anaerobic ammonium oxidation. *Systematic and Applied Microbiology*, v. 23, p. 93–106, 2000.

SCHMID, M.C., RISGAARD-PETERSEN, N., VAN DE VOSSENBERG, J., KUYPERS, M.M.M., LAVIK, G., PETERSEN, J., HULTH, S., THAMDRUP, B., CANFIELD, D., DALSGAARD, T., RYSGAARD, S., SEJR, M.K., STROUS, M., OP DEN CAMP, H.J.M., JETTEN, M.S.M. Anaerobic ammonium-oxidizing bacteria in marine environments: widespread occurrence but low diversity. *Environmental Microbiology*, v. 9, p. 1476–1484, 2007.

SCHMID, M., WALSH, K., WEBB, R., IRENE, W., RIJPSTRA, C., VAN DE PAS-SCHOONEN K., VERBRUGGEN, M.J., HILL, T., MOFFETT, B., FUERST, J., SCHOUTEN, S., SINNINGHE DAMSTÉ, J.S., HARRIS, J., SHAW, P., JETTEN, M., STROUS, M. *Candidatus* "Scalindua brodae", sp. nov., *Candidatus* "Scalindua wagneri", sp., nov., two new species of anaerobic ammonium oxidizing bacteria. *Syst Appl Microbiol*. v. 26, p. 529–538, 2003.

SCHMIDELL, W., REGINATTO, V.S. Remoção de Nitrogênio. In: V Curso de Tratamento Biológico de Resíduos. Florianópolis-SC, 2005.

SCHMIDT, I., BOCK, E., JETTEN, M.S.M. Ammonia oxidation by *Nitrosomonas eutropha* with NO_2 as oxidant is not inhibited by acetylene. *Microbiology*, v. 147, 2247–2253, 2001.

SCHMIDT, I., SLIEKERS, O., SCHMID, M., BOCK, E., FUERST, J., KUENEN, J.G., JETTEN, M.S.M., STROUS, M. New concepts of microbial treatment processes for the nitrogen removal in wastewater. *FEMS Microbiology*, v. 27, p. 481–492, 2003.

SCHMIDT, J.E., BATSTONE, D.J., ANGELIDAKI, I. Improved nitrogen removal in upflow anaerobic sludge blanket (UASB) reactors by incorporation of Anammox bacteria into the granular sludge. *Water Science and Technology*, v. 49, p. 69–76, 2004.

SEYFRIED, C.F., HIPPEN, A., HELMER, C., KUNST, S., ROSENWINKEL, K.H. One-stage deammonification: nitrogen elimination at low costs. *Water Science and Technology*, v. 1, p. 71–80, 2001.

SHIMAMURA, M., NISHIYAMA, T., SHIGETOMO, H., TOYOMOTO, T., KAWAHARA, Y., FURUKAWA, K., FUJII, T. Isolation of a multiheme protein with features of a hydrazine-oxidizing enzyme from an anaerobic ammonium-oxidizing enrichment culture. *Applied and Environmental Microbiology*, v. 73, p. 1065–1072, 2007.

SIEGRIST, H., REITHAAR, S., KOCH, G., LAIS, P. Nitrogen loss in a nitrifying rotating contactor treating ammonium-rich wastewater without organic carbon. *Water Science and Technology*, v. 38, p. 241–248, 1998.

SINNINGHE DAMSTÉ, J.S., RIJPSTRA, W.I.C., GEENEVASEN, J.A.J., STROUS, M., JETTEN, M.S. Structural identification of ladderane and other membrane lipids of planctomycetes capable of anaerobic ammonium oxidation (anammox). *FEBS Journal*, v. 272, p. 4270–4283, 2005.

SINNINGHE DAMSTÉ, J.S., STROUS, M., RIJPSTRA, W.I.C., HOPMANS, E.C., GEENEVASEN, J.A.J., VAN DUIN, A.C.T., VAN NIFTRIK, L.A., JETTEN, M.S.M. Linearly concatenated cyclobutane lipids form a dense bacterial membrane. *Nature*, v. 419, p. 708–712, 2002.

SLIEKERS, A.O., DERWORT, N., CAMPOS-GOMEZ, J.L., STROUS, M., KUENEN, M., JETTEN, M.S.M. Completely autotrophic nitrogen removal over nitrite in a single reactor. *Water Research*, v. 36, p. 2475–2482, 2002.

SLIEKERS, A.O., THIRD, K.A., ABMA, W., KUENEN, J.G., JETTEN, M.S.M. CANON and Anammox in a gas-lift reactor. *FEMS Microbiol. Lett.*, v. 218, p. 339–344, 2003.

STROUS, M. Microbiology and application of anaerobic ammonium oxidation. PhD Thesis, 144 p., 2000.

STROUS, M., FUERST, J.A., KRAMER, E.H.M., LOGEMANN, S., MUYZER, G., Van De Pas-Schoonen, K.T., Webb, R., Kuenen, J.G., Jetten, M.S.M. Missing lithotroph identified as new Planctomycete. *Nature*, v. 400, p. 446–449, 1999a.

STROUS, M., HEIJNEN, J.J., KUENEN, J.G., JETTEN, M.S.M. The sequencing batch reactor as a powerful tool for the study of slowly growing anaerobic ammonium-oxidizing microorganisms. *Applied Microbiology and Biotechnology*, v. 50, p. 589–596, 1998.

STROUS, M., KUENEN, J.G., FUERST, J.A., WAGNER, M., JETTEN, M.S.M. The Anammox case – a new experimental manifesto for microbiological eco-physiology. *Antonie van Leeuwenhoek*, v. 81, p. 693–702, 2002.

STROUS, M., KUENEN, J.G., JETTEN, M.S.M. Key physiology of anaerobic ammonium oxidation. *Applied and Environmental Microbiology*, v. 65, p. 3248–3250, 1999b.

STROUS, M., PELLETIER, E., MANGENOT, S., RATTEI, T., LEHNER, A., TAYLOR, M.W., HORN, M., DAIMS, H., BARTOL-MAVEL, D., WINCKER, P., BARBE, V., FONKNECHTEN, N., VALLENET, D., SEGURENS, B., SCHENOWITZ-TRUONG, C., MEDIGUE, C., COLLINGRO, A., SNEL, B., DUTILH, B.E., OP DEN CAMP, H.J.M., VAN DER DRIFT, C., CIRPUS, I., VAN DE PAS-SCHOONEN, K.T., HARHANGI, H.R., VAN NIFTRIK, L., SCHMID, M., KELTJENS, J., VAN DE VOSSENBERG, J., KARTAL, B., MEIER, H., FISHMAN, D., HUYNEN, M.A., MEWES, H.-W., WEISSENBACH, J., JETTEN, M.S.M., WAGNER, M., LE PASLIER, D. Deciphering the evolution of an Anammox bacterium from a community genome. *Nature*, v. 440, p. 790–794, 2006.

STROUS, M., VAN GERVEN, E., KUENEN, J.G., JETTEN, M.S.M. Effects of aerobic and microaerobic conditions on anaerobic ammonium-oxidizing (Anammox) sludge. *Applied and Environmental Microbiology*, v. 63, p. 2446–2448, 1997.

SZATKOWSKA, B., CEMA, G., PLAZA, E., TRELA, J., HULTMAN, B. A one-stage system with partial nitritation and Anammox processes in the moving-bed biofilm reactor. *Water Science and Technology*, v. 55, p. 19–26, 2007.

SZATKOWSKA, B., PLAZA, E. Temperature as a factor influencing the Anammox process performance. *Water and Environmental Management Series*. Young Researchers, p. 51–58, 2006.

TANG, C., ZHENG, P., WANG, C., MAHMOOD, Q. Suppression of anaerobic ammonium oxidizers under high organic content in high-rate Anammox UASB reactor. *Bioresource Technology*, v. 101, p. 1762–1768, 2010.

THIRD, K.A., SLIEKERS, O.A., KUENEN, J.G., JETTEN, M.S.M. The CANON System (Completely Autotrophic Nitrogen-removal Over Nitrite) under ammonium limitation: interaction and competition between three groups of bacteria. *Systematic and Applied Microbiology*, v. 24, p. 588–596, 2001.

TILCHE, A., BACILIERI, E., BORTONE, G., MALASPINA, F., PICCININI, S., STANTE, L. Biological phosphorus and nitrogen removal in a full scale sequencing batch reactor treating piggery wastewater. *Water Science and Technology*, v. 40, p. 199–206, 1999.

TOH, S.K., ASHBOLT, N.J. Adaptation of anaerobic ammonium-oxidising consortium to synthetic coke-ovens wastewater. *Applied Microbiology and Biotechnology*, v. 59, p. 344–352, 2002.

TOH, S.K., WEBB, R.I., ASHBOT, N.J. Enrichment of autotrophic anaerobic ammonium-oxidizing consortia from various wastewaters. *Microbial Ecology*., v. 43, p. 154–167, 2002.

TRIGO, C., CAMPOS, J.M., GARRIDO, J.M., MENDEZ, R. Start-up of the Anammox process in a membrane bioreactor. *Journal of Biotechnology*, v. 126, p. 475–487, 2006.

TSUSHIMA, I., KINDAICHI, T., OKABE, S. Quantification of anaerobic ammonium-oxidizing bacteria in enrichment cultures by real-time PCR. *Water Research*, v. 41, p. 785–794, 2007b.

TSUSHIMA, I., OGASAWARA, Y., KINDAICHI, T., SATOH, H., OKABE, S. Development of high-rate anaerobic ammonium-oxidizing (Anammox) biofilm reactors. *Water Research*, v. 41, p. 1623–1634, 2007a.

TURK, O., MAVINIK, D.S. Selective inhibition: a novel concept for removing nitrogen from highly nitrogenous wastes. *Environmental Technology Letters*, v. 8, 419–426, 1987.

VAN DE GRAAF, A.A., DE BRUIJN, P., ROBERTSON, L.A., JETTEN, M.S.M., KUENEN, J.G. Autotrophic growth of anaerobic ammonium-oxidizing microorganisms in a fluidized bed reactor. *Microbiology*, v. 142, p. 2187–2196, 1996.

VAN DE GRAAF, A.A., DE BRUIJN, P., ROBERTSON, L.A., JETTEN, M.S.M., KUENEN, J.G. Metabolic pathway of anaerobic ammonium oxidation on the basis of 15N studies in a fluidized bed reactor. *Microbiology*, v. 143, p. 2415–2421, 1997.

VAN DE GRAAF, A.A., MULDER, A., DE BRUIJN, P., JETTEN, M.S.M., ROBERTSON, L.A., KUENEN, J.G. Anaerobic oxidation of ammonium is a biologically mediated process. *Applied and Environmental Microbiology*, v. 61, p. 1246–1251, 1995.

VAN DE GRAAF, A.A., MULDER, A., SLIJKHUIS, H., ROBERTSON, L.A., KUENEN, J.G. Anoxic ammonium oxidation. *Proceedings of Europeans Congress on Biotechnology*, 5th, p. 388–391, 1990.

VAN DER STAR W., Growth and Metabolism of Anammox Bacteria. Ph.D Thesis. Delft University of Technology, 2008.

VAN DER STAR, W.R., ABMA, W.R., BLOMMERS, D., MULDER, J.W., TOKUTOMI, T., STROUS, M., PICIOREANU, C., VAN LOOSDRECHT, M.C.M. Startup of reactors for anoxic ammonium oxidation: experiences from the first full-scale anammox reactor in Rotterdam. *Water Research*, v. 41, p. 4149–4163, 2007.

VAN DER STAR, W.R.L., VAN DE GRAAF, M.J., KARTAL, B., PICIOREANU, C., JETTEN, M.S.M., VAN LOOSDRECHT, M.C.M. Response of anaerobic ammonium-oxidizing bacteria to hydroxylamine. *Applied and Environmental Microbiology*, v. 74, p. 4417–4426, 2008.

VAN DONGEN, L.G.J.M., JETTEN, M.S.M., VAN LOOSDRECHT, M.C.M. The combined SHARON/Anammox process. London: IWA Publishing, 2001a.

VAN DONGEN, U., JETTEN, M.S.M., VAN LOOSDRECHT, M.C.M. The Sharon-Anammox process for treatment of ammonium rich wastewater. *Water Science and Technology*, v. 44, p. 153–160, 2001b.

VAN HULLE, S.W.H., VAN DEN BROECK, S., MAERTENS, J., VILLEZ, K., SCHELSTRAETE, G., VOLCKE, E.I.P., VANROLLEGHEM, P.A. Construction, start-up and operation of a continuously aerated laboratory-scale SHARON reactor in view of coupling with an Anammox reactor. *Water SA*, v. 31, p. 327–334, 2005.

VAN LOOSDRECHT, M.C.M. Innovative nitrogen removal. In: Biological Wastewater Treatment. Principles, Modelling and Design. Henze, M., van Loosdrecht, M.C.M., Ekama, G., Brdjanovic, D. (eds). London: IWA, p. 139–153, 2008.

VAN LOOSDRECHT, M.C.M., HAO, X.-D., JETTEN, M.S.M., ABMA, W. Use of Anammox in urban wastewater treatment. *Water Science and Technology: Water Supply*, v. 4, p. 87–94, 2004.

VAN LOOSDRECHT, M.C.M., SALEM, S. Biological treatment of sludge digester liquid. Proceedings of IWA Specialist Conference Nutrient Management in Wastewater Treatment Process and Recycle Steams, p. 13–22. Krakow, Poland, 2005.

VAN NIFTRIK, L.A., FUERST, J.A., SINNINGUE DAMSTÉ, J.S., KUENEN, J.G., JETTEN, M.S.M., STROUS, M. The anammoxosome: an intracytoplasmic compartment in anammox bacteria. *FEMS Microbiology Letters*, v. 233, p. 7–13 2004.

VAN NIFTRIK, L.A., GEERTS, W.J.C., VAN DONSELAAR, E.G., HUMBEL, B.M., WEBB, R.I., FUERST, J.A., VERKLEIJ, A.J., JETTEN, M.S.M., STROUS, M. Linking ultrastructure and function in four genera of anaerobic ammonium-oxidizing bacteria: cell plan, glycogen storage and localization of cytochrome c proteins. *The Journal of Bacteriology*, v. 190, p. 708–717, 2008.

VÁZQUEZ-PADÍN, J., FERNÁDEZ, I. FIGUEROA, M., MOSQUERA-CORRAL, A., CAMPOS. J.L., MÉNDEZ, R. Applications of Anammox based processes to treat anaerobic digester supernatant at room temperature. *Bioresource Technology*, p. 2988–2994, 2009.

VERSTRAETE, W., PHILIPS, S. Nitrification-denitrification processes and technologies in new contexts. *Environ. Poll.*, v. 102, p. 717–726, 1998.

VILLAVERDE, S., FDZ-POLANCO, F., GARCIA, P.A. Nitrifying biofilm acclimation to free ammonia in submerged biofilters. Start-up influence. *Water Research*, v. 34, p. 602–610, 2000.

VILLAVERDE, S., GARCÍA-ENCINA, P.A., FDZ-POLANCO, F. Influence of pH over nitrifying biofilm activity in submerged biofilters. *Water Research*, v. 31, p. 1180–1186, 1997.
VOLCKE, E.I.P., VAN LOOSDRECHT, M.C.M., VANROLLEGHEM, P.A. Controlling the nitrite:ammonium ratio in a SHARON reactor in view of its coupling with an Anammox process. *Water Science and Technology*, v. 53, p. 45–54, 2006.
WANG, T., ZHANG, H., YANG, F., LIU, S., FU, Z., CHEN, H. Start-up of the Anammox process from the conventional activated sludge in a membrane bioreactor. *Bioresource Technology*, v. 100, p. 2501–2506, 2009.
WETT, B. Development and implementation of a robust deammonification. *Water Science and Technology*, v. 56, p. 81–88, 2007.
WETT, B. Solved upscaling problems for implementing deammonification of rejection water. *Water Science and Technology*, v. 53, p. 121–128, 2006.
WIESMANN, U. Biological nitrogen removal from wastewater. In: Advanced in Biochemical Engineering/Biotechnology. Fiechter, A. (ed.)., v. 51, p. 113–154, 1994.
WONG-CHONG, G.M., LOEHR, R.C. Kinetics of microbial nitrification – nitrite-nitrogen oxidation. *Water Research*, v. 12, p. 605–609, 1978.
WYFFELS, S., BOECKX, P., PYNAERT, K., VERSTRAETE, W., VAN CLEEMPUT, O. Sustained nitrite accumulation in a membrane-assisted bioreactor (MBR) for the treatment of ammonium-rich wastewater. *Journal of Chemical Technology and Biotechnology*, v. 78, p. 412–419, 2003.
WYFFELS, S., BOECKX, P., PYNAERT, K., ZHANG, D., VAN CLEEMPUT, O., CHEN, G., VERSTRAETE, W. Nitrogen removal from sludge reject water by a two-stage oxygen limited autotrophic nitrification denitrification process. *Water Science and Technology*, v. 49, p. 57–64, 2004.
XUE, Y., YANG, F.L., LIU, S.T., FU, Z.M. The influence of controlling factors on the start-up and operation for partial nitrification in membrane bioreactor. *Bioresource Technology*, v. 100, p. 1055–1060, 2009.
YAMAMOTO, T., TAKAKI, K., KOYAMA, T., FURUKAWA, K. Long-term stability of partial nitritation of swine wastewater digester liquor and its subsequent treatment by Anammox. *Bioresource Technology*, v. 99, p. 6419–6425, 2008.
YANG, L. Investigation of nitrite build-up within an enriched nitrification process. PhD Dissertation, School of Civil Engineering, Purdue University, West Lafayette, Indiana, 1990.
YANG, L., ALLEMAN, J.E. Investigation of batchwise nitrite build-up by an enriched nitrification culture. *Water Science and Technology*, v. 26, p. 997–1005, 1992.

Chapter 6
Molecular Biology Techniques Applied to the Study of Microbial Diversity of Wastewater Treatment Systems

João Paulo Bassin, Márcia Dezotti, and Alexandre Rosado

6.1 Introduction

6.1.1 Microbial Diversity

These are extraordinary times for the field of microbiology given the rapid development in scientific knowledge and the existence of an enormous biodiversity, which is only just beginning to be characterized. Microorganisms have evolved over approximately four billion years, and two billion years ago, they were the only forms of life on earth. They are able to exploit a wide spectrum of energy sources and to live in almost all habitats. In the course of their existence, all of the basic biochemistry of life has evolved, and all of the other living forms developed from the ancestral microorganisms. They represent the richest repertoire of chemical and molecular diversity in nature. In addition, they comprise the foundation of basic environmental processes, such as biogeochemical cycles and food chains, and also maintain vital and complex relations between themselves and with higher organisms (ROSADO and DUARTE 2002).

Diversity is a critical subject which involves all levels of biological organization, from molecular to global. In general, biological diversity programs have emphasized studies on plants and animals, and little attention has been given to microorganisms. The diversity of microorganisms is as vast as it is unknown (ROSADO and DUARTE 2002). The study of microbial diversity is also important with regard to

J.P. Bassin (✉) • M. Dezotti
Chemical Engineering Program, COPPE, Federal University of Rio de Janeiro, Rio de Janeiro, Rio de Janeiro, Brazil
e-mail: jbassin@peq.coppe.ufrj.br; mdezotti@peq.coppe.ufrj.br

A. Rosado
Institute of Microbiology Prof. Paulo de Goés, Federal University of Rio de Janeiro, Rio de Janeiro, Brazil

M. Dezotti et al., *Advanced Biological Processes for Wastewater Treatment*,
DOI 10.1007/978-3-319-58835-3_6

advances in biotechnology. The new technologies developed, mainly those related to nucleic acid analysis, bioinformatics, analytical chemistry, and the sampling and characterization of ecosystems, have placed microbial diversity in a prime position in the scientific arena.

The detection and identification of bacteria were traditionally carried out based on their main forms of obtaining carbon and energy, their nutritional demands, and the culture medium for their growth, besides direct observation by microscopy (KENNEDY 1999; HERBERT 1990).

The tradition methods to describe microorganisms are not normally directly applicable to microbial communities; the microorganisms are generally described and classified by their phenotype. The phenotype is a very broad term which encompasses the observable traits of the cell, such as the morphology, physiological activity, structure of the cell components, and, in some cases, the ecological niche which the cell occupies. Unfortunately, these traits provide limited information regarding the evolutionary relationships of microorganisms, which, theoretically, should serve as a base for any classification system (HUGENHOLTZ and PACE 1996). In the case of microbial systematics based on phenotype, many tests to determine phenotypic traits require information on cellular function or growth. In this context, plating and cultivation in culture media is required in order to obtain pure cultures of the microorganisms or for direct observation by microscopy to establish their phenotype (HUGENHOLTZ and PACE 1996).

Conventional microbiological techniques based on the isolation of pure cultures and on morphological, metabolic, biochemical, and genetic tests have provided extensive information with respect to the diversity of microbial communities both in natural and engineered systems. However, this information is limited and thus needs to be further refined (ZAK et al. 1994). In most cases information is not available in relation to the physiological needs (nutritional and physicochemical) of the microorganisms under study, and the complexity of the syntrophic and symbiotic relationships, abundant in nature, contributes to the difficulties associated with obtaining pure cultures of most microorganisms which live in a wide diversity of environments (SANZ and KÖCHLING 2007).

The limitations of traditional techniques for the detection and identification of microorganisms are even greater when the aim is to study the diversity of microorganisms associated with a certain environment. Bacteria display a higher diversity than any other group of organisms, and each culture medium is, to a greater or lesser extent, selective to certain microbial groups, favoring the growth of some of these groups, while the development of others, also present in the original sample, is inhibited. Even on using a medium which is selective toward a certain target organism, some strains in a non-cultivatable state in the environment will be excluded from the analysis (COUTINHO et al. 1999). Thus, in many cases, microorganisms isolated using conventional methods which employ a culture medium may not represent those which are actually present in larger numbers.

According to PACE (1996), only a tiny proportion of the microorganisms present in the environment, equivalent to 1%, can be cultivated using standard cultivation and plating techniques, which highlights the limitations of conventional methods. Moreover, cell counts obtained by microscopy represent only a quantitative evaluation of the microbial population and provide little information regarding the diversity of organisms in the sample (PICKUP 1991). In addition, it is important to note that knowledge regarding prokaryotic organisms is limited. Their small size and often the absence of distinguishable phenotypic traits means that most of these organisms cannot be cultivated (PACE 1997).

As alternatives to the traditional methods and seeking to overcome the scarcity of information provided by these methodologies, several techniques have been developed, notably those based on nucleic acids. In this regard, molecular biology has arisen as an alternative approach to overcome the intrinsic limitations of conventional techniques in the study of the diversity of microorganisms. The limitations of traditional methods, together with the technological advances in the area of molecular biology, have led to molecular techniques being widely used for the study of microbial diversity (VAN ELSAS et al. 1998). Thus, the elucidation of the microbial world in various biological systems has become more interesting.

6.1.2 Basic Concepts of Genetics

Prior to describing the principles and concepts of the main molecular biology techniques employed to study microbial diversity, it is worth looking at the origin of these tools. Thus, a brief summary of the fundamental concepts of genetics will be outlined, encompassing information related to the nucleic acids (deoxyribonucleic acid, DNA, and ribonucleic acid, RNA), their composition, and the important stages of duplication/replication, transcription, and translation involving these biomolecules which carry genetic information.

All cells of organisms, from bacteria to humans, contain one or more sets of a collection of DNA characteristic of the species. This DNA profile is known as the genome of living beings, where all of the genetic information required for their existence is contained. This information held in the DNA is expressed in cells and is perpetuated in the progeny. Thus, DNA can be considered to have the function of storing and transmitting genetic information (genetic code).

The genome can be subdivided into chromosomes, each containing one continuous and very long molecule of DNA. The chromosomes, in turn, are composed of millions of regions called genes (segments of DNA that codify the information required for the production of a certain polypeptide or a segment of RNA), which are distributed to each daughter cell at the moment of cell division. Thus, the cell needs to make a copy of its genes in order to divide them equally between its daughter cells. The position of the genes on the chromosome is known as the locus.

DNA is a long, unbranched polymer, composed of subunits called deoxyribonucleotides, grouped into four types. Each type subdivides into smaller molecules that include nitrogen bases (dNTPs) which can be purines (adenine-A and guanine-G) or pyrimidines (thymine-T or cytosine-C), a pentose sugar (deoxyribose) and a phosphate group. The bases A and T and the bases G and C are complementary and are linked to the pentose by one bond on the carbon 1′ of the latter, and the phosphate is associated with the carbon 5′ of the same pentose and the carbon 3′ of the pentose

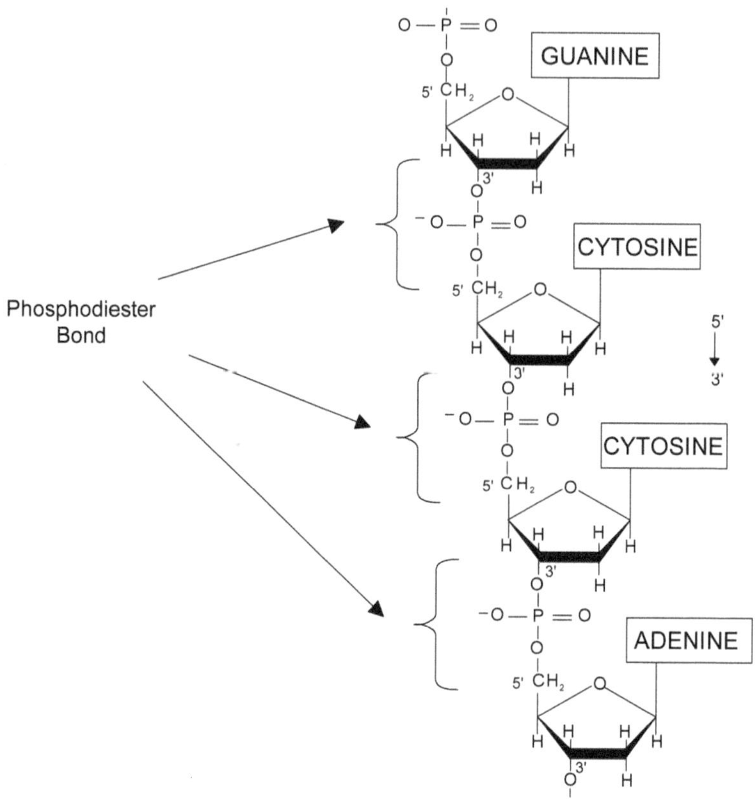

Fig. 6.1 Schematic representation of phosphodiester bond between two nucleotides (adapted from LEHNINGER 1985)

of the adjacent nucleotide. This latter bond is called the phosphodiester bond and constitutes the strand of nucleotides. This type of bond confers polarity on the DNA molecule which plays a role in the replication and transcription, called the $5' \rightarrow 3'$ direction, as shown in Fig. 6.1. The direction of the 3′ and 5′ phosphodiester bonds of one strand is the opposite to that of the other strand. Thus, these strands are commonly referred to as antiparallel. As a result, at each extremity of the molecule, one of the polynucleotides ends in a 3′ and the other in a 5′.

DNA is comprised of two strands of nucleotides, and the association between them is carried out by the pairing of nitrogen bases. As previously mentioned, adenine associates with thymine and cytosine with guanine. This configuration is achieved through hydrogen (H) bridge bonds, as shown in Fig. 6.2, which are responsible for the stability of the helix. There are two possibilities for a bridge between adenine and thymine (involving the nitrogen of one base and the oxygen of the complementary base) and three between cytosine and guanine (involving two nitrogen with oxygen and one nitrogen with nitrogen). This scenario is universal in living organisms, since the structure of DNA is the same for bacteria, fungi, plants, and animals. In fact, the differentiation lies in the sequence in which the nucleotides are associated.

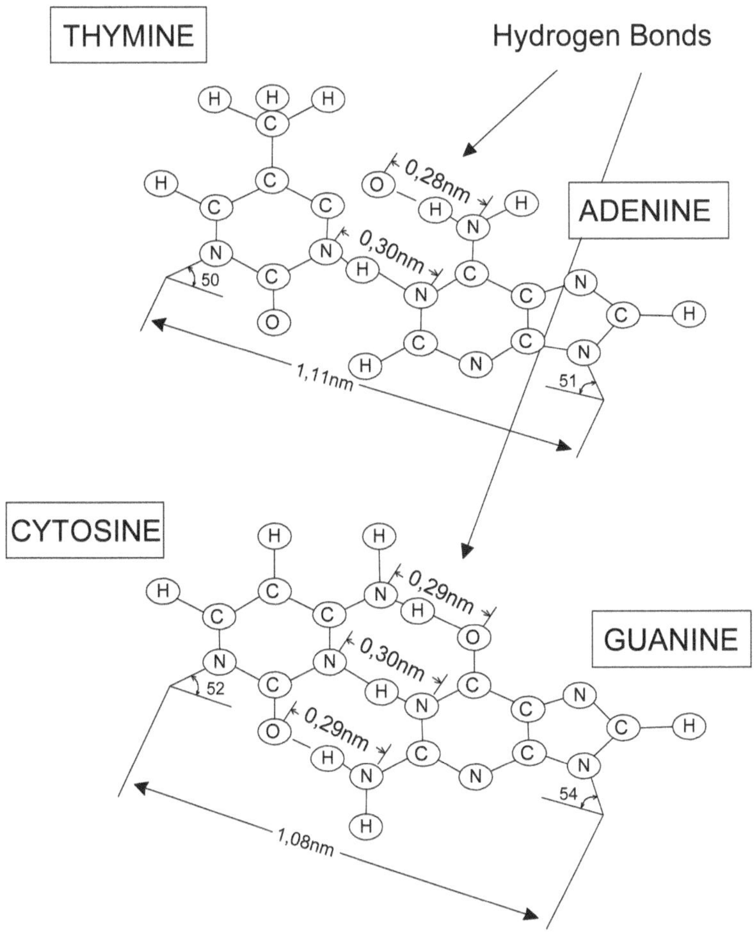

Fig. 6.2 Hydrogen bridge bonds between adenine and thymine or cytosine and guanine (adapted from LEHNINGER 1985)

The denaturation through the rupture of the hydrogen bridges can be complete or partial and occurs earlier in the case of the A = T bonds, which have two hydrogen bridges in contrast to the C ≡ G bonds, which are more resistant, since they have three hydrogen bridges. The partial denaturation allows the zones rich in AT and the zones rich in CG to be identified.

Biochemical analysis of DNA preparations of different species shows that, although the composition of nucleotides is very varied, there is a general quantitative rule that the number of adenine bases is equal to the number of thymine bases (A = T) and likewise for guanine and cytosine (G = C). The constructed model revealed that the effective number of hydrogen bridges which can be formed between G and C or between A and T was greater than for any other combination.

The double-helix model for DNA, proposed by Watson and Crick in the 1950s, perfectly explains the quantitative biochemistry.

With the discovery of the double-helix structure of DNA, the transfer of genetic information between cells can be explained. Since each strand of DNA contains a sequence of nucleotides which is exactly complementary to the nucleotide sequence of the other strand, both strands carry the same genetic information. Thus, if we designate the two strands as A and A′, strand A can serve as a template for creating a new A′ and vice versa. Therefore, the genetic information can be copied through a process in which strand A is separated from strand A′ and each one serves as a template for the production of a new complementary strand. This process is known as semiconservative replication, which enables the perpetuation of the DNA molecule through cell division (mitosis or meiosis).

Based on this understanding, the scientists could conclude that with the base pairing mechanism, it becomes evident that DNA carries information by way of the linear sequence of its nucleotides. Each nucleotide can be considered as one letter of an alphabet of four letters, used to "write" biological messages. Organisms differ from each other because their respective DNA molecules carry different nucleotide sequences (ALBERTS et al. 1994).

The enzyme which catalyzes the formation of the 5′ → 3′ phosphodiester bond between the new nucleotide and the strand being formed is called DNA polymerase. After the formation of this bond, the DNA polymerase moves along the old nucleotide of the template strand, repositioning itself to promote the binding of a new dNTP to the growing strand. It is worth noting that DNA polymerase, responsible for the polymerization of the DNA molecule, can only catalyze the growth of a new polynucleotide in the 5′ → 3′ direction, considering that it is only able to add nucleotides at the free 3′-OH end of a growing strand. This mechanism ensures that the new nucleotide provides the energy required for its own incorporation, that is, the energy comes from the breaking of the linkages between the phosphates bound to the carbon 5′ of the triphosphate of the nucleotide to be incorporated and not by the end of the growing strand.

The addition of a nucleotide to a strand via DNA polymerase only occurs when it is correctly aligned with the corresponding nucleotide of the template strand. If this is not the case, the DNA polymerase removes the mismatched nucleotide from the free 3′-OH end, a property known as 3′ → 5′ exonucleotide activity, in the opposite direction to the growth of the strand. This property is in fact a correction mechanism which enables DNA polymerase to verify the last nucleotide incorporated, correcting its own incorporation errors as it moves along the template DNA molecule. The capacity for auto-correction of DNA polymerase is crucial to ensure a high degree of fidelity in the DNA replication which, in turn, provides genetic stability over generations.

If the errors which occur during the DNA replication process are not corrected by the DNA polymerase enzyme, they will perpetuate in the form of mutations. Furthermore, DNA is continually damaged by physical and chemical external agents, and although cells have sophisticated repair mechanisms, some damage will remain and will also be expressed in the form of mutations. However, low mutation rates are important for evolution, bearing in mind that new alleles (alternative forms of the gene which occupy the same place on homologous chromosomes) are formed from mutations.

The other known nucleic acid is RNA (ribonucleic acid), formed from the transcription of DNA (reproduction of a DNA strand in a complementary RNA sequence). The transcription process is catalyzed by the enzyme RNA polymerase and begins in certain regions of the DNA known as promoters, this being the site where the RNA polymerase binds (initiation stage). The RNA polymerase unwinds the DNA, initiating the synthesis of RNA which, as in the case of DNA, occurs in the $5' \rightarrow 3'$ direction (elongation stage). Finally, the termination of the polynucleotide strands of the synthesized RNA occurs.

In contrast to DNA, RNA is comprised of a single strand of nucleotides which are linked through phosphates ($5' \rightarrow 3'$ direction). Furthermore, RNA is smaller and less stable molecule in comparison with DNA. The bases which constitute RNA are adenine, uracil, cytosine, and guanine. RNA can be observed in the cell nucleus and cytoplasm.

From the functional and structural point of view, three major varieties of RNA can be identified: transfer RNA (tRNA), messenger RNA (mRNA), and ribosomal RNA (rRNA). The main function of the different types of RNA is the synthesis of proteins (a process called translation), which are the final compounds of the gene expression of a genetic code stored in DNA. Proteins have structural and catalytic functions.

The messenger RNA carries genetic information directly from the DNA, present in the nucleus, transporting it to the ribosomal sites. It enables the transcription of the genetic code to be carried out. The mRNA molecule is much larger than the protein which it forms, considering that three nucleotides are required to codify one amino acid. In prokaryotic cells, in particular, mRNA molecules can be even larger, since in bacteria a long mRNA molecule can be translated starting from different locations, giving rise to more than one protein, depending on the location at which the translation was initiated.

Ribosomal RNA represents the locus of protein synthesis in the cytoplasm. It is found free in the cytoplasm or associated with rough endoplasmic reticulum. When combined with proteins, it forms particles called ribosomes which are easily observed by electron microscopy. When bound to messenger RNA filaments, ribosomes form polyribosomes, where protein synthesis occurs. Ribosomal RNA is much more abundant than other types of RNA, accounting for 80% of the cellular RNA. Ribosomes are formed of two subunits, one large and one small, with different structural and functional characteristics. The subunits bind in a reversible way at the beginning of the synthesis of the protein molecule and separate when the protein is completed. The large subunit of ribosomes in eukaryotic cells contains three types of RNA, with sedimentation coefficients of 28S, 5.8S, and 5S. On the other hand, the large subunit of ribosomes in prokaryotic cells contains two types of RNA, one with a sedimentation coefficient of 23S and the other 5S. The small subunit is comprised of only one type of RNA (18S in eukaryotic cells and 16S in prokaryotic cells). As will be described in Sect. 6.2.1, this is the subunit most commonly used in studies which employ molecular biology techniques for the identification and quantification of microorganisms.

tRNA presents the smallest molecule of the three types of ribonucleic acid, with a structure in the form of a cloverleaf. Its function is to bind specific amino acids required for protein synthesis and transfer them to the correct positions on the polypeptide chains under formation in the complexes formed by ribosomes and messenger RNA (polyribosomes). Thus, one property of tRNA is that it can combine with amino acids and recognize certain locations of the mRNA molecule comprised of a sequence of three bases. These sequences, which are distinct for each amino acid, are called codons. The correspondence between a nucleotide triplet and an amino acid is known as the genetic code. The sequence of three bases in the tRNA molecule, which recognizes the codon, is called the anticodon. For each amino acid there is at least one tRNA. It should be noted that tRNA can recognize more than one codon and thus the code is known as degenerate. This characteristic is important to studies conducted in the area of genetic engineering, for instance, the amplification of gene sequences starting from degenerate primers, which are based on a sequence of proteins.

6.2 Principles and Concepts of Molecular Biology Techniques Applied to the Study of Microbial Diversity

6.2.1 Introduction to Molecular Biology Techniques

The best form of biological classification is based on phylogenetic or evolutionary relationships. Genotypic information, that is, the sequences of semantic molecules (genes or transcripts, DNA, RNA, and proteins), constitutes the history of evolution, and the determination of the DNA sequence not only enables the evolutionary relationship between sequenced molecules to be measured but also provides an insight into the evolution of organisms which gave origin to the generation of these sequences.

The comparison of organisms to determine their evolutionary relationships can be carried out by comparing the nucleotide sequences of their genomes. However, this information is difficult to obtain and to interpret and generally is not a practical approach. Nevertheless, genotypes have been compared in a limited way using DNA-DNA hybridization to determine the degree of kinship between cultivated microorganisms (STACKEBRANDT et al. 1993). Studies on DNA reassociation have also been carried out to estimate the complexity of microbial communities. In this analysis DNA extracted from a certain community is denatured and then reassociated. The rate of reassociation is a measure of the magnitude of the number of individual genotypes in the environmental sample (TORSVIK et al. 1990). The conclusion drawn from these studies is that natural microbial ecosystems are extremely complex. A small sample of soil, for instance, can contain thousands of different genotypes. Evolutionary relationships can also be inferred through comparative analysis of the individual gene sequences of the genome (ZUCKERKANDL and PAULING 1965). The evaluation of the phylotype of an organism provides considerable information regarding its characteristics. Some properties of an organism can be predicted based on the properties of related organisms. It is expected that all of the representatives of a certain phylogenetic group have the same properties which occur frequently in the group.

Given the difficulty associated with using the complete primary structure of DNA to carry out these comparisons, protein-codifying gene sequences, amino acid sequences, and genes which have a sufficient degree of conservation for the species analyzed (e.g., ribosomal RNA genes) are compared (HILLIS et al. 1996).

In this context, molecular biology techniques are frequently based on ribonucleic acids (RNAs) of small ribosomal subunits (16S rRNA for prokaryotes) or their corresponding genes (rDNAs), considered to be the biopolymers most suitable for biodiversity studies. The sequences which codify for 16S ribosomal RNA have been used most extensively to classify phylogenetically the diversity of life (WOESE 1987; WOESE et al. 1990).

This molecule in particular was selected because of its universality and its abundance in all living beings (10^3–10^5 ribosomes per cell) and also because it derives from a common ancestor and is genetically stable and its size is compatible with amplification by polymerase chain reaction (PCR) (described in Sect. 6.2.2). Besides these characteristics, the rRNA molecules have regions which are extremely well conserved among all organisms which share that species of rRNA, since, given its rigid structure, needed to maintain its function, the rate of mutation during the evolutionary process is very low when compared with other genes (ROSADO et al. 1997). At the same time it has regions which are highly variable and the degree of variation in these regions can vary from one taxon to another. The presence of these variable regions offers great possibilities for the design of domain-specific, genus-specific, or even species- or strain-specific probes (WOESE 1987; ROSADO et al. 1996).

These specific characteristics of the rRNA molecule allow the comparison of organisms within a single domain and also enable strains within the same species to be differentiated and the microbial diversity to be phylogenetically classified. Furthermore, the gene sequence is sufficiently long for the data generated to be statistically relevant, and it can be easily sequenced using current technology. In this context, differences in the sequences of ribosomal nucleic acids have led to an expansion in knowledge in the area of phylogeny (construction of phylogenetic trees) and quantitative schematic representations of the evolutionary diversity (HILLIS et al. 1996; WOESE 1987; OLSEN et al. 1994). The construction of phylogenetic trees, in turn, is the most incisive way to infer phylogenetic relationships from molecular sequencing data.

On analyzing information gathered using techniques based on the RNA molecule, it was possible, for the first time, to evaluate the biodiversity of a natural habitat in a complete and relatively simple manner (SANZ and KÖCHLING 2007). One significant advantage of using information on rRNA sequencing is its availability in databases (RDP, GenBank, EMBL) which, in most cases, can be accessed free of charge, allowing the comparison of new sequences obtained with others already recorded in the databanks (COUTINHO et al. 1999).

Studies on the molecular diversity of microorganisms should focus on defined functional groups. The analysis of ribosomal sequences obtained directly from environmental samples can reveal new species, break established paradigms, and even lead to the restructuring of the taxonomy of functional groups. For example, the sequencing of fragments of 16S rDNA of a group of ammonium-oxidizing bacteria amplified from soil and sediment samples revealed that the vast majority of sequences were new, indicating that the three cultivated genera of this group (*Nitrosomonas*, *Nitrosospira*, and *Nitrosococcus*) were not dominant in any of the samples analyzed (STEPHEN et al. 1996). The results of this study redefined this functional group, previously considered to be of low diversity, and highlighted the importance of the molecular approach in studies to evaluate and monitor microorganisms in the natural environment (ROSADO and DUARTE 2002).

The perfecting of molecular biology techniques together with advances in the area of bioinformatics has led to specialization within environmental microbiology to such an extent that an area called molecular microbial ecology has emerged (ROSADO et al. 1997).

In environmental biotechnology, bioreactors destined for wastewater treatment processes can be considered as one of the most elaborate niches created with the main objective of cleaning up water (OKABE and KAMAGATA 2010). These engineered systems have been applied for more than a century in order to minimize the impact of anthropogenic activity on the environment. Over this period, the design of bioreactors was gradually changed to improve its performance in terms of pollutant (e.g., organic matter and nutrients) removal efficiency, leading to a more stable operation of the treatment system.

Despite the extensive use of bioreactors for more than a century in the wastewater treatment sector and all of the innovations related to these biological systems, many plants still face a variety of problems such as bulking sludge, excess foaming, and unsatisfactory nitrification and phosphate removal (MARTINS et al. 2004; SEVIOUR et al. 2000). The reasons for these operational problems are diverse and can be attributed to inappropriate reactor design, inadequate operation of the reactor, exposure to inhibitors, and shock loads of organic compounds, toxic substances, and salt. However, it should be noted that most problems related to functioning and performance of the bioreactors are associated with changes in the structure of the microbial communities.

Biological wastewater treatment processes harbor a great variety of microorganisms. In the activated sludge process, for instance, which is one of the most applied technologies, several types of microorganisms are present, such as bacteria, protozoans, fungi, micrometazoans, and algae. Of the whole microbial population, 95% correspond to bacteria (JENKINS et al. 1993), which play a crucial role in the biodegradation of pollutants.

Despite the importance of biological processes for the treatment of wastewaters, information on the microbial ecology of these systems is scarce. For a long time, biological processes have been regarded as a "black box," without complete elucidation of the microbiological phenomena taking place in bioreactors. This was mainly due to the lack of analytical techniques capable of determining the identity of microorganisms playing a key role in the biodegradation processes. Therefore, in most cases, the interest has been centered entirely on the removal efficiency of certain pollutants, and the main objective is to adequate the wastewater to the requirements established by the environmental legislation. In this context, the powerful metabolic potential of the microbial communities was exploited without a detailed knowledge of the organisms responsible for the treatment.

In the 1900s, research on the microbiology of the biological wastewater treatment processes had many methodological limitations. Methods which were dependent on cultivation, such as plate counts or most probable number (MPN), were the preferred choice for the detection and quantification of bacteria in wastewater treatment systems. In fact, despite the limitations of these techniques, they are still used in some cases for the quality control of effluents, particularly with regard to pathogens and several indicator organisms.

The introduction of different culture-independent molecular techniques and new insights into microbiology (notably in relation to wastewater treatment) in the past decade have aided the improvement of the design, operation, control, and hence the performance of new generations of bioreactors. The molecular methods allowed to identify the most important microorganisms in the treatment system (particularly the prokaryotes) and contributed to gain an understanding of the mechanisms of these processes. Therefore, it became possible to determine precisely the composition, structure, function, and dynamics of complex microbial communities present in biological wastewater treatment plants.

Microbial ecology and biotechnology are hence intrinsically related. The former generates a theoretical foundation for carrying out biotechnological processes, whereas in environmental biotechnology, the concepts and tools of microbial ecology are applied in order to better manage the processes. When managed appropriately, a wide range of benefits can be obtained.

On contemplating the possibility of identifying the microorganisms present in a certain treatment system, a question may arise: why is it important to determine which types of bacteria are present in each functional group of a certain wastewater treatment plant? The answer to this question may be given taking into consideration that the identification of the key microorganisms that are involved in degradation processes taking place in biological treatment reactors, be it related to organic matter, nutrients (nitrogen and phosphorus), or even micropollutant and pathogen removal, is of great importance since these organisms are responsible for the maximum utilization of a certain substrate. Thus the identification of the organisms playing important role in the treatment is essential for obtaining high removal efficiency of pollutants, which is the main objective of all wastewater treatment plants. Furthermore, molecular tools can aid the identification of bacteria not yet cultivated, some of which are important for the formation of floc structures and others responsible for problems occurring during the operation of the bioreactors, such as bulking sludge (commonly attributed to filamentous bacteria) and the excessive formation of foam (WAGNER and LOY 2002). Therefore, the importance of the identification and analysis of both beneficial and harmful microorganisms in the systems is clear, as the efficiency and robustness of a wastewater treatment plant depend on the composition and activity of its entire microbial community.

In summary, the possibility of relating the biological system performance to the composition of the microbial community present in treatment systems and the functions of each bacterial population, individually, is edging ever closer to reality. The ways in which this information will be used to gain better control over biological processes are being revealed continuously as knowledge on this topic is gained by microbiologists and environmental engineers.

Several functional groups of bacteria involved in the most commonly employed treatment processes have already been clearly identified and described. These groups are represented, predominantly, by the bacteria involved in nitrification and to some extent by those involved in denitrification, by several bacteria involved in enhanced biological phosphorous removal (EBPR) and by most of the bacteria which cause problems related the sludge sedimentation (bulking) and the formation of foam and scum. In each functional group, for example, nitrifiers, a limited number of phylogenetic lineages (<10) are found in nitrifying plants, in general, with few dominant populations (3–5) being present in one plant (NIELSEN et al. 2009a).

Besides the importance of identifying key microorganisms associated with the degradation of a variety of pollutants in wastewater treatment systems, attention should be given to the appearance of new and interesting sustainable solutions. These include the recovery of nutrients (e.g., P) from wastewaters or the conversion of the components present in wastewaters into compounds which can be used as, for instance, bioplastics (polyhydroxyalkanoates, PHAs). The conversion of the components in wastewater into energy through the production of methane via anaerobic digestion has been used for decades, and these processes have begun to be developed together with other energy-generating processes such as microbial fuel cells (NIELSEN et al. 2009a). In order to reach these objectives, it is clear that the management of these complex engineered biological systems requires knowledge on the microbial populations and the factors which affect their activity. This knowledge does not relate only to the identification, which needs to be reliable and precise, but also to the ecophysiology, ecology, and population dynamics of the microbial communities involved.

It should be emphasized that the composition of the microbial community at a certain treatment plant is strongly dependent on the composition of the wastewater, the design of the process, and the mode of operation of the plant. Clearly, certain functional groups are dominant only when specific processes are included in the treatment plant design (for instance, nitrogen removal or EBPR). However, little is known regarding the factors which control the composition of a microbial community. Thus, studies on the microbiology of wastewater treatment plants are important in order to observe and record factors which may be decisive for the presence of different species.

An activated sludge system aimed solely at the removal of organic matter requires only aerobic tanks (besides the settling tank). However, when denitrification/EBPR stages are also included, anoxic/anaerobic tanks are also required. The presence of anoxic/anaerobic tanks substantially modifies the structure of the microbial community.

The sludge age (sludge retention time) is another determinant parameter for the selection of the dominant bacterial populations in the bioreactor. Low values for this parameter (less than 5–10 days) may prevent the occurrence of nitrification due to the low growth rates of the nitrifying microorganisms. On the other hand, a high value for sludge age (more than 20–30 days) is crucial for obtaining complete N and P removal in temperate climates. In plants which operate at high temperatures (over 40 °C), microbial communities which are less common are generally selected (NIELSEN et al. 2009a).

As mentioned previously, the composition of the influent wastewater is another decisive factor for the bacterial growth, and it has a notable influence on the determination of the dominant populations in certain systems. Industrial and domestic wastewaters differ considerably. The former may contain many substances which adversely affect the microbial development in biological treatment processes (organic/inorganic substances which have an inhibitory effect on microorganisms). Moreover, they may be lacking in important nutrients, such as P, N, and other micronutrients. Depending on their origin, industrial wastewaters can contain high salt concentrations, which affect the microbial ecosystem. On the other hand, domestic wastewaters provide a more balanced relationship between organic compounds and nutrients and also contain a higher number of microorganisms. These microorganisms entering the bioreactor with the influent wastewater can also affect the composition of the microbial community.

Another important factor to be taken into consideration, which has also been little explored and is poorly understood, is the reactor configuration used to carry out a specific process, such as nitrification. The investigation of whether different reactor configurations, operating in continuous or batch modes, receiving the same wastewater, lead to development and growth of different microorganisms is a research topic which should be highlighted. Table 6.1 summarizes some factors which can determine the microbial community in wastewater treatment systems.

Studies describing the complete composition of the community of microorganisms in wastewater treatment plants have been published. The work of JURETSCHKO et al. (2002), who studied the activated sludge process for the treatment of industrial wastewater aimed at removing C and N, and a study by KONG et al. (2007), who investigated the biological removal of N and P from a mixture of domestic and industrial wastewaters, are examples. Several studies have also been carried out on specific populations in several (laboratory and full-scale) treatment systems, considering, for instance, only nitrifying and/or denitrifying microorganisms, polyphosphate-accumulating organisms, or filamentous bacteria. In Sect. 6.3.2 some studies are described, each related to specific microorganisms in different reactor configurations. Table 6.2 shows the most common species and genera of bacteria found in wastewater treatment plants.

To highlight the importance of the molecular approach to studies regarding the biological nitrogen removal process, particularly the nitrification step, for decades the main microorganisms responsible for this process were considered to be *Nitrosomonas* and *Nitrobacter*. However, with the application of molecular techniques, it became evident that other microorganisms are also important. The same

Table 6.1 Overview of important factors which determine the structure of microbial communities in wastewater treatment systems (adapted from NIELSEN et al. 2009a)

Factors	Description of factors
Process performance	• Removal of C, removal of C and nitrification • Removal of C and N (nitrification/denitrification) • Removal of C and N and EBPR • Chemical precipitation of P • Sludge age
Plant operation	• Oxygen concentration • Cell retention time • Addition of chemical agents (Fe/Al salts, polymers) • External addition of C (methanol) • Biomass concentration
Type of treatment plant	• Activated sludge (continuous or batch regime) • Biofilters (percolation filters) • Membrane bioreactors • Moving bed biofilm reactors • Aerobic granular sludge reactors
Wastewater composition	• Industrial/domestic • Soluble fractions/particulates (C, N and P) • Specific organic compounds • Micronutrients • Toxic substances • Salinity • Alkalinity • pH

occurred for the microorganisms responsible to biological phosphorus removal, which are able to accumulate polyphosphate. *Acinetobacter* spp. were known as examples of such organisms. However, several other microorganisms were identified with the use of molecular microbial ecology to be the main responsible for the bio-P removal process (e.g., polyphosphate-accumulating organisms).

Another example of information which has been obtained from molecular tools is related to ammonium-oxidizing bacteria (AOB) in wastewater treatment systems. The studies suggest that different plants have different AOB populations and different levels of species richness. OKABE et al. (1999), for instance, observed that a laboratory-scale biofilm reactor for the treatment of domestic wastewater was dominated by *N. europaea*. SCHRAMM et al. (1998) and JURETSCHKO et al. (1998) observed that the populations of AOB in laboratory and full-scale plants were dominated by *Nitrosospira* and *N. mobilis*, respectively. However, based on the gene *amoA*, it has been observed that the *Nitrosospira* species are not important AOBs in the full-scale treatment systems (WAGNER and LOY 2002).

It is important to mention that the knowledge and information acquired with the use of molecular techniques can, in some cases, be minimal and their usefulness for engineers and technicians responsible for the design and operation of bioreactors is not well defined. Thus, it should be considered how knowledge obtained applying molecular techniques can be used to improve the performance of treatment systems. Questions regarding how important it is to know the exact composition of the

Table 6.2 Microorganisms commonly observed in wastewater treatment systems (adapted from NIELSEN et al. 2009a)

Functional group	Populations commonly reported
Nitrifiers	
Ammonium-oxidizing bacteria (AOB)	Genus *Nitrosomonas* (*N. europaea*, *N. eutropha*, *N. mobilis*, and *N. oligotropha* (class *β-Proteobacteria*) Genus *Nitrosospira* (class *β-Proteobacteria*)
Nitrite-oxidizing bacteria (NOB)	Genus *Nitrospira* (sublineage 1 and 2) (phylum *Nitrospirae*) Genus *Nitrobacter* (class *α-Proteobacteria*)
Anammox bacteria	Lineage *Brocadia*, *Kuenenia*, *Scalindua* and *Anammoxoglobus* (phylum *Planctomicetos*)
Denitrifying bacteria	Genus *Candidatus* Accumulibacter (class *β-Proteobacteria*) Genus *Azoarcus* (class *β-Proteobacteria*) Genus *Curvibacter* (class *β-Proteobacteria*) Genus *Thauera* (class *β-Proteobacteria*) Genus *Zoogloea* (class *β-Proteobacteria*)
Polyphosphate-accumulating organisms (PAOs)	Genus *Candidatus* Accumulibacter (class *β-Proteobacteria*) Genus *Tetrasphaera* (phylum *Actinobacteria*)
Glycogen-accumulating organisms (GAOs)	Genus *Candidatus* Competibacter (class *γ-Proteobacteria*) Genus *Defluviicoccus* (classe *α-Proteobacteria*)
Filamentous bacteria	Species of class *α-Proteobacteria* Genus *Sphaerotilus* (class *β-Proteobacteria*) Genus *Thiothrix* (*Theiothrix* spp and type 021N) (class *γ -Proteobacteria*) *Candidatus* Microthrix parvicella (phylum *Actinobacteria*) Genus *Skermania* (phylum *Actinobacteria*) Genus *Gordonia* (phylum *Actinobacteria*) Genus *Rhodococcus* (phylum *Actinobacteria*) Genus *Dietzia* (phylum *Actinobacteria*) Species of phylum and class *Chloroflexi* Genus *Haliscomenobacter* (phylum *Bacteroidetes*)

microbial community in order to improve the performance of the process need to be answered, in order to take advantage of the enormous bank of information provided by the molecular tools. Further information regarding the relationship between a certain microbial community and the operation and stability of biological treatment systems is still needed. The future perspective is that, based on the reliable identification of microbial populations by way of molecular techniques, these questions will be answered.

Clearly, it is not necessary to have specific information on the phylogenetic position or taxonomic classification of each organism acting in a certain biological system in order to design a wastewater treatment plant. However, knowledge gained from molecular techniques can aid a better understanding of the biological processes and break certain paradigms which have been established over time (SANZ and KÖCHLING 2007). The use of conventional methods combined with molecular tools should also be encouraged, since it has been demonstrated that both methodologies leave gaps when used individually, but these can be filled when they are used concomitantly (MUYZER and SMALLA 1998).

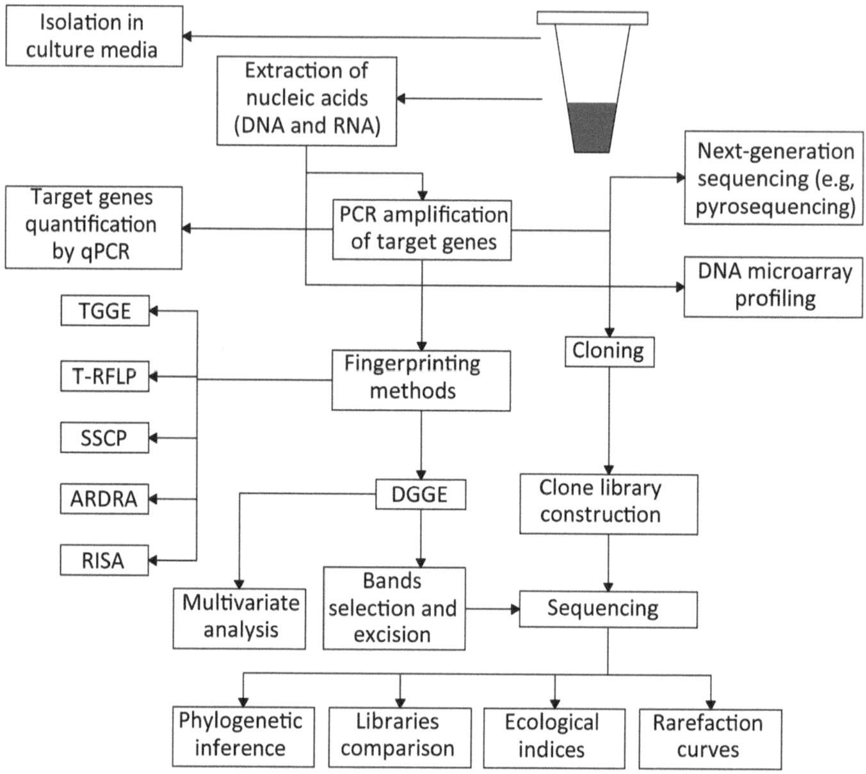

Fig. 6.3 Schematic representation of several molecular techniques used for the characterization of microorganisms from different environments (adapted from ANDREOTE et al. 2009)

The schematic diagram in Fig. 6.3 shows some molecular techniques used to study the genetic diversity of environmental samples of microorganisms. In general, the strategy used to collect the samples and the method employed for DNA/RNA extraction are important factors which determine the success of the experiment.

To complement the information given in Fig. 6.3, Fig. 6.4 provides an illustrative summary of the main procedures used in molecular microbial ecology, many of them used to assess the microbial community of wastewater treatment bioreactors.

In both natural and engineered environments (e.g., biological wastewater treatment reactors), there is a high diversity of microorganisms which are present in different numbers. This high diversity represents the variety of different populations which are organized into different communities. In general, these are structured in a similar way, being comprised of many populations represented by a few individuals and a few dominant populations with many individuals.

On carrying out a survey of a certain microbial community to determine the species and the number of individuals present in each species, it will be difficult to sample all of the populations present. Thus, in order to obtain data which are closer to the reality, some steps must be respected before the sampling. The first step

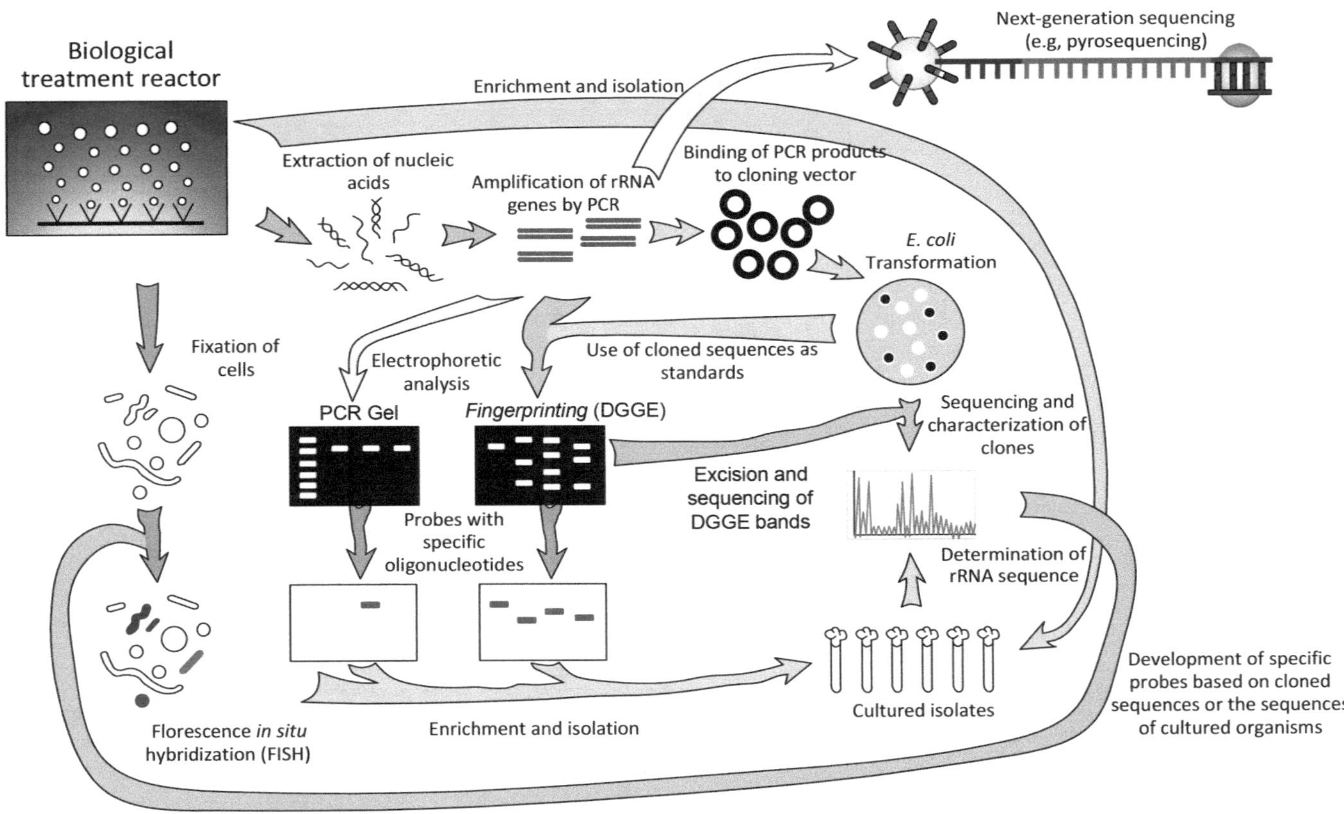

Fig. 6.4 Illustration of the main techniques used in molecular microbial ecology of engineered wastewater treatment systems (adapted from HEAD et al. 1998)

before beginning any study is to formulate a hypothesis to be proved. The objectives to prove this hypothesis can then be outlined after which the methodology can be defined. This last step is very important since not only the analysis to be carried out needs to be defined but the entire sampling process to be conducted. Issues such as the sample number and frequency have to be carefully addressed since they provide the statistical support for the study. If the aim is to describe the community, the samples must be collected randomly from the environment under investigation (e.g., biological reactors for wastewater treatment) since different populations can be distributed in the form of aggregates (in the case of biofilms), uniformly or even randomly.

Once the sampling strategy has been defined, the collection can be carried out. The highest degree of care must be taken to avoid contamination at this stage so that the sample obtained faithfully represents the number or activity of the populations of the environment under investigation. Specific equipment can be used depending on the sample to be collected. The preservation of samples until processing is also of great importance. They should be kept in a water bath with ice at a temperature of between 0 and 4 °C and sheltered from light. In some cases the freezing of samples can be harmful. However, for molecular analysis, freezing the sample at −20 °C immediately after collection is recommended. The preservation time also has to be appropriate. Several studies have demonstrated that the earlier a sample is processed the closer the results will be to the reality. Thus, ideally, the sample should be processed immediately after collection.

The processing, and also the analysis, will be dependent on the type of sample and the microbial group to be studied. However, in all cases the samples should be homogenized prior to processing. The sample may be diluted or concentrated depending on the environment sampled. Both qualitative and quantitative analyses need to be carried out.

With the development of molecular microbial ecology, investment in DNA extraction methods for different types of organisms from different samples has become very important and consists of an essential step in research using molecular techniques (ROSADO et al. 1996). The choice of the best method for the isolation of DNA or RNA from different environments is dependent on the type of nucleic acid to be isolated and the type of environment from which it will be extracted. Different methods are available for the extraction of nucleic acids from samples originating from different sources, and each case should be considered individually.

Several protocols have been described for the extraction of nucleic acids from different environments (OGRAM et al. 1987; SMALLA et al. 1993; VAN ELSAS et al. 1997). The extraction has been carried out based on two approaches: direct lysis in situ (OGRAM et al. 1987) or cell extraction followed by cell lysis (HOLBEN et al. 1988). These methods originate from pioneering studies which served as a starting point for the development and perfecting of other simpler and faster protocols (TREVORS and VAN ELSAS 1995).

Most techniques currently used for the extraction of nucleic acids from an environment involve direct DNA and/or RNA extraction. This methodology enables a greater total DNA yield to be obtained (OGRAM 2000). The identification of the most efficient method for the study of bacterial diversity should take into consideration the extraction of nucleic acids from highly differentiated cell types, such as bacterial spores and Gram-positive and Gram-negative cells (NIEMI et al. 2001). These methods, in general, involve the use of mechanical disaggregation of the cells. In this case, the main problem is the excessive fragmentation of the nucleic material obtained. Fragmentation inhibits studies on the hybridization of the total DNA of a community and can lead to artifacts on applying the PCR technique (NIEMI et al. 2001).

The extraction of DNA from water samples basically involves four different stages. In the first method, the lysis of microbial cells is carried out by chemical methods using, for instance, SDS and phenol (OELMÜLLER et al. 1990). The second method involves mechanical cell lysis (FERRIS et al. 1996). In the other methods the DNA is extracted from the samples using heat generated in the thermocycler during PCR. KIRCHMAN et al. (2001) proposed that a small fragment of the membrane used in the sample filtrations is placed in the amplification tube to serve as a template in the reaction. ØVREÅS et al. (1997) used a concentrated suspension of non-lysed cells as a template. According to the authors only 3% of the cells were lost using this methodology.

In this context the selection of an extraction protocol will be dependent on the material to be used as well as the application of the nucleic acid, which will require different degrees of purity. Several commercial kits for DNA extraction are available on the market, each one containing a specific protocol to be followed. For each specific sample the best method available should be selected according to sample origin.

After extracting the genetic material, target genes can be amplified by PCR (in the case of DNA) or reverse transcription polymerase chain reaction (RT-PCR) (in the case of RNA). The target DNA molecules can also be quantified during amplification by means of quantitative real-time PCR (qPCR), which provides the number of molecules produced per cycle (Fig. 6.3).

The PCR or RT-PCR product can be analyzed by molecular fingerprinting methods, such as denaturing gradient gel electrophoresis (DGGE) and temperature gradient gel electrophoresis (TGGE) (Sect. 6.2.3), terminal restriction fragment length polymorphism (T-RFLP), amplified fragment length polymorphism (AFLP), ribosomal intergenic spacer analysis (RISA), and single-stranded conformation polymorphism (SSCP) (Sect. 6.2.6), and/or can be cloned and sequenced (Sect. 6.2.4) for identification of the microorganisms. These sequences can be used for phylogenetic inference and to characterize the diversity and richness of microbial populations by means of several ecological indices. Furthermore, the obtained sequences can be very useful for designing oligonucleotide probes to be applied in fluorescence in situ hybridization (FISH—Sect. 6.2.5), a technique which does not require the extraction of DNA (Fig. 6.4).

The identity of PCR products can also be directly revealed by next-generation high-throughput sequencing technologies. These powerful miniaturized and parallelized platforms can generate hundreds of mega- to gigabases of nucleotide sequence data. Recently developed technologies can also analyze single molecules and do not require previous PCR amplification. Further details on the next-generation sequencing approaches are given in Sect. 6.2.7.

DNA microarray (Sect. 6.2.6.7) is also a powerful molecular technique which may be used to determine the expression of thousands of genes from a certain biological sample in a single experiment. It usually makes use of cDNA obtained from RT-PCR of the RNA extracted from the sample. Thus, information regarding the degradation potential of the microbial communities from wastewater bioreactors can be obtained.

The schematic representation in Fig. 6.4 highlights the importance of combining conventional techniques of culture and isolation with molecular methods, particularly when aiming to better understand the microbial communities present in a certain environment. The representative diagrams of each technique shown in Fig. 6.4 will be better understood as the reader progresses through this chapter.

There follows a description of the main molecular biology techniques used both for the identification and quantification of microorganisms in various biological systems and to study the dynamics of microbial populations under different environmental conditions. Some techniques applied with less frequency, but with a considerable potential for use, will also be described.

6.2.2 PCR

The first molecular techniques employed after the total extraction of nucleic acids from microbial communities of environmental samples were DNA hybridization, the determination of the percentage of GC content in the communities, DNA reassociation, and RFLP. The major limitation of these techniques is the need for a large quantity of relatively pure DNA (FRANKLIN et al. 1999). The molecular techniques currently employed to study microbial diversity in a wide variety of biological systems are based on the principle of amplifying a target sequence through the polymerase chain reaction (PCR). Described by MULLIS and FALOONA (1987), PCR is a powerful tool which revolutionized molecular biology practices, enabling the development of highly sensitive and specific protocols for the detection and/or quantification of microorganisms, plasmids, transposons, or genes (ROSADO et al. 1996).

The polymerase chain reaction can be carried out based on an understanding of the DNA duplication mechanism and is based on several factors associated with the mechanism, such as the fact that DNA serves as a template for its own duplication and that the enzyme DNA polymerase catalyzes the formation of a new strand of DNA. Through PCR it is possible to amplify selectively in vitro a single DNA or RNA molecule millions of times within only a few hours, using the basic elements of the natural DNA replication process. This makes it possible to detect and analyze specific gene sequences of a sample, without cloning or the need for hybridization (OSTE 1988; VAN DER ZEE and HUIS 1997). Disadvantages of this technique include the need for prior knowledge regarding the target DNA sequences which need to be copied and infidelity in the DNA replication (STRACHAN and READ 2002).

In summary, the method is based on the capacity of specific primers for a certain DNA region of an organism to anneal only to the desired sequence via the complementarity of bases. The primers, which are able to locate a single site in a highly complex genome, are short oligonucleotides generally of 10 or 20 base pairs. They promote a perfect hybridization with the opposite filaments of the target sequence (base pairs of the DNA strand under study) and trigger the synthesis of the complementary DNA sequence with the aid of the enzyme DNA polymerase. This enzyme is responsible for the extension of the target DNA, recognizing the complex formed by the primer and the strand of the template DNA, which results in the simultaneous copying of both directions of the DNA segment flanked by the two annealed primers (OSTE 1988).

In order for the PCR procedure to reproduce, in the laboratory, the natural DNA replication process, the DNA polymerase needs to tolerate the thermal cycles used during this procedure. Thus, this enzyme is commonly extracted from *Thermus aquaticus*, a bacterial species which survives at high temperatures (SAIKI et al. 1988). When extracted from this microorganism, the enzyme is referred to as Taq DNA polymerase.

As a result of the PCR, parts of the gene, generally the exons of the DNA, can be rapidly amplified using known specific primers. The PCR procedure involves a succession of three stages (representing one cycle), which are determined by different temperatures:

- Thermal denaturation of the double-stranded DNA of the sample which will serve as a template, generating two single strands. This stage involves rapid incubation (30 s to 1 min) at a temperature of 90–95 °C.
- Hybridization of oligonucleotide primers or specific annealing by complementarity with a target DNA sequence by cooling. The primers flank the DNA ends of interest on the two strands resulting from the previous step (denaturation). The annealing occurs at position 5′ of the single strand of the template DNA. The ideal temperatures for this stage vary between 45 and 60 °C, and it lasts approximately 1 min. The temperature is selected according to the size and composition of oligonucleotide primers.
- Amplification or elongation of oligonucleotides with the aid of the enzyme Taq DNA polymerase, carried out at 70–72 °C, for a period of 1–3 min. This stage is characterized by the extension or polymerization and is initiated when the primer has bound to the complementary segments of the template strand. The DNA polymerase binds the nucleotides together, completing the single strand and transforming it into a double strand, promoting its extension and producing an exact copy of the target sequence.

In the following cycles, a chain reaction subsequently occurs in which the recently synthesized DNA strands are separated from the original strands in the denaturation stage and each strand again serves as a template in the annealing and extension stages for the formation of new strands, which guarantees an exponential increase. Figure 6.5 shows a schematic representation of the principle of the polymerase chain reaction.

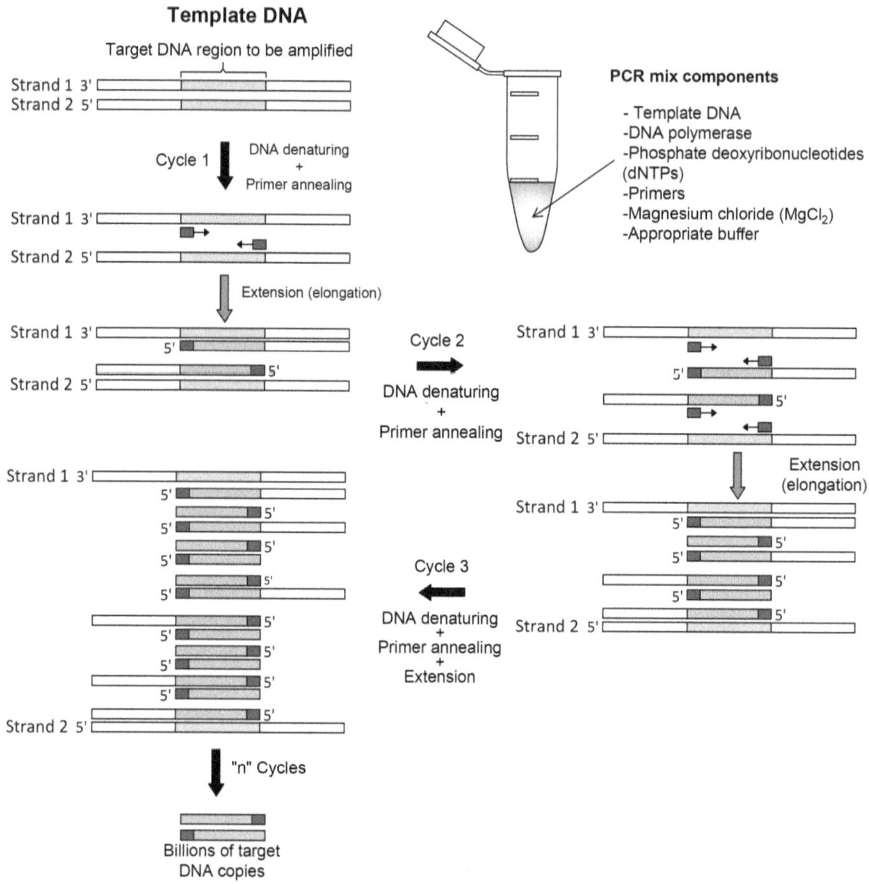

Fig. 6.5 Principle of the PCR in which small specific DNA sequences (primers) are used to link target DNA strands denatured by heating. The extension of the molecule for the formation of a complementary strand is carried out through the annealing of primers, via the thermostable enzyme DNA polymerase

The denaturation, annealing, and extension processes can be carried out in a cyclic form, due to the thermostability of the enzyme DNA polymerase, until the target sequence is found in detectable quantities. These processes are usually repeated for 30–40 cycles. Since all of the components are present in the sample from the beginning of the reaction, the procedure used to carry out the PCR can be automated and conducted on an instrument known as a thermocycler, programmed by adjusting the time, temperature, and number of cycles, which are mainly dependent on the objective to be accomplished (OSTE 1988; VAN DER ZEE and HUIS 1997).

Before the first cycle in the thermocycler is started, the oligonucleotide primers, DNA polymerase, phosphate deoxyribonucleotides (dNTPs—dATP, dTTP, dCTP, dGTP), magnesium chloride ($MgCl_2$), and an appropriate buffer are mixed in a PCR tube. Some commercial buffers for PCR already contain $MgCl_2$. This component

Table 6.3 Some components of the PCR reaction and their most common concentrations (adapted from LORENZ 2012)

Component	Final concentration
Template DNA	10–100 ng
PCR buffer	One-tenth of reaction volume (buffer 10× concentrate)
$MgCl_2$	1.5 mM
dNTPs (dATP, dCTP, dGTP, dTTP)	200 μM
Primers (forward and reverse)	20 μM = 20 pmol/μL
Enzyme (Taq DNA polymerase)	2.5 units
Sterile nuclease-free water	Until completing the final volume (50 μL)

provides Mg^{2+} divalent cations which act as a cofactor for type II enzymes such as restriction endonucleases and polymerases. The reaction volume generally varies from 25 to 100 μL, and the most common concentrations (for a 50 μL reaction) are given in Table 6.3. These conditions allow the amplification of most samples, but can be modified for each new PCR reaction as required. The temperature of each stage, the duration of each cycle, and the number of cycles must be adjusted in order to increase the reaction yield. The annealing temperature determines the degree of specificity of the reaction, with higher temperatures increasing the specificity of the primer/template DNA.

Theoretically, considering that the sequence flanked by the primers doubles with each cycle (Fig. 6.6), after 30 cycles a theoretical factor of one billion copies of the target sequence can be reached.

The quantity of DNA amplified during the polymerase chain reaction (PCR) can be analyzed by electrophoresis in agarose or polyacrylamide gel, a technique used to separate and characterize the DNA fragments obtained. Thus, it can be observed whether or not the amplification occurred. Electrophoresis is a basic procedure in all analysis carried out on nucleic acids. This technique is based on the fact that the nucleic acid molecules have a negative electric charge in aqueous solution, which is originated from the ionization of the phosphate groups present in the nucleic acid molecules. The negatively charged molecules can migrate in a solid support (agarose or polyacrylamide) submitted to an electric field. The nucleic acids migrate in the direction of the positive pole and separate according to their molecular weight. Small molecules tend to migrate at higher velocities compared to large molecules.

Acrylamide gel, in particular, is commonly used for DNA fragments of less than 1 kb. The gel can later be stained with dyes (e.g., ethidium bromide and SYBR® Safe or SYBR® Green) and visualized directly or using light (HIRATA and HIRATA 1991). Figure 6.7 shows an agarose gel, visualized by staining with SYBR® Safe in a transilluminator and the bands resulting from the PCR product. In this case the primers used were Bac341f (5′-CCTACGGGAGGCAGCAG-3′) and Bac907r (5′-CCC CGT CAA TTC CTT TGA GTT-3′) (MUYZER et al. 1993), yielding a PCR fragment of around 550 base pairs.

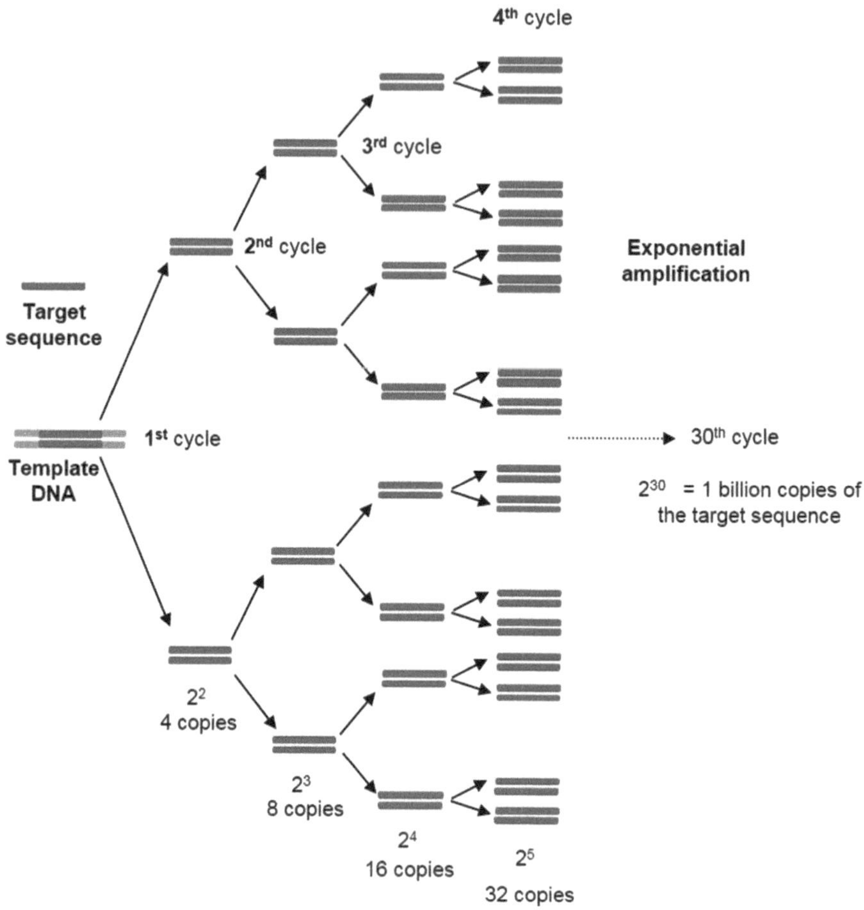

Fig. 6.6 Schematic diagram illustrating the duplication of the sequence flanked by the oligonucleotide primers

The primers binding to the region of the template DNA are important for the enzyme DNA polymerase to perform its polymerization function, since amplification of the DNA will only occur if there is hybridization of the primer with a DNA segment of the sample. The efficiency and specificity of the amplification must be taken into account when the primer is designed. In addition, it is crucial that there is no complementarity between the two 3′ ends of the primers to avoid the formation of dimers which would reduce the formation of the target sequence.

The primer design generally represents the most critical parameter in the development of PCR protocols, bearing in mind that in many cases it determines the success of the DNA fragment amplification. There is no region which guarantees the amplification of the desired segment, although some strategies can be used in the designing of the primer pairs. Examples of these strategies include that the primer length should be around 15–30 base pairs and the content of GC residues should be

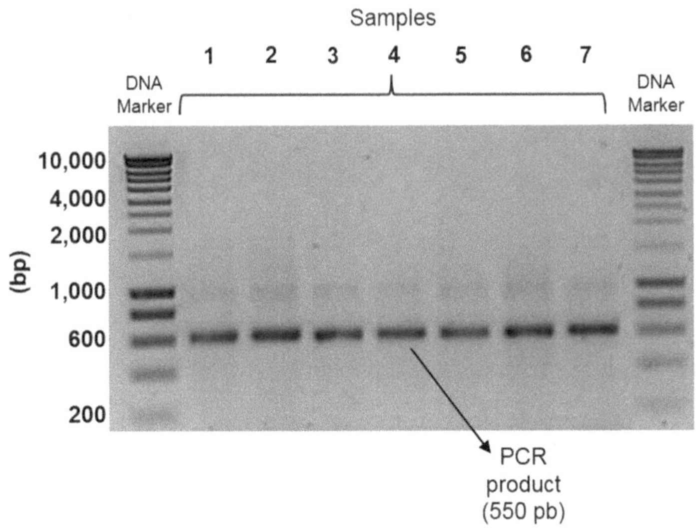

Fig. 6.7 Electrophoresis in agarose gel of PCR products

around 50%. Besides that, these nucleotides must be randomly distributed in the primer. The designing of the primers, when carried out with care, can allow savings in terms of time and costs, considering that they consist of the most expensive components in the PCR technique.

One fault which has been observed in the PCR technique is the preferential amplification of sequences caused by the nonspecific pairing of some regions of the DNA molecule, which hinders the binding of the PCR primers with the target region, generating nonspecific products. Another factor which needs to be taken into consideration in PCR studies on complex samples is the preferential amplification caused by the different sizes and concentrations of the genes in the sample. One way to prevent the selective amplification in these samples is to create conditions which facilitate the denaturation of the DNA strands, with the addition of alkalis, cosolvents, or acetamide to the reaction (REYSENBACH et al. 1992).

The choice of the DNA region located between two primers which is specifically and selectively amplified in an exponential manner strongly influences the degree of specificity of the PCR procedure and is dependent on the purpose of the test. The specificity is dependent on the degree of homology between the target sequence and the DNA of different genera and species.

As described above, highly conserved regions of a universal gene are normally chosen as target to carry out the PCR test for the identification of microorganisms. This strategy is based on the use of universal sequences as targets for amplification, as, for instance, the sequences of ribosomal RNA (rRNA). The rRNA genes are well conserved both functionally and with regard to their sequence, besides being associated with regions of moderate variation (WOESE 1987), which makes the comparison of rRNA sequences a very useful tool not only for identification but also for understanding the phylogenetic and evolutionary relationships of the organisms

(RODÍCIO and MENDOZA 2004). In fact, the first applications of techniques based on nucleic acids in the study of microbial ecology were related to the phylogenetic relationships between microorganisms determined through the analysis of the sequence of 16S rRNA (MACRAE 2000). Furthermore, other techniques were developed based on the use of these markers, such as in situ hybridization and ARDRA (KOZDRÓJ and VAN ELSAS 2000), which will be described in Sects. 6.2.5 and 6.2.6, respectively.

The small subunit of rRNA, that is, 16S, is a molecule which is widely used for phylogenetic analysis in prokaryotes. The 16S rRNA is a polyribonucleotide of approximately 1500 nucleotides, codified by the rrs (rDNA) gene. The 16S rRNA, as with any single-chain sequence, folds into a secondary structure which intercalates single chains and double chains, as shown in Fig. 6.8 (RODÍCIO and MENDOZA 2004).

23S rRNA has also been gaining attention since it has a higher number of variable regions, allowing phylogenetic analysis at the species and subspecies levels, and thus it can be used to complement 16S rRNA (CHRISTENSEN et al. 1998). Although 23S rRNA has approximately 3000 nucleotides, presenting therefore twice as much information and greater accuracy in relation to phylogenetic inferences in comparison with 16S rRNA, the latter has become a reference because it can be sequenced more easily. However, this does not diminish the importance of the use of 23S rRNA as a supplement for the data generated from 16S rRNA in studies on closely related organisms (STAHL 1997).

The spacing region between the ribosomal genes 16S and 23S of rRNA has been applied not only for the identification of species, but it has also been the target of many studies seeking to characterize and develop molecular markers in prokaryotes and eukaryotes (JENSEN et al. 1993). The space between genes 16S and 23S of rDNA is known as the internal transcribed spacer (ITS), and it has a higher degree of variability compared with regions 16S and 23S. These variations have been used to differentiate bacterial species and for phylogenetic analysis (LEBLOND-BOURGET et al. 1996).

In summary, interest in the conserved region of rRNA originates from two factors. The first relates to the principle that the three genes which codify the ribosomal subunits are contained in a single locus, which occurs several times in one genome. The second factor is associated with the presence of variable regions flanked by codifying regions, known as intergenic spacer (IGS) and internal transcribed spacer (ITS) regions, both of which are very useful for the differentiation of species and genera (EDEL 1998). In general, the use of well-conserved regions enables the identification of a set of related microorganisms and not an individual genus or species. The use of highly variable regions provides greater specificity, which allows the identification of a particular genus or species. The greater the variability of the region the more discriminatory the PCR will be (MITCHELL et al. 1993).

Besides the use of universal primers, PCR can also be carried out with functional genes encoding enzymes which catalyze specific reactions. If the process to be monitored involves a characteristic biochemical reaction, such as the biological degradation of a certain organic compound in a wastewater treatment reactor, the gene which encodes for the enzyme involved in the degradation can be amplified by

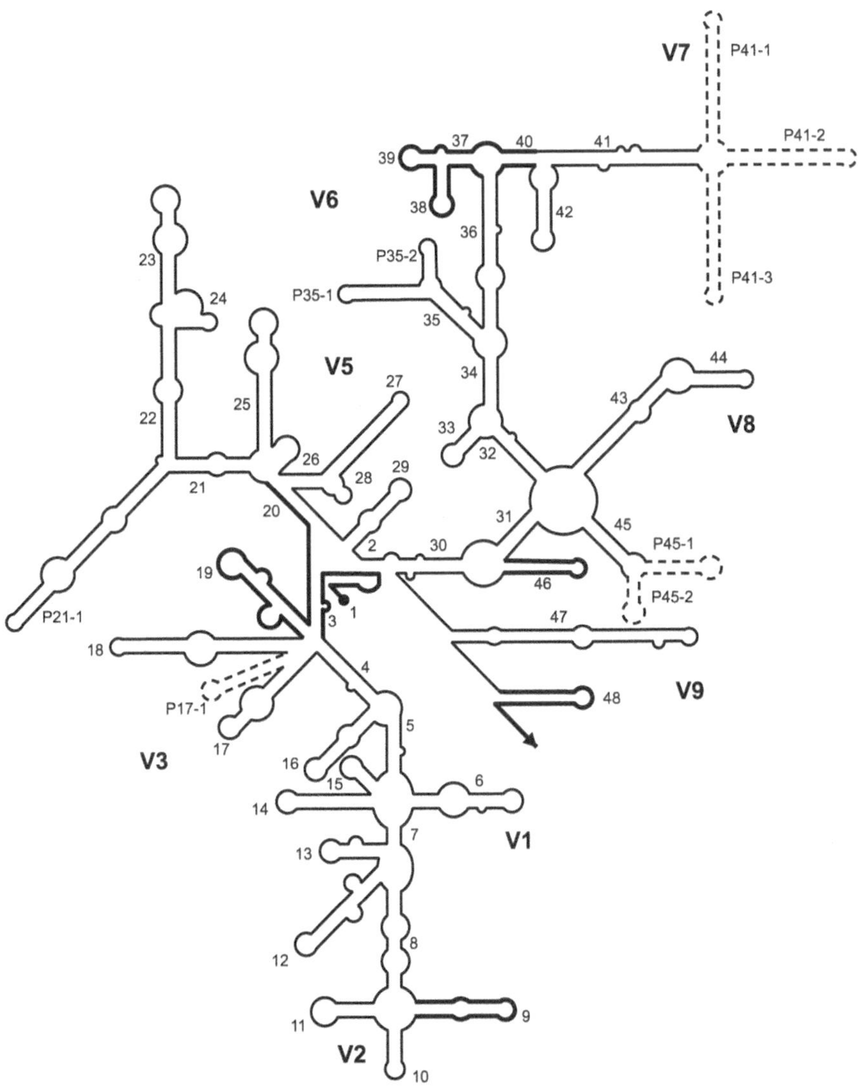

Fig. 6.8 Schematic representation of the secondary structure of 16S rRNA. The conserved regions are represented by the *bold lines* and the variable regions (V1–V9) by *fine lines* (adapted from RODÍCIO and MENDOZA 2004)

PCR and the products submitted to fingerprinting methods (e.g., DGGE) and/or sequencing. In this PCR approach where the gene of interest is targeted, unimportant bacteria are excluded from the analysis, and classification at lineage level is possible. However, information regarding the overall profile of bacterial community is not provided. Furthermore, microorganisms which may contribute to the process functioning but do not have the genes of the target enzyme are not detected.

The PCR technique is still evolving and innovations are frequently incorporated into this procedure. Some operational problems are prone to occur, for instance, extreme sensitivity and contamination, which leads to false-positive results, and samples containing substances which inhibit the amplification, giving false negatives.

Since PCR technique can be used to amplify DNA from a small number of molecules, the contamination of any component of the mixture with exogenous DNA can lead to the simultaneous amplification of another species of DNA. In addition, the presence of a PCR product and primers from previous amplifications can also act as a source of contamination. The inclusion of a negative control in each PCR experiment is one way to detect contamination in the reaction. This control should contain the same components used in the other samples, except the target DNA molecule.

Another point to be noted is the importance of also including a positive control to observe the occurrence of false negatives, which can originate from problems with the thermocycler, with a reagent used or with the enzyme Taq DNA polymerase. The false negatives can also originate from the choice of primers with a high concentration of pairs of guanine-cytosine nucleotides (GC greater than 50%). This bond is comprised of a triple-hydrogen bridge, characterizing a stronger bond, and thus it is more difficult to denature, requiring a longer time to reach the required temperature to break this bond. These factors can lead to incomplete denaturation, reducing the possibility for hybridization and consequently causing a decrease in the PCR product yield (HIRATA and HIRATA 1991).

To avoid the occurrence the false positives, the reagents used must be carefully selected and dosed, in order to maintain their quality and purity, together with standardized concentrations and equilibrated pH, parameters which should remain constant from one reaction to another. Disequilibrium in the PCR system can lead to the appearance of dimers, which can be avoided by maintaining standardized procedures in carrying out all of the manual stages, as well as using special initial annealing techniques in the thermocycler program. False-positive results can also originate from contamination by amplicons, which can be avoided by following some specific procedures for the appropriate handling of samples (DEGRAVE et al. 1994).

Other modifications to the PCR technique aimed at improving the sensitivity and specificity have been proposed. One of these is known as nested PCR, where initial amplifications are carried out with a pair of primers without a high degree of specificity (e.g., universal primers). The product amplified by PCR is then submitted to a second round of amplification with another set of primers located more internally in relation to the first pair, increasing the sensitivity and specificity of the method (DIEFFENBACH et al. 1993).

Reverse transcription polymerase chain reaction (RT-PCR) technique is another variant of PCR, often used to qualitatively detect RNA expression through formation of a double-stranded complementary DNA (cDNA) from single-stranded RNA (e.g., messenger RNA—mRNA). The reverse transcription of the RNA of interest into its DNA complement is mediated by the enzyme reverse transcriptase (KAWASAKI 1990). Subsequently, the synthesized cDNA is amplified by means of traditional PCR.

There is also in situ PCR, which combines the high sensitivity and specificity of the PCR with the capacity of in situ hybridization to detect small quantities of nucleic acids inside cells, which allows rare RNA and DNA sequences to be located.

Finally, an important PCR variant is the quantitative real-time PCR (also referred to as q-PCR), which allows for simultaneous amplification and detection of targeted DNA molecules during PCR by means of fluorescent dyes. The quantity of initial target DNA molecules can be estimated by way of comparison with an internal standard (CROSS 1995; LOGAN et al. 2009). q-PCR can also be used to quantify RNA. In this case, the RT-PCR and qPCR are combined giving rise to the technique often known as RT-qPCR or even real-time RT-PCR. This approach is very useful for analyzing gene expression.

Real-time PCR requires a thermocycler which has a fluorescence emission monitoring system. This technique has the advantage of being quantitative, since it can be used to evaluate the number of molecules produced in each cycle, enabling the monitoring of the reaction and providing more precise and faster results compared with conventional PCR, which generates only qualitative data.

The most relevant characteristics of real-time PCR are speed, specificity, sensitivity, and quantification. The method involves basically three phases: baseline phase, where the PCR products are not sufficient for fluorescence detection; exponential (log) phase, where the quantity of PCR products doubles for each cycle; and finally the plateau phase, reached when the number of products no longer increases. Characteristics such as precision and good reproducibility in the quantification of the nucleic acids associated with real-time PCR arise from the fact that the values are determined during the exponential phase of the reaction.

The instrumentation required to perform real-time PCR includes a thermocycler, equipped with an optical system for the fluorescence excitation and emission collection and a computer with software for data acquisition and the final analysis of the reaction.

The point at which the cycle in the reaction reaches the threshold of the exponential phase is called the cycle threshold (C_T), as shown in Fig. 6.9. The C_T point enables precise and reproducible quantification in the fluorescence analysis. The emission of the fluorescence compounds generates a signal which increases as a function of the quantity of PCR product. The fluorescence intensity, which reflects the quantity of product amplified, is recorded during each cycle.

There are two main approaches to the quantification in real-time PCR: nonspecific detection and specific detection. SYBR® Green is the most commonly used example of nonspecific detection, in which fluorophores (molecules which absorb and emit light at a specific wavelength) bind to the double strand of the DNA and, with the excitation by light emitted by the optical system of the thermocycler, emit green fluorescence (Fig. 6.10a).

At the beginning of the amplification, the reaction mixture contains denatured DNA, the primers, and the SYBR® Green. The non-bound SYBR® Green molecules have weak fluorescence, producing a minimal signal which is subtracted during the computer analysis. After the primers have been recognized during the polymerization process catalyzed by the enzyme DNA polymerase, the SYBR® Green binds to

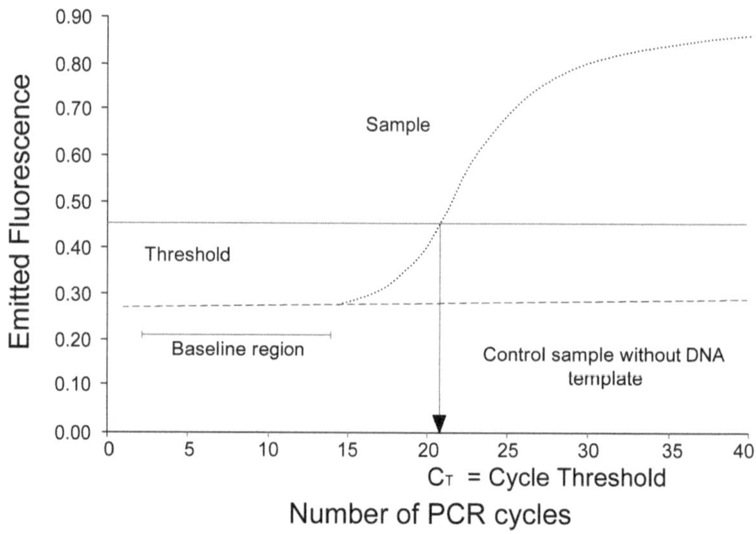

Fig. 6.9 Amplification curve of real-time PCR showing three distinct phases: baseline, log, and plateau phases

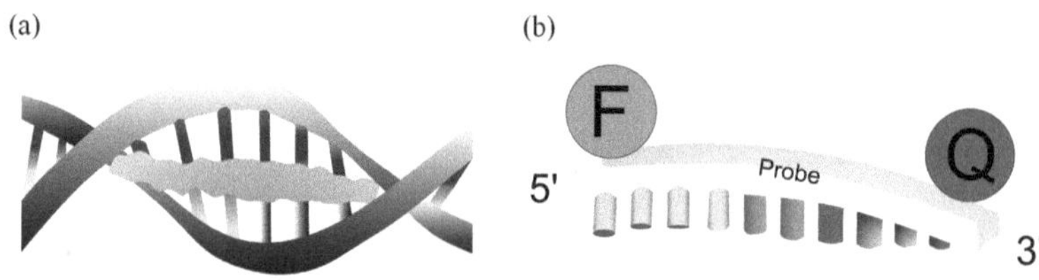

Fig. 6.10 SYBR® Green between the double strands of DNA (**a**) and TaqMan probe (**b**)

the recently synthesized double strand of DNA, enhancing the fluorescence. Thus, the reaction is monitored in continuous mode, and the fluorescence increase is observed in real time. In the subsequent cycle, in the DNA denaturation stage, the SYBR® Green molecules are released, and there is a drop in the fluorescence signal. The fluorescence detected at the end of the extension phase of each PCR cycle enables the monitoring of the increasing quantity of amplified DNA (VITZTHUM et al. 1999). The first significant increase in the quantity of PCR product can therefore be related to the initial quantity of target DNA. The higher the amount of target DNA present at the beginning of the PCR reaction the lower the number of cycles required to reach the point at which the fluorescence signal is initially detected.

SYBR® Green offers advantages such as low cost, ease of use, and sensitivity. The disadvantage is that it binds to all of the double strands of DNA which appear during the reaction, including the primer dimers and other nonspecific products, which can lead to the overestimation of concentration of the target fragment.

Besides SYBR® Green, another two methods are TaqMan and molecular beacons, both able to carry out hybridization generating energy transfer for quantification.

These latter two options consist of fluorescent probes which are highly specific for the target sequence, releasing fluorescence only when the PCR product of interest is present, i.e., when hybridization of the probe with its complementary sequence occurs. Furthermore, these specific probes allow to perform multiple reactions.

Functioning as a probe, TaqMan is a labeled DNA fragment used for the hybridization of another DNA molecule, employed to detect specific sequences of the DNA fragments amplified in the PCR. As shown in Fig. 6.10b, this probe has a fluorophore at one end and at the other end there is a quencher, that is, a molecule which accepts the energy of the fluorophore in the form of light and dissipates it in the form of light or heat.

During real-time PCR the TaqMan probe hybridizes with the target complementary single-stranded DNA sequence for amplification. In this final stage, the TaqMan is degraded due to the 5′ → 3′ exonuclease activity of the DNA polymerase, separating the quencher from the fluorescent molecule during the extension phase. With the separation of the fluorophore from the quencher, the fluorescence intensity increases, and thus during the amplification process the light emission increases exponentially. This increase in the fluorescence occurs only when the probe hybridizes and when the amplification of the target sequence is established (HEIDM et al. 1996).

Molecular beacons are nucleotides used as single-stranded probes which form a secondary structure between the 5′ and 3′ ends, known as a stem-and-loop structure. The loop contains a sequence which is complementary to the target sequence, and the stem is formed by annealing of the complementary sequences which are located at the ends. A fluorophore is bound to one end and a quencher is bound to the other, both through covalent bonds (Fig. 6.11a).

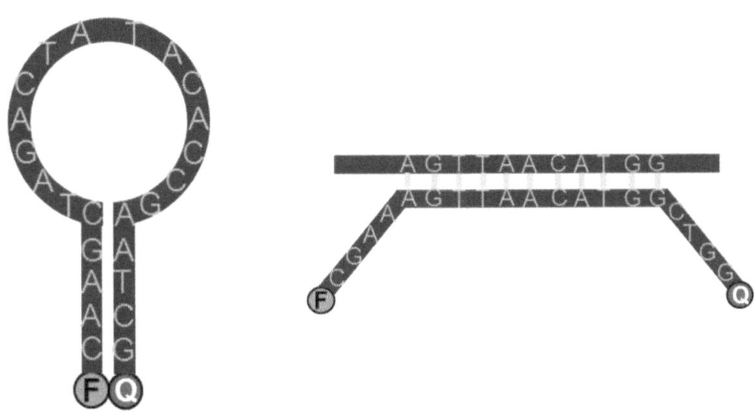

(a) (b)

Fig. 6.11 (**a**) Oligonucleotide used as probe is synthesized so as to allow the formation of a secondary structure at the 5′ and 3′ ends. (**b**) Every strand formed during the amplification is a target for the annealing of the molecular beacon, increasing the fluorescence intensity

When there are no targets, the oligonucleotide remains free in solution and does not emit fluorescence since the quencher is close to the fluorophore, capturing energy. When the molecular beacon finds its target hybridization occurs, which results in a reorganization of its structure (change in the conformation), the fluorophore is dissociated from the quencher, thus emitting fluorescence (Fig. 6.11b).

It is important to mention that the molecular beacons can be synthesized with fluorophores of different colors and thus tests which require the detection of several targets can be carried out in the same reaction. The high level of specificity means that target sequences which differ by only a single nucleotide can be discriminated.

6.2.3 DGGE

Denaturing gradient gel electrophoresis (DGGE) has been widely applied in microbial ecology to study the phylogeny of microbial communities (MUYZER et al. 1993). It consists of a molecular fingerprinting method since it is able to generate "fingerprints" or "bar codes" of microbial communities. The DGGE technique is based on the electrophoresis of DNA fragments, previously amplified by PCR in polyacrylamide gel with a linear denaturing gradient. These fragments are submitted to an environment with increasing levels of denaturing chemical agents (urea and formamide), which promote conformational changes in the molecule, reducing its migration (SIGLER et al. 2004; ERCOLINI 2004). When the denaturing agent is temperature, the technique is called temperature gradient gel electrophoresis (TGGE).

In the case of DGGE, the concentration of polyacrylamide in the gel is constant, the urea and formamide have an increasing concentration gradient, and the temperature is fixed at around 60 °C. On the other hand, in TGGE the separation of the DNA fragments is obtained with an increasing temperature gradient and a high constant concentration of urea and formamide.

The principle of dissociation is based on the fact that the electrophoretic mobility of DNA in a polyacrylamide gel is sensitive to the secondary structure of the DNA molecule, with respect to its conformation, which can be helicoidal, partially denatured, or as a single strand. The partially denatured molecules, comprised partially of a double helix and partially a single strand, move more slowly in the gel compared with double or single strands (MUYZER et al. 1998).

When the DNA is submitted to electrophoresis under increasing conditions of denaturation (chemical or thermal), the fragments remain as double strand until the conditions required for the denaturation of molecular domains, called melting domains, are reached. These domains have identical denaturation temperatures, and thus at a certain temperature there is a complete denaturation of these domains, which are intercalated along the molecule. When a domain is denatured, the conformation of the molecule changes from helicoidal to partially denatured. Under these conditions, the migration of the molecule in the gel almost ceases, and it occupies a particular position in the gel, as shown in Fig. 6.12.

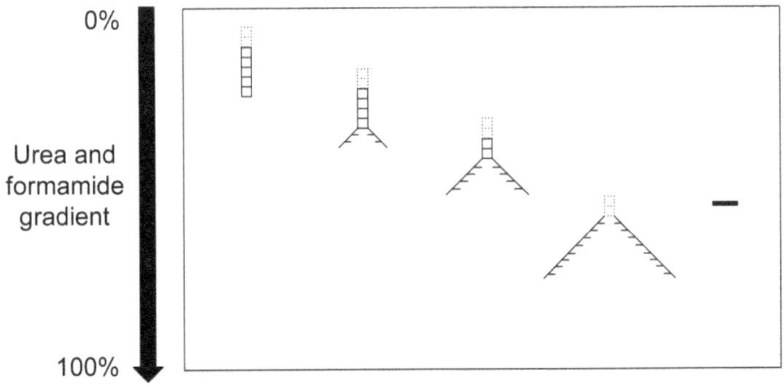

Fig. 6.12 Schematic representation of a parallel gel containing a gradient of denaturing agents (adapted from ROSADO and DUARTE 2002)

Variations in the nucleotide sequences of these domains lead to different denaturation conditions, and the molecules with different sequences stop migrating at different positions in the gel, resulting in their separation (ROSADO and DUARTE 2002).

The determination of the melting behavior of the DNA fragments is important to be determined before their analysis by DGGE/TGGE, so that the ideal gradient and duration of electrophoresis can be applied (MUYZER et al. 1998). With prior knowledge on the nucleotide sequence of one or more DNA molecules which will be analyzed in the denaturing gels, it is possible to theoretically predict the optimum gradient for the separation of different products. In this regard, computer programs can be employed, such as MELT95 (Ingeny, Leiden, the Netherlands) and several others which can predict the denaturation behavior of DNA molecules (ROSADO and DUARTE 2002).

The gradient and time period which provide the best separation of the PCR products are generally determined using a perpendicular gel and/or a parallel gel containing different gradients, where the samples are applied for varying time periods. In the perpendicular gel, there is an increasing concentration of denaturing agents located in a position perpendicular to the direction of electrophoresis. The DNA is applied in a lane which occupies almost all of the gel width. After the DNA staining, a sigmoidal curve can be observed, and on the left side of the gel (containing lower concentrations of denaturants), the PCR products have greater electrophoretic mobility, since the DNA remains as a double strand. On the other hand, on the right side of the gel (with a higher concentration of denaturants), the migration of the DNA is suddenly interrupted by the high concentration of denaturing agent (urea/formamide or temperature). In the intermediate range of denaturant concentrations, where the PCR products have different mobilities, a strong transition in the mobility occurs when the denaturant concentration reaches that required for the initiation of the separation of the double strands in a certain domain (the domain in which denaturation occurs at the lowest temperature or under the mildest chemical condition). In order to determine the best gradient for the separation of specific PCR products, various parallel gels containing different gradients can be used (MUYZER et al. 1998).

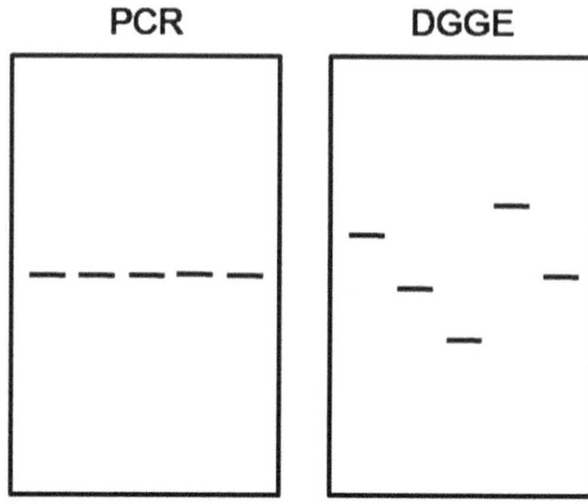

Fig. 6.13 Illustration showing PCR products with different nucleotide sequences analyzed in agarose (PCR) and through DGGE in polyacrylamide with urea/formamide denaturing gradient (adapted from ROSADO and DUARTE 2002)

The optimum time for the electrophoresis is determined through the electrophoresis in gels containing parallel gradients. Gels containing a parallel gradient have an increasing gradient of denaturing agents (chemical and physical), which increases from the beginning to the end of the gel. They are used to analyze multiple samples in the same gel. Before analyzing the samples in the parallel gel, the duration of the electrophoresis must be determined in order to obtain the maximum resolution between the different PCR products (DNA fragments). To achieve this, the individual samples are applied in a parallel gel, for the same period of time, in experiments called "time travel" (MUYZER et al. 1993).

The PCR products generally consist of fragments of the same size but with different nucleotide base sequences. Thus, they will present different electrophoretic patterns in the DGGE, that is, different migration profiles in the denaturing gel, which enables the differentiation of each microorganism (MUYZER et al. 1993; COUTINHO et al. 1999; ERCOLINI 2004). In this way, by providing a "bar code" pattern for the microbial community of the samples studied, DGGE complements the PCR information, since this latter technique only provides a single band which is not very descriptive, as discussed previously in Sect. 6.2.2.

As in the case of the PCR, the bands obtained by DGGE/TGGE can be visualized after staining with ethidium bromide, silver, or SYBR® Green. Figure 6.13 shows a schematic diagram of a situation in which fragments of the same size, obtained after amplification by PCR and visualized in agarose gel, can be separated by DGGE since they have different nucleotide sequences. It should be noted that the migratory behavior of the DNA fragment in DGGE is not only determined by the nucleotide composition (G + C content) but also by the interactions between nucleotides in the molecule (BRESLAUER et al. 1986; SUGIMOTO et al. 1996).

As a result, different DNA sequences, originating from different bacteria, denature in different concentrations of denaturant, which results in a pattern of bands where, theoretically, each band represents a specific population of organisms in the same sample. Thus, an overall view of the predominant species in the sample analyzed can be obtained. In DGGE or in TGGE, it is possible to detect approximately

50% of sequence variations in DNA fragments of up to 500 base pairs (MYERS et al. 1985). This percentage can be increased to almost 100% through the attachment of a segment rich in GC (GC clamp) to one side of the DNA fragment.

This clamp, containing 30–50 bp, when attached to the 5′ end of a primer, is amplified by PCR together with the DNA and introduced into the amplified DNA, acting as a domain with high resistance to denaturation, hindering the disassociation of the double strand of the DNA into single strands (SHEFFIELD et al. 1989). Thus, the occurrence of a loss in the definition of the electrophoretic patterns is avoided (SHEFFIELD et al. 1989; ERCOLINI 2004). The addition of this clamp increases the capacity of the DGGE to detect changes in a single base in the DNA (SHEFFIELD et al. 1989).

Some authors have used psoralen as a chemical clamp, which consists of a photoactive compound which acts as a base intercalator, marking one of the primers. This specific type of clamp has the advantage of not altering the length of the primers, although it has the drawback of not allowing the direct reamplification of PCR products. Besides that, it increases the DNA degradation rate on exposure to UV light (ROSADO and DUARTE 2002). Bipolar clamps, which are those attached to the two ends of the PCR products, have been recommended in order to increase the sensitivity of the gels with temperature as the denaturant in TGGE (GILLE et al. 1998).

Once generated, the fingerprints can be sent to the database in which the similarity between the different banding patterns can be accessed to determine the structural resemblances and differences which exist between different environments or treatments. Figure 6.14 shows the DGGE gel obtained from the PCR product amplified using primers specific to 16S rRNA, representative of the microbial community of two nitrifying reactors (SBR_1 and SBR_2) submitted to different salinity regimes (BASSIN et al. 2012). Analysis of the gel was carried out using the software program BioNumerics (Applied Maths). A dendrogram is shown in the upper part of Fig. 6.14 in which the similarity between the banding patterns can be observed. This actually represents the similarity between the microbial communities present in the samples. Besides considering the number of different bands to estimate the diversity of the microbial community, some authors consider the intensity of each band as indicative of the abundance of a specific population represented by the band (NÜBEL et al. 1999). However, this interpretation should be done with caution.

According to MUYZER et al. (1993), DGGE is a technique which is not only useful to characterize microbial communities with a high degree of complexity but also to infer the phylogeny of the community members, test the purity of microbial lineages, and monitor the isolation of microorganisms from different samples. Since it adopts band patterns, DGGE also allows the monitoring of the dynamics of specific populations as a function of environmental impacts (the introduction of stress-causing agents such as chemical products or other pollutants) and also the variations in the conditions of the feeding and the operating system (for instance in the case of treatment bioreactors). This enables the study, in a relatively rapid and simple manner, of the spatial-temporal variability of the microbial populations. Furthermore, in view of the different PCR primers available, DGGE can also be used to investigate an infinite number of phylogenies or specific target organisms.

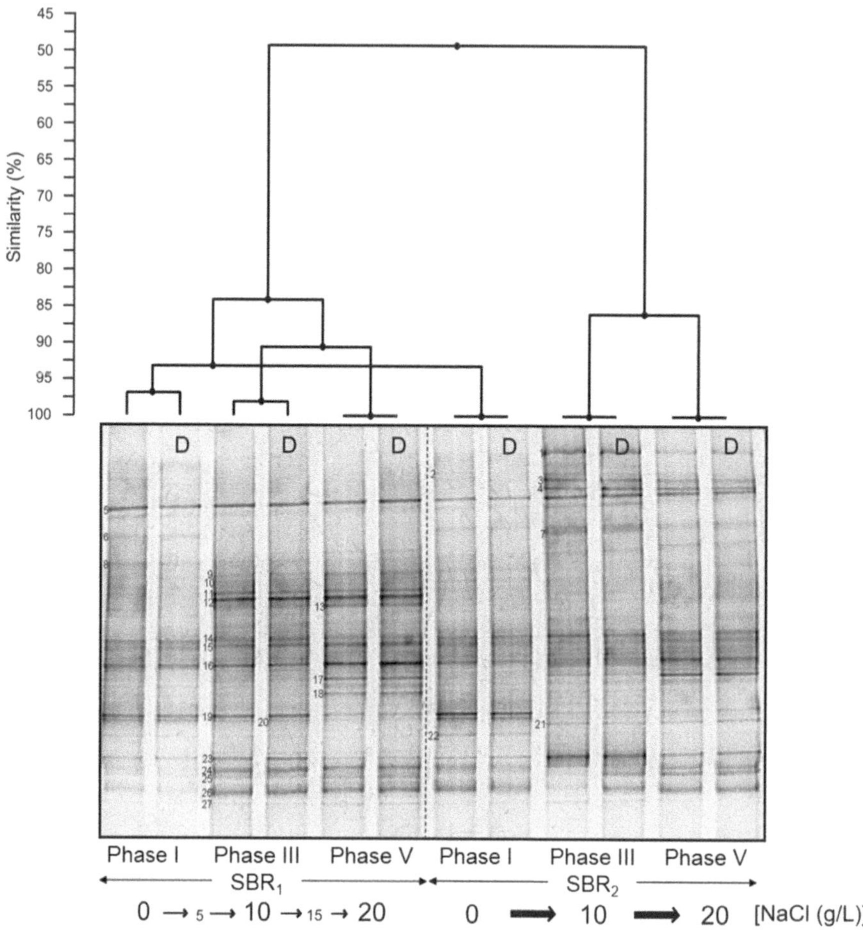

Fig. 6.14 DGGE of PCR products amplified with primers specific to the 16S rRNA gene. An analysis of the grouping showing the similarities between the different fingerprints is presented in the form of a dendrogram. D indicates the duplicate for each sample, while the numbers indicate individual bands which were excised from the gel for sequencing (BASSIN et al. 2012)

The DGGE technique also allows the excision of bands in the gels for the sequencing of DNA fragments contained in the bands. The fingerprinting obtained with the DGGE technique, when associated with band sequencing and phylogenetic analysis, has been widely employed in studies on the structural diversity of microbial communities and the phylogenetic placement of their members (MUYZER and SMALLA 1998; MUYZER et al. 1996). The sequencing of the bands present in the gel can be submitted to a global databank, thus allowing comparison with the results of different research groups worldwide. Concomitantly, the sequences obtained can be reanalyzed as the databank contents increase, providing the phylogenetic updating of the results. The DNA present in the gel can also be transferred to nylon membranes and hybridized with specific probes for the determination of certain microbial

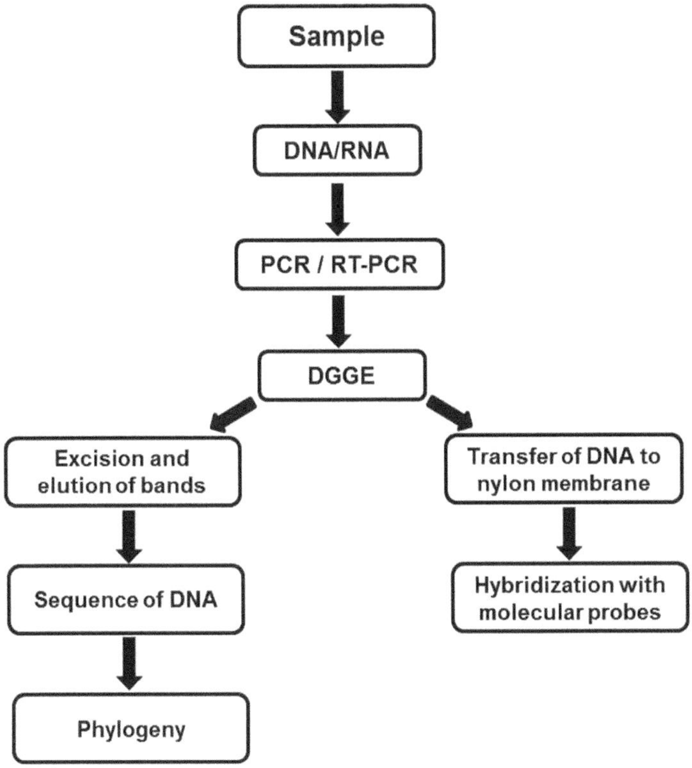

Fig. 6.15 Schematic representation of the stages and possibilities in DGGE analysis of DNA extracted from different samples

groups. Figure 6.15 shows a diagram of the different stages and possibilities which exist when working with gels containing denaturation gradients.

Another attractive feature of DGGE is the possibility to compare and analyze different samples in a single gel, in contrast to cloning. However, a drawback associated with this technique is that very long sequences may not be separated efficiently. The PCR product should not surpass 500 bp, which can limit the information required for the inference of the phylogeny (MUYZER and SMALLA 1998).

There are other aspects which should be taken into account in studies on microbial communities using techniques based on PCR-DGGE. Care should be taken regarding the fact that some samples contain several distinct species even though the G + C% contents are very similar, hindering the analysis of certain groups of organisms, referred as operational taxonomic units (OTUs). JONES and THIES (2007) proposed a complementary method for DGGE, coupling bidimensional analysis in polyacrylamide gel in order to separate PCR fragments generated by microbial communities, to avoid the superposition of the fragments.

With regard to the environmental microbiology, DGGE has been widely employed as a tool to obtain the profile of complex microbial populations without the need for their cultivation (MUYZER and SMALLA 1998). 16S rRNA is the region most commonly used in DGGE, although the use of other specific primers has been recently proposed, particularly for analysis involving closely related microorganisms. In fact, it has been demonstrated that other regions of the bacterial genome, when compared with the region 16S rRNA, allow better results to be obtained in the differentiation of organisms of the same species (ERCOLINI 2004; YERGEAU et al. 2005; TACÃO et al. 2005).

6.2.4 Cloning and Sequencing

The cloning and sequencing of the gene which codifies for 16S rRNA is considered to be the most powerful method for the exploration of microbial diversity and is widely employed in the field of molecular microbial ecology (MUYZER 1999).

Cloning basically involves cleaving a certain gene sequence of the genome and inserting it into another DNA known as the vector, which is normally a plasmid (small DNA fragments of approximately 3 kb which can be naturally found in bacteria and are resistant to a certain antibiotic). This cleavage process followed by insertion became possible as the knowledge regarding restriction enzymes and the ligases evolved.

It should be noted that besides the control genes (resistance to antibiotics), the vectors have a multiple cloning site with several sites of enzyme restriction and initiating sequences. This allows the rapid sequencing of the cloned fragment. Fragments of up to 5 kb can be inserted into plasmids. After the cleavage and insertion of the fragment (or gene) into the vector, this can be placed into bacterial cells employing several methods such as the use of calcium chloride, electroporation, and thermal shock. Once in the cells the plasmids can multiply and produce many plasmids which contain the cloned fragment. Thus, the cloning enables an organism to contain and express a gene which is not naturally present in its structure. Figure 6.16 represents, schematically, the cloning process.

As mentioned above, gene fragments which codify for 16S rRNA (or other functional genes) present in the genomic DNA of complex samples can be amplified through the polymerase chain reaction (PCR). A genomic library derived from the amplification of these samples is produced using the cloning method. The cloning stage allows the separation of the different copies of the target 16S rDNA sequence present in the DNA extracted from microbial communities. Once the cloning has been carried out, the clone library for the 16S rRNA gene can be accessed in different ways, such as hybridization of colonies with specific probes to determine the gene, PCR with specific primers to confirm the insertion of the cloned PCR products, or sequencing to identify the fragments. After the identification, a phylogenetic evaluation of the microbial community representative of the original sample can be carried out with the aid of phylogeny programs (OLSEN et al. 1994; COLE et al. 2003). Another way to carry out this type of molecular characterization is cloning and sequencing of cDNA transcribed from 16S rRNA with the use of the enzyme reverse transcriptase (AMANN et al. 1995).

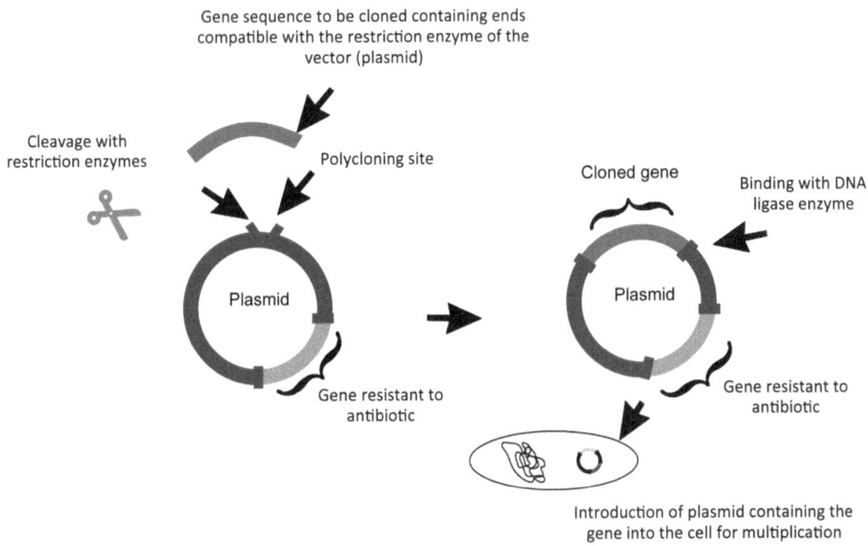

Fig. 6.16 Schematic representation of the cloning stages

While the amplicons generated from pure bacterial cultures can be directly sequenced, in the case of genomic DNA extracted from microbial communities the cloning stage needs to be included to separate the different copies of 16S rRNA, since a mixture of DNA from different sources cannot be directly sequenced.

In general, the cloning and construction of an rRNA gene library has been applied in combination with other techniques in wastewater treatment studies. The cloning of the complete gene allows the obtainment of more precise phylogenetic information in comparison with other molecular techniques such as FISH or DGGE, providing very precise taxonomic studies and phylogenetic trees of high resolution. The identification of microorganisms which have still not been cultivated or identified is also possible applying the cloning technique. In addition, it should be mentioned that cloning has potential for use as a tool for the designing of new specific primers for PCR and molecular probes for the detection and/or quantification of certain microbial groups by fluorescence in situ hybridization (SANZ and KÖCHLING 2007).

The stage of selecting clones from a library is crucial for the process of metagenomic sequencing. The selection methodologies can be based on sequence or on function. The analysis through sequence is based on the use of conserved DNA regions to design probes or PCR primers employed in the selection of clones which contain a certain sequence of interest (SCHLOSS and HANDELSMAN 2003) and is not dependent on the expression of a certain characteristic by the host. Furthermore, it is based on reliable analysis methods, such as PCR, hybridization of colonies, and microarray. As a disadvantage, the method is not selective for complete genes, and thus only genes with partial sequences of the chosen genes can be selected (KNIETSCH et al. 2003). In the case of analysis by function, this is dependent on the capacity of testing a large number of clones in the library for the production of the desired enzyme or metabolite. Therefore, it is necessary to standardize the tests on large scale, bearing in mind that most tests are not suitable for a simple and rapid evaluation of a large quantity of clones. Also, there are some cases where the clone selection stage is not carried out, such as the random sequencing of DNA, in which the whole metagenomic library, normally of small inserts, is sequenced (PEIXOTO et al. 2008).

The creation of gene libraries allows the identification of a substantial part of the microbial species present in a sample, although a complete panorama of the microbial habitat, besides being very time consuming, is laborious to obtain. The fact is that several clones need to be sequenced in order to ensure that most of the individual species in the sample have been taken into account. It should be remarked that cloning is not a quantitative technique and PCR can favor certain species due to the differences in the accessibility of the target DNA (SANZ and KÖCHLING 2007).

The fact that the cloning method is laborious and time-consuming makes it impractical when dealing with a large number of samples. This hinders the monitoring of changes in microbial communities, particularly when several sample points are required. On the other hand, when time and effort are not limiting factors, cloning encompasses the majority of microorganisms, including minority groups, which are difficult to detect with fingerprinting methods.

Another point which is worth noting is the problems associated with the metagenomic libraries related to the low frequency with which the desired clone is found (SCHLOSS and HANDELSMAN 2003). In this context, it is common to adopt sample enrichment strategies to increase the chance of recovering a clone with the desired characteristic (PEIXOTO et al. 2008). One enrichment strategy is the creation of a library starting with an environment where the desired characteristic is present in large quantities. Another enrichment method consists of the addition of a substrate such that the microorganisms which have the desired characteristic can multiply to the detriment of the others. Also, a stable-isotope probe (SIP) can be used, in which a substrate marked with ^{13}C is added to the sample. Microorganisms which can utilize the substrate will incorporate ^{13}C into their DNA. On extracting the nucleic acid of the sample, it is then possible to separate the marked from the non-marked material by ultracentrifugation with a density gradient (RADAJEWSKI and MURRELL 2002).

Despite their broad application in the study of microbial communities in natural habitats, cloning techniques are less widespread in research on wastewater treatment processes. This lack of popularity may be due to the need for specialized personnel and equipment, which are not always available in engineering or chemistry departments. This technology can involve considerable complexity, which often restricts it to a support tool, even though it has enormous potential for application in the area of wastewater treatment (SANZ and KÖCHLING 2007).

After the cloning step, sequencing procedures are used to determine the nitrogen base sequence corresponding to a fragment or gene. There are several methodologies for DNA chain sequencing. These include the chemical method and the dideoxy method with radiation or fluorescent labeling (based on the Sanger method).

One of the most commonly used methods is automated dideoxy with fluorescent labeling, in which the sequence under study is amplified using mixtures, each containing one of the dNTPs (A, T, C, or G). In these mixtures, besides the normal nucleotide, dideoxynucleotides (A, T, C, or G) are present. These do not have hydroxyls in the pentose, which hinders the binding of a new nucleotide to them. At this point in the amplification, the DNA chain growth is interrupted. Bearing in mind that there is competition between the dideoxy and the normal dNTP, the polymerization is interrupted at different times. In addition, it is possible to obtain different-sized fragments. There are two ways to carry out this reaction. In the first, a primer containing a fluorescent label (tetramethylrhodamine) is placed into four tubes, each containing one of the dideoxy nucleotides and the other normal nucleotides. The second procedure is carried out by adding one of the four nucleotides containing a different fluorescent label (fluorescein; NDB-4-chloro, 7-nitrobenzene, 2-oxal-diazol, tetramethylrhodamine, and Texas Red) and the appropriate dideoxynucleotide in four separate reactions. Thus, for each sequence studied, four polymerase chain reactions (PCRs) analyzed by gel electrophoresis are carried out, in which the fragments are identified through their molecular size by means of fluorescent labeling.

In the second procedure, one analytical run can be carried out with all samples in a single gel slot, since the fragments have different labels. The final result is organized sequentially on a computer, which provides the product which corresponds to the correct sequence of the bases comprising the gene or the expressed sequence. Automated sequencing is able to provide information on sequences of up to 1 kb per reaction, allowing the accumulation of large quantities of information.

6.2.5 FISH

Fluorescence in situ hybridization (FISH) is one of the simplest and fastest techniques for the identification, localization, and quantification of species of microorganisms present in a wide variety of samples. This technique has also emerged as an alternative to the traditional methods for the identification of microorganisms which are based on isolation through cultivation in appropriate culture media. Basically, in situ hybridization promotes the interaction of knowledge on cell biology, cytogenetics, and molecular genetics (SPECTOR et al. 1998). Regarding the microbiology of wastewater treatment systems, the FISH technique has been shown to be very useful, allowing to detect microorganisms with high detection sensitivity.

The FISH technique is based on the existence of known and specific rRNA sequences of an organism, which allow complementary sequences to be designed. This specific sequence is known as a probe. The fluorescence probe is commonly a short sequence (15–30 nucleotides) of single-stranded DNA, bound to fluorophores which recognize complementary sequences of rRNA (16S and 23S subunits in the case of bacteria) in previously fixed cells (permeable to the probe). As a consequence, an in situ hybridization occurs (DNA-RNA combination). Thus, both the

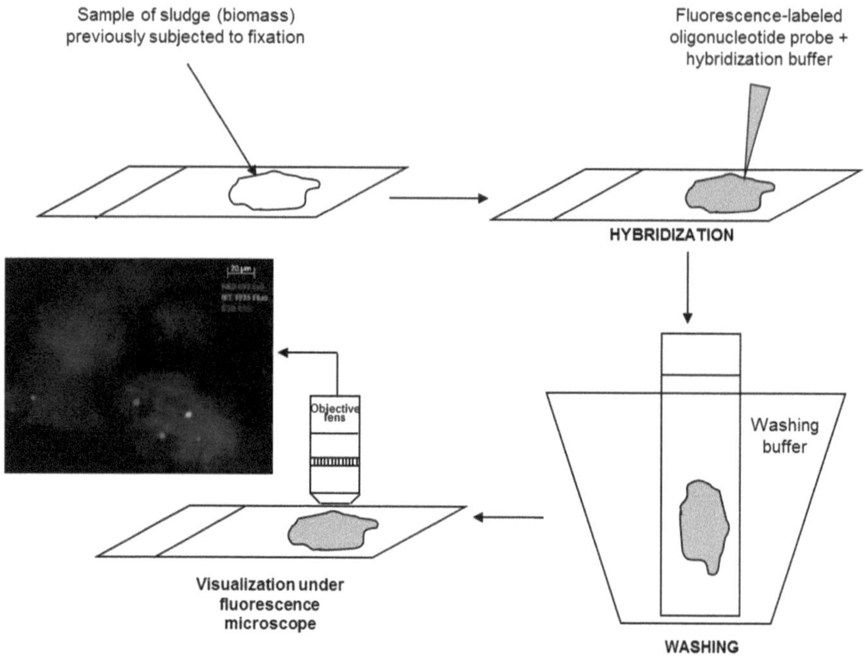

Fig. 6.17 Simplified schematic representation of the steps involved in the FISH technique for identification of microorganisms in a sludge sample (adapted from NIELSEN 2009)

location and the length of the target sequence, which, in fact, correspond to different microorganisms representative of different taxonomic levels, can be accessed by fluorescence microscopy (AMANN et al. 1990). Figure 6.17 shows schematically the procedure involved in the characterization of samples by FISH.

No probe will hybridize with cells which do not contain the target sequence. On the other hand, cells containing the desired sequence will retain the hybridized probe, and, due to the high number of ribosomes within the active cells, they become fluorescently labeled.

The FISH methodology becomes even simpler and faster when the required probes are available, allowing the direct visualization of the non-cultivated microorganisms. One of the advantages of this technique is the possibility to observe the spatial distribution of the microorganisms in the different environments in which they are found. The quantification of the specific microbial groups is also possible, which is not the case with conventional techniques (most probable number, plate counts) or with other molecular techniques. There is no need for highly trained or specialized personnel, since basic knowledge of microscopy and some laboratory experience are sufficient.

The DNA probe can be labeled directly through the incorporation of a fluorescence-labeled nucleotide precursor or indirectly through the incorporation of a nucleotide containing a reporter molecule (biotin or digoxigenin). After the incorporation

Table 6.4 Fluorescent dyes most commonly used to label oligonucleotide probes in FISH analysis (adapted from NIELSEN et al. 2009a)

Fluorochrome	Color	Maximum excitation λ (nm)	Maximum emission λ (nm)	Extinction coefficient (M^{-1} cm^{-1})
Cy3	Red	552	565	150,000
Cy5	Red[a]	649	670	250,000
SYBR Green	Green	494	520	73,000
DAPI	Blue	350	456	27,000
FLUOS	Green	494	523	74,000
TAMRA	Red	543	575	65,000
Alexa-488	Green	493	517	71,000
Alexa-546	Red	562	573	104,000
Alexa-350	Blue	343	441	19,000

[a]Emission in the infrared region, which requires a digital camera which detects infrared light and image analysis to enable the observation

in DNA, this reporter molecule is bound, through affinity, to a fluorescence-labeled molecule. Table 6.4 shows the fluorescent dyes most commonly used to label oligonucleotides in fluorescence in situ hybridization analysis.

The specificity of the probes allows the detection/identification at any taxonomic level required, from the domain to a resolution which enables the differentiation of very specific phylogenetic groups, such as species. In general, a probe aimed at the domain of bacteria is used in combination with other more specific probes.

One point to be considered relates to the fact that, although the FISH technique has allowed a better understanding of the dominant microbial communities in different biological systems, this methodology does not describe the aspects related to the function of microorganisms, but only to the phylogeny of the microbial communities.

The main disadvantage of this technique is the limited availability of specific probes for certain groups of bacteria. Although it is possible, in theory, to develop the most appropriate probe for each application in virtue of the growth of databases for the rRNA sequence (16/18S and 23/28S rRNA), in some cases it is almost impossible to develop a probe which detects specifically certain microbial groups which share the same metabolic properties, for instance, those involved in nitrification and sulfate reduction.

Some factors which hinder the use of the FISH technique are poor cell permeability, insufficient ribosome content, and inaccessibility of the ribosomes. Another point which should be noted is the fact that the FISH technique can produce false-positive (autofluorescence of the sample) and false-negative results, and both methodological and environmental factors can influence the performance of this technique. The choice of the probe and the fluorochrome, the operating conditions and the protocols used, the hybridization temperature, the fact that some microorganisms possess autofluorescence, the specific type of ecosystem, and the physiological state of the target cells can significantly influence the efficiency of this molecular tool (OUVERNEY and FUHRMAN 1997; DAIMS et al. 1999; MOTER

Table 6.5 Oligonucleotide probes aimed at the 16S rRNA gene regions used to detect ammonia-oxidizing bacteria (AOB) in nitrifying activated sludge and biofilm systems (adapted from DAIMS et al. 2009; LIPSKI et al. 2001)

Probe	Target organisms	Sequence	Reference
Nso1225	Ammonia-oxidizing bacteria (subclass *β-Proteobacteria*)	CGC CAT TGT ATT ACG TGT GA	MOBARRY et al. (1996)
Nso190	Ammonia-oxidizing bacteria (subclass *β-Proteobacteria*)	CGA TCC CCT GCT TTT CTC C	MOBARRY et al. (1996)
Nsm156	*Nitrosomonas* spp., *Nitrosococcus mobilis*	TAT TAG CAC ATC TTT CGA T	MOBARRY et al. (1996)
Nsv443	*Nitrosospira* spp.	CCG TGA CCG TTT CGT TCC G	MOBARRY et al. (1996)
NSM1B	*Nitrosomonas*	TCT GTC GGT ACC GTC AT	HOVANEC and DELONG (1996)
NEU	Halophilic and halotolerant *Nitrosomonas* spp.	CCC CTC TGC TGC ACT CTA	WAGNER et al. (1995)
Nmv	*Nitrosococcus mobilis*	TCC TCA GAG ACT ACG CGG	JURETSCHKO et al. (1998)
Cluster 6a192	*Nitrosomonas oligotropha* (*Cluster* 6a)	CTT TCG ATC CCC TAC TTT CC	ADAMCZYK et al. (2003)

and GOBEL 2000). Therefore, all these factors need to be taken into consideration in order to obtain reliable and comparable information.

It is important to note that the selection and improvement of the hybridization conditions for a new probe is a laborious and relatively complicated process, which requires experience and dedication, since the results are often unsatisfactory. The quantification can be laborious and subjective (manual counting) or complex (image analysis). The structural analysis of the aggregates (granular sludge and biofilms) requires a confocal microscope and an image analysis environment, which increases the costs and the need for specialist personnel (SANZ and KÖCHLING 2007).

Additionally, some prior knowledge regarding the microorganisms expected to be present in the sample is generally required for the successful application of this method, which leads to the need to use other techniques. For the identification and quantification of a particular species, a specific probe needs to be available, or its 16S rRNA sequence must be known (if a probe has not been previously published) (SANZ and KÖCHLING 2007).

The use of oligonucleotide probes aimed at the 16S rRNA region represents a revolution in microbial ecology, both for carrying out research and for practical applications. In studies regarding biological wastewater treatment, in particular, the hybridization techniques are by far the most commonly used.

Tables 6.5, 6.6, 6.7, 6.8, 6.9, and 6.10 provide information on some of the probes most commonly used for the identification of several bacteria found in wastewater treatment plants. The bacteria are divided into the following functional groups: nitrifiers (ammonia-oxidizing bacteria, AOB, and nitrite-oxidizing bacteria, NOB), anammox bacteria, denitrifiers, polyphosphate-accumulating organisms (PAOs), and glycogen-accumulating organisms (GAOs). This set of microorganisms can be found in aerobic or aerobic/anoxic/anaerobic activated sludge or biofilm systems.

Table 6.6 Oligonucleotide probes aimed at the 16S rRNA gene regions used to detect nitrite-oxidizing bacteria (NOB) in nitrifying activated sludge and biofilm systems (adapted from DAIMS et al. 2009; LIPSKI et al. 2001)

Probe	Target organisms	Sequence	Reference
NIT2	*Nitrobacter*	CGG GTT AGC GCA CCG CCT	WAGNER et al. (1996)
NIT3	*Nitrobacter*	CCT GTG CTC CAT GCT CCG	WAGNER et al. (1996)
Nb1000	*Nitrobacter*	TGC GAC CGG TCA TGG	MOBARRY et al. (1996)
NBAC2	*Nitrobacter*	GCT CCG AAG AGA AGG TCA CA	HOVANEC and DELONG (1996)
Ntspa662	*Nitrospira*	GGA ATT CCG CGC TCC TCT	DAIMS et al. (2001)
Ntspa1026	*Nitrospira*—sublineages I and II	AGC ACG CTG GTA TTG CTA	JURETSCHKO et al. (1998)
Ntspa1431	*Nitrospira*—sublineage I	TTG GCT TGG GCG ACT TCA	MAIXNER et al. (2006)
Ntspa1151	*Nitrospira*—sublineage II	TTC TCC TGG GCA GTC TCT CC	MAIXNER et al. (2006)
Nsr1156	*Nitrospira*—sublineage II	CCC GTT CTC CTG GGC AGT	SCHRAMM et al. (1998)
Nspmar62	*Nitrospira marina*	GCC CCG GAT TCT CGT TCG	FOESEL et al. (2008)

Table 6.7 Oligonucleotide probes aimed at the 16S rRNA gene regions used to detect anammox bacteria in activated sludge and biofilm systems

Probe	Target organisms	Sequence	Reference
Pla46	All *Planctomycetes*	GAC TTG CAT GCC TAA TCC	NEEF et al. (1998)
Amx368	All anammox bacteria	CCT TTC GGG CAT TGC GAA	SCHMID et al. (2003)
Amx820	*Brocadia anammoxidans Candidatus Kuenenia stuttgartiensis*	AAA ACC CCT CTA CTT AGT GCC C	SCHMID et al. (2000)
Kst157	*K. Candidatus Kuenenia stuttgartiensis*	GTT CCG ATT GCT CGA AAC	SCHMID et al. (2001)
Amx1015	*Brocadia anammoxidans*	GAT ACC GTT CGT CGC CCT	SCHMID et al. (2000)
Bfu613	*Candidatus Brocadia fulgida*	GGA TGC CGT TCT TCC GTT AAG CGG	KARTAL et al. (2008)
Apr820	*Candidatus Anammoxoglobus propionicus*	AAA CCC CTC TAC CGA GTG CCC	KARTAL et al. (2007)
BS820	*Candidatus Scalindua wagneri*	TAA TTC CCT CTA CTT AGT GCC C	KUYPERS et al. (2003)
Scabr1114	*Scalindua brodae*	CCC GCT GGT AAC TAA AAA CAA G	SCHMID et al. (2003)

Table 6.8 Oligonucleotide probes aimed at the 16S rRNA gene regions for the identification of potential denitrifying organisms (adapted from NIELSEN and HANSEN 2009)

Probe	Target organisms	Sequence	Reference
PAR651	Genus *Paracoccus*	ACC TCT CTC GAA CTC CAG	NEEF et al. (1996)
G_Rb	*Rhodobacter*, *Roseobacter*	GTC AGT ATC GAG CCA GTG AG	GIULIANO et al. (1999)
HyphoCII-654	*Hyphomicrobium denitrificans*, *H. methylovorum*, *H. facilis*	CCC ACC TCT ATC GGA CTC	LAYTON et al. (2000)
Curvi997	*Curvibacter*	CTC TGG TAA CTT CCG TAC	THOMSEN et al. (2004)
PAOmix	Most of the *Accumulibacter*	PAO462, PAO651 and PAO846	CROCETTI et al. (2000)
AZA645	Most members of *Azoarcus cluster*	GCC GTA CTC TAG CCG TGC	HESS et al. (1997)
THAU646	*Thauera*	TCT GCC GTA CTC TAG CCT T	LAJOIE et al. (2000)
ACI208	*Acidovorax* spp.	CGC GCA AGG CCT TGC	AMANN et al. (1996)
ZRA23a	Most members of the *Zoogloea* lineage	CTG CCG TAC TCT AGT TAT	ROSSELLÓ-MORA et al. (1995)
AT1458	*Azoarcus-Thauera* cluster	GAA TCT CAC CGT GGT AAG CGC	RABUS et al. (1999)
Pae997	Most of the *Pseudomonas* spp.	GCT GGC CTA GCC TTC	AMANN et al. (1996)

6.2.6 Alternative Methods Applied to the Study of Microbial Diversity

6.2.6.1 Terminal Restriction Fragment Length Polymorphism (T-RFLP)

An alternative method to DGGE to investigate the microbial diversity of different environments is terminal restriction fragment length polymorphism (T-RFLP). T-RFLP consists of a quantitative molecular method for the rapid analysis of complex microbial communities, with a high number of species (LIU et al. 1997). This technique is also based on PCR, although the procedure differs from the techniques of PCR/DGGE and PCR/cloning.

In T-RFLP, the 16S rRNA gene is amplified with universal primers, one of which is labeled with fluorochromes, which allow labeling of the PCR products. These are digested with restriction enzymes, also known as restriction endonucleases, which are "molecular scissors" able to recognize specific sequences of bases in double-stranded DNA and cleave both the double-stranded fragments at specific points (STRYER 1996). Assuming that each species present in the sample shows differences in the amplified gene sequence, the terminal restriction fragment will have a different length and can thus be separated through gel electrophoresis.

Table 6.9 Oligonucleotide probes aimed at the 16 rRNA gene regions for the identification of potential phosphate-accumulating organisms (PAOs) (adapted from NIELSEN et al. 2009b; LIPSKI et al. 2001)

Probe	Target organisms	Sequence	Reference
PAO462	*Accumulibacter*	CCGTCATCTACWCAGGGTATTTAAC	CROCETTI et al. (2000)
PA0651	*Accumulibacter*	CCCTCTGCCAAACTCCAG	CROCETTI et al. (2000)
PAO846	*Accumulibacter*	GTTAGCTACGGCACTAAAAGG	CROCETTI et al. (2000)
PAOmix	*Accumulibacter*	PAO462, PAO651 and PAO846	CROCETTI et al. (2000)
S-G-RHX-0991-a-A-18	*Candidatus Accumulibacter phosphatis*	GCTCTCTTGCGAGCACTC	HESSELMANN et al. (1999)
RHC439	*Rhodocyclus/Accumulibacter*	CNATTTCTTCCCCGCCGA	HESSELMANN et al. (1999)
RHC175a	*Rhodocyclaceae*	TGCTCACAGAATATGCGG	HESSELMANN et al. (1999)
PAO462b	*Rhodocyclus tenuis*	CCGTCATCTRCWCAGGGTATTAAC	ZILLES et al. (2002)
PAO846b	*Rhodocyclus tenuis*	GTTAGCTACGGYACTAAAAGG	ZILLES et al. (2002)
actino1011	*Tetrasphaera japonica*	TTGCGGGGCACCCATCTCT	LIU et al. (2001)
HGC69a	Gram-positive bacteria with high G + C—*Actinobacteria*	TATAGTTACCACCGCCGT	ROLLER et al. (1994)
Actino221	Potential PAOs—*Actinobacteria*	CGCAGGTCCATCCCAGAC	KONG et al. (2005)
Actino658	Potential PAOs—*Actinobacteria*	TCCGGTCTCCCCTACCAT	KONG et al. (2005)
MP2	*Microlunatus phosphovorus*	GAGCAAGCTCTTCTGAACCG	KAWAHARASAKI et al. (1998)

Table 6.10 Oligonucleotide probes aimed at the 16 rRNA gene regions for the identification of potential glycogen-accumulating organisms (GAOs) (adapted from NIELSEN et al. 2009a, b)

Probe	Target organisms	Sequence	Reference
GAM1019	*Gammaproteobacteria*	GGTTCCTTGCGGCACCTC	NIELSEN et al. (1999)
GAM1278	*Gammaproteobacteria*	ACGAGCGGCTTTTTGGGATT	NIELSEN et al. (1999)
GAOQ431	*Gammaproteobacteria*	TCCCCGCCTAAAGGGCTT	CROCETTI et al. (2002)
GAOQ989	*Competibacter* (some)	TTCCCCGGATGTCAAGGC	CROCETTI et al. (2002)
GB	*Competibacter* (most)	CGATCCTCTAGCCCACT	KONG et al. (2002)
TFO_DF218	Organisms related to *Defluviicoccus*	GAAGCCTTTGCCCCTCAG	WONG et al. (2004)
TFO_DF618	Organisms related to *Defluviicoccus*	GCCTCACTTGTCTAACCG	WONG et al. (2004)

In addition, it is possible to sequence and identify the terminal restriction fragments (T-RFs) generated by comparing them with a sequence database. The length of these fragments can also be determined and their quantity can be estimated. The different fragments obtained are distinguished by laser-induced fluorescence detection. The fluorescence data are converted into electropherograms.

Based on the terminal location of the fluorescent marker in the PCR products, it is possible to ensure that only terminal fragments are measured. The abundance of different species in the microbial communities is estimated through the determination of the number of terminal fragments, observed with the digestion of the total rDNA of the communities amplified by PCR.

Therefore, the T-RFLP pattern obtained function as a true fingerprint of the microbial communities and is a set of the number of fragments with the same length. The observation of different peaks is indicative of the presence of fragments of different lengths, and the relative abundance of each fragment is reflected by the area under the peaks on the electropherogram. The samples are run in polyacrylamide gel. Each fluorescent band corresponds to a single population, and the intensity of each band is directly proportional to the quantity of the PCR product, which provides an approximate indication of the abundance of the population in the community (REIS JR et al. 2002). This characteristic should be observed with caution, in the same way as the band intensity of a gel pattern in DGGE (SANZ and KÖCHLING 2007). Bearing in mind that differences in the sequence will generate amplicons of different lengths, it is possible to carry out clustering of groups of microorganisms which are phylogenetically different (LIU et al. 1997).

The advantage of the T-RFLP technique is its capacity to detect even rare members of the microbial community. However, some studies have indicated that pseudo-terminal restriction fragments can be formed, causing an overestimation of the microbial diversity (EGERT and FRIEDRICH 2003). Moreover, ENGEBRETSON and MOYER (2003) have demonstrated that the T-RFLP technique is very useful for estimating the diversity of communities characterized by low or medium diversity, but not for complex microbial populations.

On carrying out RFLP studies with complete bacterial DNA (whole genome), a huge quantity of fragments is generated. Thus, the differences in the fragment lengths cannot be visualized directly in the gel, since innumerous fragments resulting from the treatment with the restriction enzyme produce a continuous "drag" effect in the gel.

In order to detect the RFLP markers, the gel fragments are transferred to a nylon or nitrocellulose membrane through a process known as Southern blot. The polymorphic fragments can be visualized through hybridization with DNA probes which have sequences homologous to the DNA immobilized on the membrane (FERREIRA and GRATTAPAGLIA 1998). This method has greater resolution when the target gene is present in multiple copies and is highly conserved (MELO et al. 2002).

The T-RFLP technique has been shown to be a useful molecular tool and is constantly being statistically improved. Recent studies have described methods to improve the statistical analysis of the T-RFLP profiles, enabling a better interpretation of the data generated and allowing researchers to select more appropriate methodological and statistical options for each particular study. These new methodologies include the introduction of procedures aimed at distinguishing between signals and noise, the alignment of the T-RFLP peaks, and the use of multivariate statistical methods able to detect changes in the microbial communities due to specific reasons (SCHUTTE et al. 2008).

The T-RFLP technique can also be combined with real-time PCR, a molecular method known as real-time T-RFLP. This approach was proposed by YU et al. (2005) and can be applied to the simultaneous determination of the diversity and abundance of the microorganisms present in a complex microbial community. Validated using a model microbial community containing three specific microbial lineages, it has been observed that the real-time T-RFLP technique is an efficient molecular tool for greater discrimination of microbial communities in various natural or engineered systems, providing a means to carry out quantitative molecular fingerprinting.

6.2.6.2 Amplified Fragment Length Polymorphism (AFLP)

The AFLP method is a combination of PCR and RFLP based on restriction pattern analysis (RFLP) followed by amplification via PCR of DNA fragments selected from the restriction fragments obtained. The restriction is carried out using two enzymes which produce DNA fragments containing different ends. Short oligonucleotides (adaptors) are coupled to these ends of known sequences, which will serve as a template for the PCR. The selective amplification is carried out using two primers which are complementary to the adaptors, containing some bases complementary to the restriction ends (REIS JR et al. 2002).

6.2.6.3 Single-Stranded Conformation Polymorphism (SSCP)

The single-stranded conformation polymorphism (SSCP) technique is used to detect small alterations in the DNA sequence or mutations (ORITA et al. 1989). Under non-denatured conditions, DNA has a folded structure which is determined by its nucleotide sequence. Some changes in this sequence, for instance, simple nucleotide substitutions, can alter this structure and, consequently, its electrophoretic mobility in the gel. A change in the mobility can be identified in the shape of the profile generated by the single-stranded bands which stop at distinct positions in the gel, depending on the secondary structure assumed by the molecule according to its folding. Thus, the sensitivity of the SSCP technique is dependent on the way in which the mutations affect the folding of the molecule and how the folding affects the electrophoretic mobility of the mutant sequence (ROSADO and DUARTE 2002). This technique was first used to study microbial communities by LEE et al. (1996), for the analysis of complex environmental communities.

6.2.6.4 Random Amplification of Polymorphic DNA (RAPD)

The RAPD method requires small quantities of DNA and is able to reveal a high level of polymorphism. It is a rapid technique which is easily automated (FUNGARO and VIEIRA 1998). This method is considered to be simpler and cheaper than RFLP and is based on the amplification of nonspecific fragments of DNA. Primers are used as arbitrary nucleotide sequences, which generate amplification products via PCR which are analyzed. The ability to amplify specific fragments can be affected by the insertion or deletion of large fragments between the recognition regions of the primers, as well as by the quality of the DNA and other factors related to the PCR reaction (ELLSWORTH et al. 1993). In general, primers with around 10 bp are used in the RAPD method, with annealing at low temperatures, and the variations regarding the amplification of regions close to the introns can be studied with semi-random primers (DOWLING et al. 1996).

Since they are generated arbitrarily, the primer sequences allow the observation of the RAPD profiles with various amplification products, due to the presence of several sites homologous to these primers found at various points in the genome (FUNGARO and VIEIRA 1998). The variations in the DNA sequence of the fragments can be evaluated mainly regarding the variations in the number, length, and conformation. Although the analysis of fragments has a lower resolution compared with nucleotide sequences, in many cases analysis by RAPD has been used as an efficient and economically viable technique for the analysis of a large number of microbial populations (MELO et al. 2002). Molecular phenotypes, generated by RAPD, can be used to differentiate at various taxonomic levels and can even discriminate at the intraspecies level (REIS JR et al. 2002).

It should be mentioned that the RAPD technique is most commonly used for isolates in pure cultures. However, there are some studies reporting the use of RAPD-PCR along with 16S rRNA PCR combined with DGGE to monitor changes in the structure of activated sludge microbial communities during adaptation to the influent wastewater (Li et al. 2010).

6.2.6.5 Ribosomal Intergenic Spacer Analysis (RISA)

Ribosomal intergenic spacer analysis (RISA) was developed by Borneman and Triplett (1997) and was firstly applied to the study of microbial diversity in soils.

Identification of microorganisms is based on the 16S rRNA gene; other DNA regions, such as those which carry information for specific genes of interest or even specific intergenic regions, can also be used. The spacing between the 23S rRNA (large subunit of the rRNA in prokaryotes) and 16S rRNA (small subunit of the RNA in prokaryotes) genes is used as a target of PCR in RISA. This region is highly variable both in terms of its length, extending from 50 bp to over 1.5 kb, and also its nucleotide sequences.

In the RISA technique, the observed polymorphism is related to the degree of heterogeneity. The amplification products are separated in the polyacrylamide gel based on their different lengths and visualized after staining with silver. This tool has been used successfully to obtain fingerprints of microbial communities, where each band corresponds to at least one microorganism (DORIGO et al. 2005).

The RISA technique offers higher resolution than analysis of the 16S rRNA gene, considering that the spacer region is not as evolutionarily conserved as the ribosomal genes. Differences in the length of this sequence and in the base composition allow a fast monitoring method and also serve as a tool to distinguish between microorganisms at the subspecies level (JENSEN et al. 1993).

FISHER and TRIPLETT (1999) developed an automated version of the RISA technique, called ARISA, in order to access the microbial diversity in a faster and more efficient manner. The amplification by PCR of the 16S–23S region is carried out using a primer labeled by fluorescence, which enables the detection of amplicons by automated capillary electrophoresis. The total number of different fluorescence peaks obtained with the ARISA technique for a certain sample is taken as an estimate of the species diversity, and the lengths of the fragments can be compared with those present in, for instance, the GenBank database.

Despite the interesting features of the RISA technique, it has been scarcely used for assessing the microbial community of wastewater treatment systems. CIESIELSKI et al. (2013) employed RISA to monitor changes in the community of methanogenic Archaea in an anaerobic digester used for sludge treatment. Different variants of RISA (standard and automated) were tested. All provided good results. The automated approach enabled rapid analysis of the archaeal community, whereas the conventional was more favorable for identification of microorganisms.

6.2.6.6 Phospholipid Ester-Linked Fatty Acid (PLFA) Analysis

The PLFA method reveals the presence of certain microbial groups through the detection of their phospholipid ester-linked fatty acid molecules, thus enabling the identification without depending on genes and on their amplification (ROONEY-VARGA et al. 1998).

Although there are some previous works reporting the use of PLFA analysis for monitoring the microbial community structure of sediments (Salomonová et al. 2003), and freshwater reservoirs (SUSHCHIK et al. 2003), this technique has been little used to investigate microbial communities of biological wastewater treatment reactors. CHANG et al. (2011) investigated the effects of intermittent aeration on the bacterial community profile of a submerged membrane bioreactor. They observed the dominance of both monounsaturated and saturated PLFA types in activated sludge biomass, followed by branched PLFAs.

6.2.6.7 DNA Microarray

It is currently known that the data obtained through the sequencing of DNA molecules, although relevant, are limited, since it is necessary to investigate both the processes used for the transcription of the information contained in the sequences obtained and their products, i.e., the proteins. In this context, the technological strategy called DNA microarray has revolutionized the capacity to collect information in the area of functional genomics.

Representing one of the most recent advances in molecular methods, DNA microarray (also known as DNA microchip) has great potential for the elucidation of a great variety of aspects related to the microbial ecology of different environments. These aspects include the identification of variations in the structure of a microbial community present in different samples, the identification of phylogenetic groups which can be active or non-active during a certain period, or even the identification of differences between lineages isolated from different environments. In addition, this technique has boosted research on the functional genomics of different organisms (OGRAM 2000; GUSCHIN et al. 1997).

The most common use of DNA microarray is for the determination of gene expression, aimed at measuring the levels of expression of different transcripts simultaneously, which are detected by hybridization. In other words, DNA microarrays allow to obtain the complete profile of the genes which are expressed in any type of cell or even in different samples offering, in this case, important markers for the evaluation of the environment from which the samples were obtained. The capacity to determine the differential expression of thousands of genes in a single experiment greatly contributes to the increase in the attractiveness of this technique (CHEUNG et al. 1999).

The methodology allows, simultaneously, the hybridization with large gene-specific probes and the detection of the phylogenetic or metabolic characteristics. In addition, it enables the hybridization with probes specific for DNA regions which encode enzymes, providing important information with regard to the degradation characteristics of the microbial communities under study (DENNIS et al. 2003).

Microchips consist of predefined arrays of DNA molecules (fragments of genomic DNA, cDNAs, or oligonucleotides) prepared with oligonucleotide probes immobilized in an ordered manner and in specific areas (probe cells) in a polyacrylamide gel matrix attached to a solid surface, generally glass plates (microscope slides) coated with compounds which provide a positive charge (YERSHOV et al. 1996). Also, they can be prepared in positively charged nylon membranes. For use in the microchips, the oligonucleotides synthesized must be purified by gel electrophoresis or high-performance liquid chromatography. This requirement increases the control of the stringency quality, which ensures a high specificity.

In a single microchip, thousands of different oligonucleotides can be immobilized, which allows the simultaneous detection of a great variety of microorganisms in a sample. In a single chip almost all of the genes of the genome of a series of organisms can be present. Furthermore, a microchip, with washing using distilled water, can be used up to 30 times without a reduction in the hybridization signal (GUSCHIN et al. 1997).

The use of microarrays is associated with the detection and quantification of nucleic acids originating from biological samples, which are able to be hybridized, through the complementarity of bases, with the DNA fixed in the array. This in turn presents sequences similar to the genes of interest and complementary to the mRNA or complementary DNA (cDNA). The DNA or RNA samples are labeled with fluorochromes (cyanine 3, Cy3, or cyanine 5, Cy5) when glass microarrays are used and with isotope 33-P when the microarrays are produced with a nylon membrane. In both cases, a hybridization image needs to be generated. This is obtained using a laser reader (scanner) for fluorochromes or using phosphorus readers for the isotope 33-P (DALMA-WEISZHAUSZ et al. 2006; CHEUNG et al. 1999).

A typical experiment consists of comparing the levels of gene expression for two different test conditions, such as case-control study, pre- and posttreatment, or with or without a certain experimental manipulation. More specifically, the procedure using a microarray platform which is most commonly used internationally (i.e., GeneChip®, Affymetrix Inc., Santa Clara, CA, USA) includes the conversion of the total RNA obtained from the biological sample into double-stranded cDNA through a reverse transcriptase reaction. Subsequently, the cDNA serves as a template for an in vitro transcription reaction in the presence of oligonucleotides labeled with fluorophores, resulting in labeled complementary RNA (cRNA). These labeled molecules are hybridized on the microarray slide and may or may not bind to the array probes, depending on whether the sequences are complementary or not. The hybridized slide is then stained and placed in a scanner connected to software which enables the data analysis and the quantification of the fluorescence intensity at each point. The signal generated represents the binding of the cRNA of the sample to the probe in the array and is proportional to the abundance of cRNA present, up to a certain concentration of transcripts. The quantification of the signal allows the expression of thousands of genes to be compared under different experimental conditions (LOCKHART et al. 1996; CHEUNG et al. 1999). Figure 6.18 shows a simplified scheme of a DNA microarray.

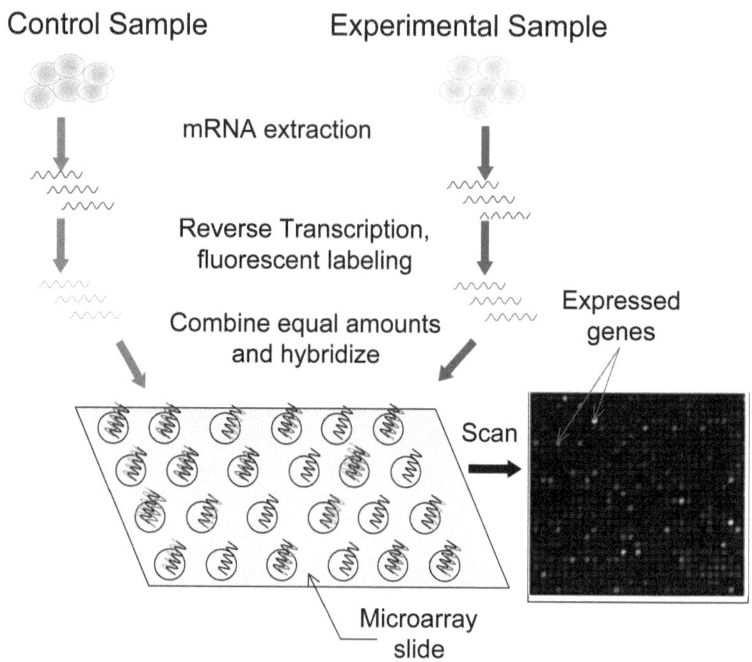

Fig. 6.18 General representation of the DNA microarray technique. The synthetic DNA probes are attached to different spots of the microarray slide and can hybridize to complementary sequences, enabling the measurement of the gene expression (adapted from http://bitesizebio.com/7206)

For comparison purposes, Table 6.11 summarizes the main characteristics of some alternative and complementary molecular methods addressed in Sect. 6.2.6. It should be noted that these methods are viable options for the study of microbial diversity, although their current state of development combined with the limited availability of references in the area of biological wastewater treatment makes them less commonly used compared with more traditional methods, such as DGGE and FISH.

6.2.7 *Next-Generation High-Throughput Sequencing Methods*

Research on biological wastewater treatment continuously grows in the world given the importance of this subject in the environmental protection context. A more comprehensive understanding of the mechanisms underlying the biological treatment process often requires knowledge of the microbial community responsible for the removal of the main classes of pollutants. For a complete overview of the ecology of the degradation processes mediated by microorganisms, several samples collected from the bioreactors should be analyzed and their identity revealed by sequencing methods.

Table 6.11 Main characteristics of some alternative and complementary methods used to study microbial diversity (adapted from SANZ and KÖCHLING 2007)

Method	Description	Advantages	Disadvantages
T-RFLP	Monitoring method which takes into consideration the 16S rRNA sequence, based on differences in the restriction behavior, (polymorphism)	– Relatively simple procedures – Makes use of intensity of fluorescence signal, allows an approx. Estimate of the species abundance	– Heterogeneity in the fragment length makes the phylogenetic analysis less reliable – Has no advantages compared with DGGE, which is the most commonly used method
RISA	Phylogenetic profiling based on the intergenic region between the 23S and 16S rRNA sequences	High sensitivity, detecting even subspecies	Small database for comparative analysis in the comparison of 16S rRNA sequences
PLFA	Profile of microorganisms obtained via characteristic fatty acid content	– Molecular characterization of the microorganisms without the involvement of genes – Complementary information for tests based on 16S	– Not considered a good option when used as the only standard method – Need for a chromatograph/mass spectrometer for the identification
DNA microarray	Hybridization method for several samples	– Appropriate for analysis of large number of samples – Parallel analysis of different parameters	– High cost of equipment – Difficult procedure to perform

Recent advances in science and technology are leading to a revision and reorientation of the molecular methodologies, allowing old and current issues to be approached under a new perspective. The greater availability of molecular approaches allows the generation of increasingly relevant data on microbial communities. This not only requires more sophisticated analysis techniques but has also contributed to an explosion in the development of promising methodologies to describe the DNA sequences of microbial cells or the microbial diversity of a certain environment in a more rapid and efficient manner.

The classic Sanger sequencing method (described in Sect. 6.2.4) was the dominant approach for DNA sequencing for the last 30 years. However, the technological advances observed in distinct areas and the need for obtaining fast and reliable sequencing results for multiple samples at limited cost and time pushed the development of many advanced and high-throughput sequencing methods by the research community (Voelkerding et al. 2009). A great variety of these technologies emerged in the 1900s and became available in the 2000s. The commercial launch of the massively parallel sequencing platforms was a breakthrough in DNA sequencing, which revolutionized the field and established a new era of genomic analysis (Tan et al. 2015).

Also referred to as "next-generation sequencing" or "second-generation sequencing," the high-throughput sequencing approaches allow obtaining massive information regarding the identity of specific microorganisms and microbial communities living in a wide range of biological systems, such as those employed for wastewater treatment. These methods make use of miniaturized and parallelized platforms which differ from each other in many aspects, such as sequencing mechanism and configuration. However, they have in common the concept of massively parallel sequencing of either clonally PCR-amplified products or single DNA molecules which are spatially separated in a flow cell. Such design is different from the classic Sanger sequencing method, based on electrophoretic separation of chain-termination products generated in individual sequencing reactions (Voelkerding et al. 2009; Bayés et al. 2012; Tan et al. 2015).

The recent sequencing technologies enable quick generation of great amount of data at a relatively lower cost compared to the Sanger method. Whereas the cost has been decreased to less than 4–0.1%, time has been reduced by a factor of 100–1000 based on daily throughput (Kircher and Kelso 2010).

The principle of the recently developed methods is based on the concept of sequencing by synthesis, whereby millions of relatively short reads are generated from amplified single DNA fragments using a repetition of cycles of nucleotide extensions mediated by polymerase or ligase (Voelkerding et al. 2009; Bayés et al. 2012). The procedure starts by placing single DNA molecules into individual reaction vessels (sequencing library preparation by random fragmentation of DNA) where amplification takes place either in liquid (emulsion PCR) or solid phase (bridge PCR on a planar solid surface) (Diehl et al. 2006; Bentley et al. 2008; Shendure and Hanlee 2008). As the DNA templates are clonally amplified, millions of DNA clones are sequenced in a microfluidic device coupled with a high-resolution imaging system. Data acquisition is performed by imaging of the fluorescently labeled nucleotides incorporated by the corresponding enzyme at each cycle. Finally, the images are converted into sequence reads (Shendure and Hanlee 2008; Morozova and Marra 2008; Bayés et al. 2012).

Given the massively parallel sequencing process, the next-generation methods are capable of generating hundreds of mega- to gigabases of nucleotide sequence data in a single run of the equipment, depending on the platform (Voelkerding et al. 2009). In comparison with old times when the limiting factor was the generation of sequencing data, the challenge in current days is the storage, handling, analysis, and interpretation of the great amount of data obtained by means of the high-throughput sequencing platforms. Such technologies provide hundreds of millions of short DNA sequence reads which should be properly handled to render reliable and high-quality results in a relatively short time span. In this context, specialized bioinformatics tools and knowledge are mandatory. This makes the analysis of the data the main obstacle to be faced in sequencing projects.

Apart from the computer resources a hardware infrastructure (e.g., computing power, disk storage capacity, and memory) needed to store and process all sequencing datasets, it is necessary to count on knowledgeable bioinformaticists capable of analyzing the large amount of sequence data released by the next-generation sequencing methods (Bayés et al. 2012; Rodríguez-Ezpeleta et al. 2012). Moreover, the need for specific software packages for data analysis makes this task even more complex, as most programs rely on considerable number of parameters which are dependent on the biological system under study, type of data, and experimental design (Rodríguez-Ezpeleta et al. 2012).

A great variety of bioinformatics tools available for analysis and interpretation of next-generation sequence data can be found in the review carried out by (Shendure and Hanlee 2008). Their functions fall within several categories, such as (1) alignment of the obtained raw sequence reads (often with 50–400 base pairs) to known (reference) microbial genomes, (2) base calling and variant calling, (3) de novo assembly from paired or unpaired reads piecing together the resulting sequence reads into longer continuous stretches of sequence (contigs), and (4) filtering, genome browsing, and annotation (Shendure and Hanlee 2008; Dolled-Filhart et al. 2013; Ekblom and Wolf 2014).

While Basic Local Alignment Search Tool (BLAST) is a largely employed alignment solution for long reads generated by conventional sequencing, the analysis of large sets of short reads provided by next-generation methods requires rapid and efficient alignment algorithms and tools (Shendure and Hanlee 2008). Moreover, the assembly of short sequencing read is also a challenge. The use of mate-paired sequences (also designated as mate pairs) can be used to facilitate de novo assembly and obtain a more complete picture of the genome (Wetzel et al. 2011).

The next-generation sequencing methods consist of a rapidly evolving field. Many technologies are commercially available, while other new methodologies and instruments are constantly being released on the market. Currently, the most used next-generation sequencers are 454 pyrosequencing (Roche Applied Science, Basel, Switzerland), Illumina/Solexa genome sequencer (Illumina, San Diego, CA, USA), SOLiD (Applied Biosystems, Foster City, CA, USA), and the Single Molecule Real-Time platform (Pacific Biosciences, Menlo Park, CA, USA). Table 6.12 presents the most well-known platforms developed in recent years and their particular features. It should be emphasized that the data here provided quickly changes and becomes out of date as a result of the advancement in the sequencing technologies. In this sense, it is recommended that the reader should always consult updated literature information. A detailed description of the high-throughput sequencing analysis and the principle behind the main methods and equipment used for this purpose can be found elsewhere (Shendure and Hanlee 2008; Bayés et al. 2012; Rodríguez-Ezpeleta et al. 2012).

Some of the next-generation platforms (HiSeq 2000 and 5500x1 SOLiD) generate short length sequences (much shorter than conventional sequencing methods) (Shendure and Hanlee 2008), which may limit bacterial identification. Furthermore, short sequence reads provide less phylogenetic information compared to full-length 16S rRNA gene sequences. Therefore, the identification of bacteria using these

Table 6.12 Examples of next-generation sequencing methods and their particular features (adapted from Morozova and Marra 2008; Voelkerding et al. 2009; Kircher and Kelso 2010; Bayés et al. 2012)

	Characteristics								
Platform/company name/website	Amplification approach—library preparation)	Sequencing mechanism	Reads per run	Read length	Output data per run	Time per run	Accuracy	Advantages	Disadvantages
ABI3730x1 (Applied Biosystems)	PCR/cloning	Automated Sanger method (dideoxy chain termination)	1536	400–900 bp	1.3 Mb	20 min–3 h	99.999%	High quality, long read length	High cost, low throughput
Genome Sequencer (GS) FLX (454 Life Sciences, Roche) http://www.454.com/	Clonal—emulsion PCR	Pyrosequencing	1 M	400–700 bp	500–700 Mb	10–24 h	99.9%	Read length, fast	Error rate with polybase more than 6, high cost, low throughput
HiSeq 2000 (Illumina, Solexa) http://www.illumina.com/	Solid-phase amplification (bridge PCR)	Sequencing by synthesis using reversible dye terminators	1000–3000 M[a]	50–100 bp	600 Gb[a]	3–10 days	98%	High throughputs	Short read assembly
5500x1 SOLiD (Life Technologies) http://www.appliedbiosystems.com/	Clonal—emulsion PCR	Massively parallel sequencing by ligation and two-base coding	1200–2400 M[a]	35–50 bp	120 Gb–300 Gb[a]	4–7 days	99.94%[b]	Accuracy	Short read assembly
PacBio RS (Pacific Biosciences) http://www.pacb.com/	No PCR amplification (single molecule)	Phospholinked Fluorescent Nucleotides	100 GB	30 pb	NA	<1	–		

[a]From two flow cells or slides
[b]Raw data

sequencers has been mainly accomplished based on 16S rRNA gene hypervariable regions, which are highly informative and may enable analysis of lower taxonomic levels (e.g., genera and species) in spite of the short length of the sequence (Nasidze et al. 2009).

The new technologies need to satisfy some main objectives, such as a significant increase in the yield without a loss in the quality of the data produced by the technique used (Rogers and Venter 2005). Hence, the accuracy of data is also of great concern in the new sequencing platforms, which are often assumed to be less accurate than the Sanger-based sequencing technologies. However, it should be taken into account that the next-generation sequencers are still evolving, and great improvements in such platforms are expected over time (Shendure and Hanlee 2008).

In general, the choice of the most appropriate sequencing platform is project specific. Sometimes, combination of different technologies can be technically and economically advantageous. This stimulates the establishment of companies which provide sequencing services on demand. Nevertheless, despite the third-party services, the construction of libraries and analysis of a great amount of data will still be dependent on the research laboratory personnel (Morozova and Marra 2008).

The application of high-throughput sequencing methods has a significant impact on the study of the microbial diversity of wastewater treatment bioreactors. The use of these powerful sequencing tools in order to get the most information possible about the microbial diversity of these biological systems is becoming an increasingly routine activity (Hu et al. 2012; Zhao et al. 2015; Ng et al. 2015). In general, genomic DNA is extracted from the reactor samples at different operating conditions, converted to a next-generation sequencing library and finally sequenced. The sequencing results obtained are then aligned to known microbial reference genomes. Besides the qualitative genomic information, the relative abundance of the sequence reads may be used to obtain quantitative data regarding individual microbial species.

With a large amount of data available, the challenge will be the extraction of useful insights from the DNA sequencing results to better understand the biological processes taking place in the engineered wastewater treatment reactors. The combination of metagenomics, metatranscriptomics, and other approaches to explore microbial interactions in biological wastewater treatment processes will be crucially important for obtaining a wealth of information regarding the functioning of such microbial systems.

6.3 Application of Molecular Biology Techniques to Wastewater Treatment

6.3.1 Introduction

Several studies reported in the literature regarding the application of molecular biology techniques to the study of the microbial diversity of wastewater treatment systems, on both laboratory and full scale, are described in this section. The main aspects of each study will be highlighted, including the most important results obtained in each case. For more details about a specific study, the reader is referred to the corresponding full version of the published articles cited here. Bearing in mind that the overall objective of this book is not restricted to molecular tools, in many cases short discussions involving the biological treatment process are provided.

6.3.2 *Application of Molecular Biology Techniques in Studies to Characterize the Microbial Communities of Wastewater Treatment Systems*

BOND et al. (1995) compared the microbial communities of two activated sludge sequencing batch reactors (SBRs): one with (SBR_1) and one without (SBR_2) phosphate removal. The reactors were fed with wastewater containing on average a COD of 370 mg/L, while the average soluble phosphate concentration was 18 mg/L. Immediately after the reactor start-up, the feed of the SBR_1 was supplemented with sodium acetate to increase the COD by 100 mg/L. This procedure was carried out to favor the development of the microorganisms responsible for the bio-P removal only in SBR_1, thus obtaining two systems with different phosphate removal capacities.

After 3 weeks of operation, it was observed that the phosphate removal in SBR_1 was greater than that in SBR_2. The values for the release of phosphate during the anaerobic phase in SBR_1 and SBR_2 were 35.4 and 5.2 mg/L, respectively. The effluent phosphate concentrations in SBR_1 and SBR_2 were 1.5 and 12.0 mg/L, respectively. These data suggested notable differences in the microbial communities present in the two reactors, particularly with respect to the biochemical transformation of phosphate (BOND et al. 1995).

Clone libraries with partial sequences of the 16S rRNA gene of the bacterial populations of the sludge were constructed. In order to identify the differences which could indicate groups or genera which are important in the biological removal of phosphorus (a process known as enhanced biological phosphorous removal, EBPR), the structures of the communities were determined through phylogenetic analysis of the partial 97 and 92 sequences of the clones of SBR_1 and SBR_2, respectively. For both reactors, the predominant bacterial group with which the clones were affiliated was the subclass β-*Proteobacteria* (28%). Bacteria belonging to this subclass have been observed by FISH as being dominant in activated sludge communities (WAGNER et al. 1993, 1994). Given their numerical dominance, it is probable that the representatives of the β subclass play an important role in the degradation of organic matter and nutrients and the formation of the floc structure (BOND et al. 1995).

Other main groups were those belonging to the subclass α-*Proteobacteria*, the phylum *Planctomycetes*, and the group related to *Flexibacter-Cytophaga-Bacteroides* (particularly in SBR_2). In addition, several clone groups not affiliated with known bacterial groups were identified in the clone libraries. *Acinetobacter* spp., reported in the past as one of the most important microorganisms in phosphate removal from activated sludge according to the methods dependent on cultivation

(WENTZEL et al. 1988), were poorly represented in the clone sequences in both libraries (approximately 2%). Differences in the community structure were observed on comparing the two reactors, in particular in relation to the *Rhodocyclus* group within the β subclass, which was extensively present in the microbial community of SBR_1. This finding may be related to the fact that this microbial group, in particular, is associated with an important function in the removal of phosphate. The same authors also report that the high species diversity in SBRs and the limited number of clones analyzed make it difficult to identify one or more bacterial species which may play a crucial role in phosphate removal (BOND et al. 1995).

SCHRAMM et al. (1998) investigated, using the FISH technique coupled with confocal laser scanning microscopy (CLSM), the distribution of ammonium-oxidizing lithoautotrophic bacteria belonging to the genera *Nitrosomonas* and *Nitrobacter* in the nitrifying biofilm of a trickling filter employed in a water recirculation system for aquiculture. The ammonium concentration varied from 0.3 to 7 mM. The hybridization of fixed cells was carried out with 16S rRNA oligonucleotide probes. It was observed that the ammonium-oxidizing bacteria (AOB) formed a dense layer of cellular aggregates in the upper part of the biofilm (aerobic part), while the nitrite-oxidizing bacteria (NOB) formed less dense colonies with lower numbers compared with the AOB. The CLSM revealed the coexistence of AOB and NOB in neighboring regions, which is related to the sequential metabolism of ammonia to nitrate, via nitrite, promoted by the joint action of these two microbial groups. Due to this spatial arrangement, the diffusion pathway of *Nitrosomonas* (AOB) to *Nitrobacter* (NOB) is very short and facilitates the efficient transfer of the intermediate nitrite. Neither species is restricted to the oxic region of the biofilm, both being detected, although in much lower numbers, in anoxic layers. Some NOB colonies, in particular, have been found in upper anoxic layers, at a depth of 100–200 μm, as well as occasionally at the bottom of the biofilm.

SNAIDR et al. (1997) investigated the structure of bacterial communities at a large wastewater treatment plant (activated sludge process) based on the 16S rRNA gene. Almost complete genes which encode the 16S rRNA subunit (rDNA) were amplified by PCR and subsequently cloned. The clones were classified by dot-blot hybridization with group-specific oligonucleotide probes. The phylogenetic affiliations of the clones were compared with the results obtained with an original sample by means of in situ hybridization, in which specific rRNA oligonucleotide probes were used. It was observed that the data were in agreement. A total of 25 16S rDNA clones were completely sequenced, 11 were almost completely sequenced (>80%) and 27 were partially sequenced.

Comparative analysis of the sequences was carried out and most of the clones examined (35) were affiliated with the β subclass of the class *Proteobacteria*. The γ and α subclasses were represented by 13 and 4 clones, respectively. Eight clones were affiliated with the ε-subclass of *Proteobacteria*, and five clones were grouped with Gram-positive bacteria characterized by DNA with a low GC content. The 16S rRNA gene of two clones showed similarity with the 16S rRNA genes of the members of the phyla *Chlamydiae* and *Planctomyces*. Oligonucleotide probes were designed and used for the enumeration of the respective bacteria. Interestingly, pathogenic representatives of the genus *Arcobacter* were present in significant numbers (4%) in the activated sludge sample examined.

BURRELL et al. (1998), in order to identify the nitrite-oxidizing bacteria found in wastewater treatment plants, operated a sequencing batch reactor (SBR) fed with a solution of inorganic salts, in which nitrite was the only source of energy. With the use of microscopy, cultivation methods, and molecular techniques, the biomass in the nitrite-oxidizing system was investigated after a period of 6 months of reactor operation. Molecular biology methods have also been used to characterize the sludge used to inoculate the SBR, and this was then compared with the biomass present inside the reactor. It was observed that the SBR biomass was comprised of a complex community with a high diversity containing Gram-negative and Gram-positive bacteria.

The methods based on cultivation in culture medium (micromanipulation, sample dilution, and plating) identified only heterotrophic bacteria, and no autotrophic microorganisms were isolated. 16S rDNA clone libraries of two microbial communities revealed that the inoculation sludge presented a complex microbial community dominated by *Proteobacteria* (29% β subclass and 18% γ subclass) and a large quantity of high GC Gram-positive bacteria (10%). In relation to the inserts of 77 clones of the library corresponding to the inoculum sludge, it was observed that most of the clone sequences were grouped into the phylum *Proteobacteria* (56%), while 4% (three clones) were grouped into the phylum *Nitrospira*, specifically *Nitrospira moscoviensis*. On the other hand, *Nitrobacter* was not identified. The inserts of 102 clones of the clone library corresponding to the biomass present in the reactor, examined through analysis with restriction enzymes, were classified into 13 different operational taxonomic units (OTU). A total of 90 clones (88%) were grouped into one particular OTU, while the remaining 12 OTUs were composed of individual clones, each one corresponding to 1% of the total number of clones. The biomass inside the SBR had a predominance of bacteria related to *Nitrospira moscoviensis*, and two clone sequences were similar to that of the genus *Nitrobacter*. Based on the results obtained, the authors concluded that the "unknown" nitrite-oxidizing bacteria (referred to in this way since they were not easily isolated through conventional cultivation techniques) present in the activated sludge systems belong to the phylum *Nitrospira* (BURRELL et al. 1998).

OKABE et al. (1999) investigated the presence and distribution of ammonium-oxidizing (AOB) and nitrite-oxidizing (NOB) bacteria in the biofilm of a rotating disk bioreactor fed with both a domestic wastewater and a synthetic medium (autotrophic nitrifying biofilm). The authors employed microsensors and FISH with 16S rRNA-specific oligonucleotide probes. The combination of these techniques allowed the microbial activity to be directly related in situ to the occurrence of populations of nitrifying bacteria. FISH analysis revealed that bacteria belonging to the genus *Nitrosomonas* represented the numerically dominant AOB in both biofilms. However, bacteria belonging to the genus *Nitrobacter* were not detected. *Nitrospira* was the main genus representative of the NOB in both biofilms, mainly found in the intermediate part of the biofilm. *Cells of Nitrospira* lineages were aggregated in irregular shapes comprised of microcolonies which were grouped around the AOB.

While most of the AOB were present throughout the biofilm, the NOB were restricted to the zones where nitrite oxidation would actively occur, located in the internal parts of the biofilm. Through measurements taken with microelectrodes, it was possible to observe that the active zone of ammonium oxidation was located in the external part of the biofilm, while the active zone of nitrite oxidation was located directly below the ammonium oxidation zone, confirming the observations obtained by FISH.

AOI et al. (2000) applied the FISH technique with 16S rRNA-specific oligonucleotide probes to investigate the microbial ecology of nitrifying bacteria in several types of wastewater treatment processes. They also evaluated the dynamic response of microorganisms in different types of biofilms and suspended flocs, as well as the distribution of ammonium-oxidizing bacteria (AOB) and other heterotrophic bacteria. Samples from three types of nitrifying reactors were analyzed: (1) biofilm from a fluidized bed reactor fed with an influent rich in ammonia without the addition of organic compounds or micronutrients, (2) activated sludge from a reactor in which the inorganic influent contained micronutrients and had a low rate of ammonia oxidation, and (3) the biofilm of another fluidized bed reactor fed with organic wastewater with a high concentration of ammonia and organic compounds.

The same authors observed that the nitrifying bacteria exhibited different forms of organization according to the operational conditions imposed, particularly in relation to the composition and concentration of the substrate. The AOB were found to be dominant in inorganic wastewaters rich in ammonium. In the case of organic wastewater, both heterotrophic bacteria and AOB were distributed in different positions in the biofilm. The samples of reactor 1 revealed a predominance of bacteria of the genus *Nitrosomonas*, as well as *Nitrosococcus mobilis*. In reactor 2, members of the genus *Nitrosomonas* and *N. mobilis* were detected as well as other AOB such as *Nitrosospira*, *Nitrosovibrio*, and *Nitrosolobus*. These microbial groups only appeared in reactor 2. The authors argue that this result is probably related to the lower ammonia oxidation rates (0.3 kgN-NH_4^+/(m^3 day)) observed in this system compared with the other two reactors (1 and 3). *Nitrosomonas* and *N. mobilis* exhibited greater activity in the presence of high ammonia concentrations, while for *Nitrosospira*, *Nitrosovibrio*, and *Nitrosolobus*, the activity was greater when the ammonia concentration was low, indicating the strong dependence of the microbial ecology on the operation conditions. This makes the observation and control of ammonium-oxidizing microbial groups of great importance in the maintenance of stable nitrifying activity (AOI et al. 2000).

In reactor 3, fed with organic influent, heterotrophic bacteria were detected in external region of the biofilm, whose thickness was 100 μm. AOB were also found, particularly in the interior of the biofilm. According to the authors, the AOB and heterotrophic bacteria tend to be located in different positions in the biofilm due to the fact that the heterotrophic population has a higher growth rate than the autotrophic population, and thus, initially, these bacteria occupy the external region of the biofilm in which the dissolved oxygen (DO) and substrate concentrations are high enough to allow their growth. On the other hand, the AOB are able to spread both individually and in the form of groups within the interior of the biofilm due to their low growth rate, even though in this region the environment is unfavorable for growth in terms of the DO concentration. As in the case of reactor 1, *Nitrosomonas* was the predominant genus of AOB in reactor 3.

AOI et al. (2000) also evaluated the dynamic response of microbial communities to changes in the influent composition, particularly in relation to a gradual reduction in the TOC/N-NH_4^+ ratio. It was observed that with a decrease in this ratio, the AOB, which were previously only located in the interior of the biofilm, extended their area of growth to the more external part, which had previously been occupied exclusively by heterotrophic bacteria. The decrease in the concentration of organic compounds limited the growth of heterotrophic microorganisms, which consequently allowed the development of the nitrifying consortium (particularly AOB) in the external part of the biofilm. When the TOC/N-NH_4^+ ratio was 0 (no organic carbon in the influent), the AOB were still present in large numbers throughout the biofilm, while the heterotrophic bacteria had practically disappeared. After a certain period of time under this condition, the nitrifying activity increased, and the biofilm thickness, which at the beginning of this regime without organic carbon had been decreasing, remained stable. The results indicated that effective control of the microbial ecology can be obtained by controlling the operating conditions of the reactor, highlighting the possibility of obtaining dense and highly active biofilms containing a high diversity of nitrifying bacteria with a gradual reduction in the TOC/N-NH_4^+ ratio.

In order to evaluate the relation between excessive foaming in wastewater treatment plants and actinomycetes, microorganisms containing mycolic acid (family *Mycolata*) and generally considered to cause this phenomenon, DAVENPORT et al. (2000) developed a specific probe for these bacteria as well as a protocol to permeabilize these microorganisms and a method for their statistical quantification. The quantitative FISH technique was used to investigate the relation between the production of foam and the concentration of actinomycetes in an activated sludge system, completely mixed, with a volume of 20 m^3. The authors observed that the formation of foam occurred when the number of actinomycetes belonging to the family *Mycolata* exceeded a certain value, which was estimated to be around 2×10^6 cells/mL or 4×10^{12} cells/m^2.

The nitrification process can occur in different reactors with different configurations, although the way in which the nitrifying microbial communities act in the different configurations is unknown. Recent studies on ammonium-oxidizing bacteria (AOB) in wastewater treatment systems have suggested that different plants support different populations and different levels of species richness. In this context, ROWAN et al. (2003) compared the diversity and the community structure of AOB, belonging to the subclass β-*Proteobacteria*, in two reactor configurations operated in full scale: an aerated biological filter (ABF) and two trickling filters (TF) in series (primary and secondary). The reactors were fed with the same mixed wastewater, originating mainly from an industrial source (70% industrial/30% domestic), previously treated in a high-rate activated sludge system and then mixed with domestic wastewater (2:1 ratio—domestic/pretreated industrial). Different molecular techniques were employed to carry out the study. PCR was used for the amplification of the 16S rRNA gene fragments, using primers selective for AOB (β-*Proteobacteria*). The use of DGGE allowed the analysis of the fragments amplified by PCR, and the profiles obtained were analyzed to compare the dominant AOB populations in the reactors. In addition, a selection of the 16S rRNA gene fragments amplified by PCR and cloned was sequenced to determine the phylogenetic affiliation of the sequences representative of each species of dominant AOB.

The results obtained from the visual comparison of the DGGE profiles shown by the 16S rRNA fragments originating from the ABF and TF revealed distinct populations in different sections of each reactor and also differences between the reactors. The ABR was comprised of three interlinked layers, each one associated with a different biodegradation process: denitrification (anoxic), organic matter removal (oxic), and nitrification (oxic). A change in the aeration level appeared to have an effect on the bacterial population present in the reactor. Also, significant differences were observed in the results obtained by DGGE for the different TF depths and the primary and secondary TF. More bands were detected in the sample from the bottom of the primary filter and in the secondary filter compared with the top of the primary filter (ROWAN et al. 2003).

Results obtained by the same authors revealed that although both reactors (ABF and TF) were fed with the same wastewater, they harbored different bacterial populations. ABF had a lower diversity of AOB when compared with TF. The sample originating from the latter presented some bands on the DGGE gel which did not appear in the gel run with samples originating from the ABF. However, both reactors appeared to have a predominant population in common (ROWAN et al. 2003).

Clone libraries of the 16S rRNA gene of AOB were constructed from the samples of each of the two reactors, and those which had a greater diversity based on the DGGE profiles were selected. The nucleotide sequences were determined for each type of clone from the clone libraries and were compared with the GenBank database. The detailed analysis of the sequences recovered from the ABF and TF revealed that all of them derived from non-cultivated β-*Proteobacteria* or γ-*Proteobacteria*. In addition, most of the sequences (97% from the ABF, 80% from the top of the TF, and 85% from the bottom of the TF) showed the strongest similarity with the AOB (β-*Proteobacteria*). The remainder of the sequences recovered was related to *Thauera* spp. and *Dechlorimonas agitatus* (ROWAN et al. 2003).

The sequences obtained from the nitrification unit of the ABF and the top and bottom of the TF were identified as belonging to the genus *Nitrosomonas*. The clone library of the ABF had a preponderance of *Nitrosomonas mobilis*, while the TF samples showed greater diversity. The predominant clones of this system belonged to *N. mobilis* and *Nitrosomonas* spp. The sequences of *N. mobilis* were found in all of the reactors, but they were more abundant in the clone library of the ABF (90% of the sequences) than that of the TF (33% from the top of the secondary filter, 27% from the bottom of the secondary filter) (ROWAN et al. 2003).

PERSSON et al. (2002) investigated the composition of a biofilm adhered to the plastic supports (surface area of 240 m^2/m^3) of a full-scale nitrifying trickling filter (NTF) treating municipal wastewater. Fluorescence in situ hybridization (FISH) with 16S rRNA-targeted oligonucleotide probes in combination with confocal laser scanning microscopy was used to characterize the microbial community. In all of the samples taken from different depths of the NTF, it was observed that the ammonium-oxidizing bacteria (AOB), with the exception of two marine species of the genus *Nitrosococcus*, belonged to the subclass β-*Proteobacteria*, in particular the genus *Nitrosomonas*. It was also observed that none of the bacteria were detected with the probes specific for *Nitrosospira* spp. Furthermore, *Nitrosococcus mobilis* and *Nitrosomonas europaea* were not detected in any of the samples by specific probes.

The results showed a tendency for a decrease both in the proportion of AOB in relation to the total bacteria and in the total quantity of AOB in the different layers of the biofilm (30 ± 23% less in the more external layers compared with the inner layers). This total quantity of AOB also showed a progressive decrease with depth, that is, from the top to the bottom of the trickling filter. Measurements and simulations of the ammonium oxidation activity also showed a decrease in the activity with an increase in the depth of the NTF, which was generally operated with complete nitrification (PERSSON et al. 2002).

DIONISI et al. (2002), using the principle of competitive PCR, developed tests to quantify the ammonium-oxidizing bacteria (AOB) belonging to the species *Nitrosomonas oligotropha* and nitrite-oxidizing bacteria (NOB) belonging to the genus *Nitrospira*. The specificities of the primers used, which were designed for two target regions (*amoA* and 16S rRNA genes of *Nitrospira*), were verified by DNA sequencing. Both tests were optimized and applied to samples from an activated sludge system. The results showed that the AOB represented 0.0033% ± 0.0022% of the total bacterial population in the treatment plant. It was observed that the AOB belonging to the species *N. oligotropha* were not detected in the sample. *Nitrospira* spp. represented 0.39% ± 0.28% of the bacterial population present in the biomass. Competitive PCR analysis showed that the number of *Nitrospira* spp. cells in the sample collected from the treatment plant was 62 times greater than the number of *N. oligotropha* cells. Nevertheless, the results confirmed that the fraction of autotrophic nitrifying biomass present in the treatment plants was small, as also observed in some studies reported in the literature.

JURETSCHKO et al. (2002) evaluated the composition of the microbial community of a nitrifying-denitrifying activated sludge system treating domestic wastewater by means of molecular techniques targeting the 16S rRNA gene. The influent wastewater contained high ammonium concentrations (around 500 mg/L). Short-chain fatty acids (mainly acetate, but also butyrate and propionate) were also found in significant quantities in the influent. In order to carry out nitrification and denitrification in a single tank, the system was aerated intermittently (30 min of aeration and 15 min under anoxic conditions). During the sampling period, over 90% of the nitrogen compounds were converted to nitrogen gas through nitrification and denitrification.

Ninety-four clones of the 16S rRNA gene (almost complete length) were phylogenetically analyzed. Of these, 59% were affiliated with the class *Proteobacteria*, grouped in the β (29 clones)-, α (24 clones)-, and δ (2 clones)-subclasses. Fifteen clones were grouped among green non-sulfurous (GNS) bacteria, and 11 clones belonged to *Planctomycetes*, *Verrucomicrobia*, *Acidobacteria*, *Nitrospira*, *Bacteroidetes*, *Firmicutes*, and *Actinobacteria*, each represented by one to five clones. Interestingly, the greatest species richness measured as the number of operational taxonomic units (OTUs) was found to be within the subclass α-*Proteobacteria*, followed by *Planctomycetes*, subclass β-*Proteobacteria*, and GNS bacteria (JURETSCHKO et al. 2002).

The same authors also studied the composition of the microbial community of an activated sludge system, determined quantitatively using the FISH technique combined with confocal laser scanning microscopy. Thirty-six oligonucleotide probes, group, subgroup, and OTU specific, targeting the rRNA gene were used in this study. It was observed that 89% of all of the bacteria detected by FISH with probes specific for bacterial cells were related to specific microbial groups. Consistent with the data generated in the clone libraries constructed from the 16S rRNA gene, members of the subclass β-*Proteobacteria* were dominant in the microbial community and represented almost half of the biovolume of all of the bacteria detected by FISH. In relation to the β subclass, 98% of the cells were identified using genus- or OTU-specific probes, and a high abundance of commonly known denitrifying bacteria related to *Zoogloea* and *Azoarcus* was observed (JURETSCHKO et al. 2002).

In aerobic-anoxic systems, in which small additions of oxygen are supplied to the anoxic reactor to induce simultaneous nitrification and denitrification, ammonium oxidation in the anoxic zone may correspond to 30–50% of the total nitrification of the reactor, even when the dissolved oxygen concentration is, generally, below the detection limit. In order to investigate whether the nitrification efficiency in the aerobic/anoxic processes is related to the presence of specialized ammonium-oxidizing bacteria (AOB), PARK et al. (2002) analyzed the population of these bacteria in aerobic/anoxic Orbal processes and in conventional nitrogen removal processes. They carried out phylogenetic analysis based on the ammonia monooxygenase (*amoA*) gene. The T-RFLP analysis revealed that organisms such as *Nitrosospira* were mainly responsible for the oxidation of ammonium in the anoxic/aerobic Orbal reactor. However, populations of *Nitrosospira* and *Nitrosomonas* were not constant and showed seasonal variability. The cloning and comparison of sequences of fragments of the *amoA* gene showed that most of the AOB in the Orbal process belonged to the lineages *Nitrospira* spp. and *Nitrosomonas oligotropha*. The abundance of organisms such as *Nitrosospira* in aerobic/anoxic reactors is significant and quite surprising, since this group of AOB have not been often associated with nitrification in wastewater treatment plants.

CHEN et al. (2003) evaluated the dynamic response of cultures of nitrifying activated sludge, cultivated in batch regime, on increasing the chloride concentration up to 30,000 mg/L. The specific nitrification rate and the dominant nitrifying species were monitored. In one of the cultures, the chloride increase was carried out gradually, whereas in the other cases the increase was applied in the form of a pulse (stepwise increase). The species of ammonium-oxidizing (AOB) and nitrite-oxidizing (NOB) bacteria were examined employing the fluorescence in situ hybridization (FISH) technique, using 16S rRNA-specific oligonucleotide probes. The experiments, for both cultures, were divided into three different operational phases in terms of the chloride increase: Phase 1, with a duration of 70 days and an increase in the chloride concentration up to 10,000 mg/L; Phase 2, with a duration of 70 days and an increase in the chloride concentration from 10,000 to 20,000 mg/L; and Phase 3, with a duration of 60 days and an increase in the chloride concentration from 20,000 to 30,000 mg/L. In the culture submitted to a stepwise increase, the chloride concentration was fixed at a certain value in each phase, while for the culture submitted to a gradual increase the chloride concentration was increased by 2000 mg/L every 10 days.

It was observed that concentrations below 10,000 mgCl^-/L (Phase 1), in both experiments (i.e., with the gradual and stepwise chloride increase) did not affect the specific rate of nitrification of the microbial cultures. Actually low chloride concentration such as 2500 mgCl^-/L provoked an increase in the nitrification rate. On applying the FISH technique, it could be observed that the AOB groups in the culture submitted to increasing chloride concentrations included *N. marine*, *N. oligotropha*, *N. europaea*, *N. eutropha*, *N. halophila*, *N. mobilis*, *N. communis*, and *N. cryotolerans*. With the gradual or stepwise increase from 10,000 to 20,000 mgCl^-/L in Phase 2, there was a significant variation in the two cultures, with an initial drop in the specific rate of nitrification, reflecting the inhibition of AOB and NOB by the chloride. However, after a few days, the specific nitrification rate began to increase again, even with the increased salinity. This finding is associated with a change in the dominant AOB species. In fact, the species which are not resistant to salinity, such as *Nitrosomonas europaea* and *Nitrosomonas eutropha*, were replaced by bacteria which are resistant to saline conditions, such as *Nitrosococcus mobilis*. The specific nitrification rate of the culture submitted to a stepwise increase in the chloride concentration was lower than that reached by the culture submitted to a gradual increase, which implies that in the latter case the adaptation and development of AOB resistant to saline conditions were favored. It was also observed that *Nitrobacter* was the only dominant genus of the NOB when the chloride concentration was lower than 10,000 mg/L, while no NOB were detected when the chloride concentration was ≥18,000 mg/L. Finally, in Phase 3 (30,000 mgCl^-/L), the nitrification rate remained constant for the culture in which the chloride content was gradually increased and showed a slight increase in the case of the culture subjected to a stepwise increase (CHEN et al. 2003).

Making use of some molecular techniques, such as PCR, DGGE, cloning, and sequencing, TAL et al. (2003) investigated the microbial consortium in a moving bed bioreactor (MBBR) connected to a marine aquaculture system with recirculation. The authors aimed at obtaining more detailed information regarding the functioning of different bacterial species, both anaerobic and aerobic, in the nitrogen removal process.

After 4 months of operation of the MBBR submitted to high organic loading rate, some of the supports present in the reactor were transferred to an experimental system operated under aerobic conditions without the addition of organic load and with a hydraulic retention time similar to that of the original reactor. This system served as a source of supports submitted to low organic loads for later studies. In this way, the different biological processes for nitrogen transformation were carried out through short incubation procedures, in a batch regime, with supports submitted to low organic loads (originating from the abovementioned experimental system) as well as to high organic loads (originating from the original MBBR) (TAL et al. 2003).

Nitrosomonas cryotolerans and *Nitrospira marina* were identified as representatives of ammonia-oxidizing and nitrite-oxidizing bacteria, respectively. Heterotrophic bacteria were also detected, including *Pseudomonas* spp., *Aquaspirillum metamorphum*, and *Sphingomonas* spp. An interesting observation made by the authors was the presence of two PCR products with sequences which were very similar to those of two representatives of *Planctomycetes* spp. These organisms were detected in the system submitted to high organic loads, suggesting anaerobic ammonia oxidation (anammox) capacity. The sequencing of PCR products also revealed sequences very similar to those of sulfite-oxidizing microorganisms, such as *Sulfitobacter* spp., suggesting the establishment of anaerobic conditions which allowed the reduction of sulfate to sulfite in the MBBR. In this study, due to the methodology used for the microbial analysis (DGGE based on PCR products), it was not possible to determine the relative abundance of the different bacterial species adhered to the biofilm, although an estimate of the dominant species in the different tests (high or low organic load applied) was obtained based on the conversion processes (TAL et al. 2003).

TSUNEDA et al. (2005) investigated the formation of nitrifying granules in an aerobic upflow fluidized bed (AUFB) reactor applied for nitrification of synthetic wastewater containing 500 mgN/L of ammonium. The formation of spherical, pseudocubic, and elliptical granules, with diameters of around 346 μm, was observed in the bottom of the reactor after an operating period of 300 days. Sequencing of the gene fragments removed from the DGGE gels did not allow a perfect combination of the sequences obtained with those registered in the databases, suggesting that members of ammonium-oxidizing bacteria (AOB), which have not yet been characterized, were present in the granules. The phylogenetic analysis derived from the sequencing of a particular DGGE band revealed a nucleotide sequence with a strong similarity with *Nitrosococcus mobilis*. Other sequences derived from other bands have shown high similarity with *Nitrosomonas marina*. The authors suggested that the high ammonium loads applied to the AUFB reactor led to the predominance of the genus *Nitrosomonas*, which is consistent with other literature studies.

The analysis of the spatial distribution of the bacteria using the FISH technique revealed that the AOB were present at a depth of 100 μm from the granule surface. Groups of AOB presented diameters in the range of 10–20 μm. Using NOB-specific probes, it was observed that these bacteria were also located close to the granule surface. However, the number of NOB was well below that of AOB, result attributed by the authors to the greater specific nitrate production of the *Nitrobacter* (5.1–42 fmol/(cell h)) in comparison with the nitrite production rate of *Nitrosomonas* (0.9–42 fmol/(cell h)) (PROSSER 1989 *apud* TSUNEDA et al. 2005). Bacteria belonging to the genus *Nitrospira*, although frequently observed in nitrifying activated sludge systems (SCHRAMM et al. 1998), were not detected in the AUFB reactor using genus-specific FISH probes. The absence of *Nitrospira* was probably related to the high concentration of ammonium fed to the system (500 mg/L), a value much higher than those commonly reported for conventional activated sludge systems, which can reach 100 mgNH_4^+-N/L (TSUNEDA et al. 2005).

EGLI et al. (2003a) operated a rotating biological contactor (RBC) and observed high nitrogen removal in the tertiary treatment of a leachate containing a high concentration of ammonium (100–500 mgN-NH_4^+/L). This leachate had been previously treated in a preliminary RBC, in which the TOC removal of 88% had been achieved. In addition, 6% of the TOC had been removed through adsorption onto activated carbon applied before the first RBC. This previous treatment ensured high removal of organic compounds (such as chlorinated hydrocarbons, phenols, and anilines) in the first RBC. Thus, the effluent of the first RBC basically contained a high ammonium concentration, which was sent to the second RBC, submitted to a maximum ammonium load of 30 kgN NH_4^+/day.

The spatial structure of the microbial community in the RBC biofilm was analyzed, focusing on the presence of ammonium-oxidizing (AOB), nitrite-oxidizing (NOB) bacteria and anammox bacteria. The composition of the microbial community present in the biofilm was determined using different molecular tools. Initially, a 16S rRNA clone library was constructed for the RBC biofilm samples based on 16S rDNA fragments amplified by PCR. The sequencing revealed the presence of an unusual number of 16S rDNA sequences, with few sequences related to well-known AOB and NOB species. The DNA inserts of 26 clones were sequenced, 9 of which were different. Two different 16S rRNA sequences had 93% similarity between them, but both had at least 97% similarity with the 16S rRNA sequence of *Sphingomonas* spp. Another sequence belonged to an organism related to the genus *Staphylococcus*. Moreover, a 16S rDNA region of an AOB supposedly related to the genus *Nitrosomonas* was cloned, which was the only sequence present in the clone library related to the AOB group. Other sequences were observed, with similarity of less than 85% with the class *Sphingobacteria*, with the phylum *Actinobacteria* and with the phylum *Planctomycetes*. The 16S rRNA sequences of the anammox bacteria or NOB were not detected in the clone libraries (EGLI et al. 2003a).

From the analysis of biofilm samples using the FISH technique with known probes and by dot-blot hybridization with the same probes for total purified RNA, it was observed that four bacterial groups represented the largest populations in the biofilm samples, three of these being involved in the nitrogen conversion process according to their phylogenetic classification. The main groups of microorganisms present in the biofilm included *Nitrosomonas europaea* and *Nitrosomonas eutropha*, anammox bacteria of the type "*Candidatus Kuenenia stuttgartiensis*," NOB of the genus *Nitrospira*, and filamentous bacteria of the phylum *Bacteroidetes*. Ammonium-oxidizing bacteria corresponded to approximately 20–30% of the biomass, while members of the phylum *Cytophaga-Flavobacterium-Bacteroides* (CFB) represented 7%. According to FISH, NOB were present in relatively small amounts, representing around 5% of the microbial community. Besides those mentioned, no other AOB groups were found. According to the author, these results may be related to the high ammonium concentrations (up to 35 mM) and high salt concentrations (10 g/L), which may favor the selection of specific AOB groups resistant to the conditions imposed (i.e., *Nitrosomonas europaea/eutropha*. The presence of anammox bacteria suggested that both the operating conditions of the treatment system and the specific type of wastewater treated provided conditions appropriate for the inoculation and spontaneous enrichment of these bacteria (EGLI et al. 2003a).

Besides being investigated by the FISH technique, the relative abundance of phylogenetically similar groups was evaluated by dot-blot hybridization. Compared with the intensity of the hybridization signal obtained with the general bacterial probe (considered as 100%), the quantity of rRNA labeled with the *Nitrosomonas*-specific probe was 23%. The same relative quantity was obtained with a more general probe for AOB. The dot-blot hybridization results indicated that filamentous bacteria of the phylum *Bacteroidetes* accounted for around 40% of the microbial population. NOB corresponded to 14% of the total bacterial community. Anammox bacteria were detected in very low quantities (around 1% of the microbial population), although FISH suggested a relatively high number for this group in the biofilm. The FISH methodology allowed an analysis of the biofilm structure, clearly showing that a very high amount of aerobic nitrifying microorganisms was located in the outer layer of the biofilm. The anammox bacteria were exclusively present in a dense inner layer of the biofilm, in which oxygen diffusion was limited, while filamentous CFB bacteria were distributed throughout the biofilm (EGLI et al. 2003a).

EGLI et al. (2003b) operated continuous stirred-tank reactors (CSTR), inoculated with activated sludge, to investigate the conditions required to promote the establishment of partial nitrification of ammonium and nitrite (nitritation) under oxic conditions. The reactors were submitted to different conditions of pH, temperature, and dilution rate, fed with either synthetic medium or the supernatant of a sludge digester (both containing 50 mM of ammonium). In all of the biological systems, stable nitritation was reached within 10–20 days after inoculation. FISH analysis using 16S rRNA-specific fluorescent probes conducted on samples from different reactors showed that the nitrite-oxidizing bacteria (NOB) of the genus *Nitrospira* were only active at the beginning of the reactor operation, that is, immediately after the inoculation with activated sludge. These bacteria gradually lost their activity and were almost completely washed out of the reactors. According to the FISH and the RFLP analysis of the *amoA* gene (which encodes the subunit of the active sites of the enzyme ammonia monooxygenase), the community of ammonium-oxidizing bacteria (AOB) showed an alteration between the first 15–20 days of reactor operation, from a very diverse group of populations in the inoculation sludge, represented by members of *Nitrosomonas communis* and *Nitrosomonas oligotropha* and of *Nitrosomonas europaea-Nitrosomonas eutropha*, to a smaller group inside the reactor.

The authors also observed that the RFLP patterns in the reactors submitted to pH 7.0 and 30 °C differed considerably from those of reactors submitted to pH 7.5 and 30 °C and to pH 7.5 and 25 °C. The reactors operated at pH 7.5 and 30 °C showed the predominance of one *amoA* RFLP type from the group *N. europaea-N. eutropha*. In the reactors operated at pH 7, which presented diversified communities, transitory changes in the population occurred. The reactors maintained at pH 7.5 and 25 °C developed communities which were indistinguishable using the FISH probes applied, although they differed in terms of the *amoA* RFLP patterns.

Furthermore, under these conditions, the increase in AOB was less pronounced than at 30 °C. On the other hand, the NOB community did not stabilize at any time, showing a decline and being no longer detectable after 10 days (EGLI et al. 2003b).

Sludge samples of these systems were very similar to those of the biomass coming with the sludge digester supernatant. This finding revealed that these reactors were constantly inoculated with the biomass from the sludge digester, allowing these systems to be operated with greater dilution rates (0.75/day) when compared with the reactors fed with synthetic medium (0.2/day). The communities of the reactors fed with digester supernatant showed a high degree of similarity, although they differed considerably from those of the reactors fed with synthetic medium. Despite the performance of the different reactors being similar with regard to the parameters evaluated, the structures of the communities differed, which probably influenced the stability during disturbances (EGLI et al. 2003b).

By means of FISH and electron microscopy, DÍAZ et al. (2003) evaluated the diversity, in qualitative and quantitative terms, of the microorganisms present in anaerobic granular sludge fed with different substrates. The granules presented a structure with several layers, in which slightly or densely packed microcolonies were observed. In the external region of the granules, only bacteria were found. On the other hand, in the center of the granules, both archaea and bacteria were present. Although the cell density of the granules was high (over ten cells per gram, labeled with DAPI), only a small fraction of the cells were hybridized with the specific probes used. The authors observed significant differences in the microbial composition of the granules fed with different substrates (formate, acetate in high and low concentrations, propionate, sucrose, amide, and peptone). Active *Proteobacteria* were in low numbers in the granules fed with volatile fatty acids. Bacteria of the genus *Syntrophobacter* dominated when the granules were fed with propionate. On the other hand, methanogenic archaea, particularly species of the genus *Methanosaeta*, were predominant when the granular reactor was fed with complex substrates or acetate in low concentrations. When high concentrations of acetate or formate were used as substrate, bacteria of the genus *Methanosarcina* and members of the order Methanobacteriales were dominant, respectively.

KIM et al. (2004) used the FISH technique to analyze nitrifying microbial communities in an activated sludge reactor (ASR) and in a fixed-bed reactor (FBR) used for the treatment of wastewater generated at a pig farm. Heterotrophic oxidation and nitrification occurred simultaneously in the ASR. The COD removal and nitrification efficiencies were observed to be dependent on the applied loads. In the FBR, the nitrification efficiency was also related to the ammonium load applied to the reactor. Nitrite accumulation was observed when the concentration of free ammonia was greater than 0.2 $mgNH_3$-N/L. With the use of FISH, it was observed that the ammonium-oxidizing bacteria (probe NSO1225) and the denitrifying bacteria (probe RRP1088) were less abundant than other bacteria (probe EUB338) in the ASR. Subsequent analysis of nitrifying bacteria in the FBR showed that species of *Nitrosomonas* (probe NSM156) and *Nitrospira* (probe NSR1156) were, respectively, the dominant ammonium-oxidizing and nitrite-oxidizing bacteria in the pig farm wastewater treatment system.

LIMPIYAKORN et al. (2005) investigated the ammonium-oxidizing bacteria (AOB) in activated sludge samples originating from 12 sewage treatment systems, subjected to low ammonium concentrations. These systems have shown different ammonium removal efficiencies and were operated according to different treatment processes: anaerobic/anoxic/aerobic process (A2O), anaerobic/aerobic process (AO), and a conventional activated sludge process (AS). The samples were collected in three different seasons (summer, autumn, and winter). In this study, real-time PCR was used to determine the total number of bacteria and AOB. AOB were also studied through PCR amplification, followed by DGGE, cloning, and sequencing of the 16S rRNA genes. The total number of bacteria and of AOB was in the ranges of 1.6×10^{12}–2.4×10^{13} and 1.0×10^{9}–9.2×10^{10} cells/L, respectively. These values varied according to the season of the year and to the influent BOD/NH_4^+-N ratio. The ammonia oxidation activity per cell of AOB varied between 0 and 49.6 fmol/(cell h), a value which was dependent on the season of the year. The highest values for this parameter were observed during summer, followed by those obtained during autumn and winter.

Although periodic variations in the number of AOB were observed, this was not the case for the composition of the AOB communities, which showed very stable behavior, even with the temperature varying over the seasons. Most of the bands excised from the DGGE gel and sequenced were related to the group *Nitrosomonas* spp. The highest number of the bands analyzed has shown high sequence similarity to *Nitrosomonas oligotropha*, which was found in all of the samples. *Nitrosomonas europaea-Nitrosococcus mobilis* (frequently considered to be halophilic microorganisms able to support high salt concentrations) and unknown species of *Nitrosomonas* appeared only in the anaerobic/anoxic/aerobic (A2O) system. It was observed that the solid retention time influenced the total number of AOB, while the dissolved oxygen concentrations affected the ammonium oxidation activity per cell of AOB. It was not possible to determine a relation between the different activated sludge systems and the presence of *Nitrosospira*, which was only recovered from one sample (LIMPIYAKORN et al. 2005).

Employing the T-RFLP technique, HOSHINO et al. (2005) monitored the microbial dynamics at the time in which the denitrification process started to occur in an activated sludge reactor with intermittent aeration. The terminal restriction fragments (T-RF) which increased due to the starting of denitrification were determined and identified through analysis of the 16S rRNA sequences obtained by cloning. It was observed that the bacteria belonging to the genera *Hydrogenophaga* and *Acidovorax* were present in greater numbers after the denitrification had begun.

CORTÉS-LORENZO et al. (2005) used PCR-TGGE to evaluate the microbial diversity of a submerged filter reactor employed for the removal of ammonium and phenol from a wastewater with a high salt concentration (30 g/L) originated from a pharmaceutical plant. The influent ammonium concentration was 340 mg/L and the phenol content was equivalent to 1000 mg/L. The system operated in pre-denitrification mode and therefore consisted of two separate columns (one anoxic and the other aerobic), with a connection between them.

The spatial diversity of the microbial communities in the biofilm was analyzed by collecting samples at four different heights in the system. The TGGE profiles for the sequences of the 16S rRNA gene (V3 hypervariable region) amplified by PCR showed significant variations in the microbial diversity, mainly due to the variation in the dissolved oxygen concentration along the reactor. Several bands obtained from TGGE were reamplified and sequenced in order to reveal the composition of the microbial communities present in the biofilm. Most of the sequenced bands (10 out of 13) were related to the 16S rRNA gene of marine α-*Proteobacteria*, mainly assigned to the genus *Roseobacter*. The remainder of the sequences which were related to the α-*Proteobacteria* was assigned to the order *Rhizobiales*. Organisms phylogenetically related to the group *Roseobacter* were found in several locations and are important primary colonizers of surfaces in coastal environments, consisting of precursors for the formation of biofilms (CORTÉS-LORENZO et al. 2005). Other sequences were related to γ-*Proteobacteria*, *Nitrospira marina* (nitrite oxidizer present only in the aerobic part of the system), and the anaerobic bacteria of the family *Desulfobacteraceae*, which is responsible for the degradation of phenol (CORTÉS-LORENZO et al. 2005).

OTAWA et al. (2006) investigated in the microbial community diversity in a sequencing batch reactor (SBR_I) with intermittent aeration and in 12 plants treating wastewater comprised of animal residues (8 SBR and 4 conventional activated sludge systems). In order to investigate the seasonal variation of microbial communities, techniques such as RT-PCR followed by DGGE were used along with the construction of clone libraries for 16S rRNA and ammonia monooxygenase (*amoA*) genes. In SBR_I, the dominant bacteria were represented by uncultured *Bacteroidetes* and *Proteobacteria*. The percentages of the bands related to these two groups in relation to the total sequenced bands corresponded to 47.4% and 34.2%, respectively. The DGGE profiles showed that the bacterial communities were stable during a certain treatment cycle, although there were seasonal variations. In relation to the AOB communities, two phylotypes (*Nitrosomonas ureae-oligotropha-marina*) were dominant in SBR_I during the all seasons. Although the phylotypes of the AOB from the 13 treatment plants presented differences, the dominance of one or two groups was observed at all plants.

Sequence of DGGE bands revealed that *amoA* sequences belonging to the group *Nitrosomonas europaea-eutropha* were dominant at 11 plants, in which the influent wastewater contained high concentration of ammonium. On the other hand, at the plants where the ammonium concentration was relatively low, there was a predominance of the groups *Nitrosomonas urea-oligotropha-marina*. The authors also observed that, although the removal efficiency remained stable during the study period, the bacterial community varied during the year, indicating that the stability of the treatment efficiency does not necessarily reflect stability in the bacterial community. In fact, variations in the influent characteristics and in the reactor temperature during the process can lead to changes in the community structure of the bacteria (OTAWA et al. 2006).

FENG et al. (2007) investigated the diversity of the dominant microbial communities during the start-up of a granular sludge sequencing batch reactor, in which partial removal of nitrogen was observed via the simultaneous nitrification/denitrification (SND) process. This process is favored by the irregular distribution of dissolved oxygen and by the anoxic microenvironment. These conditions often lead to the accumulation of nitrite (YOO et al. 1999). The start-up process of the reactor was divided into three stages: the first stage (day 1–15) was characterized by low nutrient removal rates and large fluctuations; in the second stage (day 16–35), there was an increase in the ammonium concentration with the consequent accumulation of nitrite; and in the third stage (day 36–60), the accumulation of nitrite remained constant and the pollutant removal rate reached stable values.

The DGGE method was used to obtain a fingerprint of the communities of α-*Proteobacteria*, β-*Proteobacteria*, ammonium-oxidizing bacteria (AOB), and *Nitrospira* during the start-up of the reactor. The results obtained with this technique showed that the profiles for the microbial community changed during the nitrite accumulation. Substantial modifications were observed during the initial period, particularly between days 25 and 35, although the structure of the microbial community became stable after 45 days of operation. The diversity analysis based on the DGGE patterns was carried out by calculating the Shannon index (H). For the samples collected on days 25 and 35, the H value was higher for β-*Proteobacteria* compared with the α-*Proteobacteria*, although the latter had higher values for the last two samples collected (days 45 and 55). This indicates that α-*Proteobacteria* have shown a greater diversity when the stability of the SND process was reached. During the experimental period, a decrease in the H index occurred both for the β-*Proteobacteria* and for *Nitrospira*. On the other hand, this index remained almost stable for α-*Proteobacteria* and for ammonium-oxidizing bacteria, which suggests that some species of β-*Proteobacteria* are more sensitive to free ammonia (FA) than the α-*Proteobacteria*. In fact, inhibition by free ammonia can result in the elimination of some species of β-*Proteobacteria* and *Nitrospira* (FENG et al. 2007).

The use of real-time PCR, determining the 16S rDNA copies, allows the quantification of the different microorganisms present in the system. From day 25 to 35, an apparent change was observed in the size of the populations of α- and β-*Proteobacteria*, which increased from 6.7×10^{11} cells/L to 1.0×10^{12} cells/L in the first day of the SND process. AOB presented a very similar growth behavior. On the other hand, the *Nitrospira* population decreased at the beginning of the SND process and with the accumulation of nitrite. The sizes of the AOB and *Nitrospira* populations were 8.7×10^{9}–2.4×10^{10} cells/L and 1.7×10^{10}–2.1×10^{10} cells/L, respectively. The proportions of AOB and *Nitrospira* in relation to the total bacteria in the stable stage were 2.1–2.4% to 0.8–1.2%, respectively. The same authors also observed, through the Pearson correlation coefficient, a relation between the nitrifying activity, diversity, and the size of the AOB and *Nitrospira* populations. It was observed that the size of the population had a greater influence on the degradation capacity compared to the diversity (FENG et al. 2007).

LIU et al. (2007) investigated the structure of the bacterial community of two sewage treatment plants with different processes and performances. For this purpose, DGGE was used to analyze the fragments of the 16S rRNA gene amplified by nested PCR using group-specific primers. In one of the plants (plant A), an anoxic-anaerobic-aerobic process (A2O) was adopted, and in the other plant (plant B), an anoxic/aerobic process (AO) was used. The two plants showed different sludge characteristics, particularly in relation to foaming and settleability, as well as different pollutant removal efficiencies.

Samples of raw sewage, treated effluent, and activated sludge were amplified using PCR for further evaluation of the bacterial community structure by DGGE. Analysis of the community of ammonium-oxidizing bacteria and actinomycetes was also carried out to investigate the relation between the structures of specific populations and the system performance. The samples of raw sewage from both plants were located in a cluster in a dendrogram due to their similarity (Dice coefficient equivalent to 73%), which can be attributed to the similarity of the sewage constituents. Furthermore, the similarity between the influent sewage composition of the plants may have led to the similarity observed between the activated sludge samples of the two treatment systems (Dice coefficient of 60%).

The Dice coefficients obtained on comparing the similarity between the raw sewage and activated sludge samples from the two plants were very low, 27 and 29%, respectively. In fact, many of the bacterial populations of raw sewage do not appear in the activated sludge samples, suggesting that the dominant bacterial populations in the raw sewage do not play an important role in the biological treatment. Most of the microbial populations present in the raw sewage originate from sanitary wastes or soil. The bacteria from sanitary wastes are intestinal microorganisms which result from human or animal excretion and generally do not survive in unsuitable environments, such as wastewater treatment plants (MARSHALL 1980 *apud* LIU et al. 2007).

As previously described, the two plants showed high similarity in the microbial community structure of the raw sewage and activated sludge, although the microbial populations of their respective effluents differed (Dice coefficient of 25%). The difference between the sludge settleability of the two systems may be the main reason for the structural differences in bacterial communities. The sludge volume index (SVI) of plant B was 197 L/g, whereas that of plant A was 71 L/g. The relatively poor settleability of the activated sludge in plant B when compared with plant A may have led to washout of the bacteria, without the selection of one or another specific group. This may lead to a high degree of similarity between the populations of the effluent and the activated sludge (Dice coefficient of 79%). On the other hand, only some of the bacterial species were selectively washed out of plant A, which resulted in a low degree of similarity between the structures of the bacterial communities of the effluent and activated sludge (Dice coefficient of 26%) (LIU et al. 2007).

The same authors carried out a comparative study on the structure of the AOB communities in the two treatment plants. Plant A showed more bands in the DGGE in comparison with plant B. In addition, the two systems presented different AOB community structures. The greater ammonium removal obtained at plant A can be attributed to its greater richness of AOB species. The main bands of the DGGE gel related to AOB were excised and sequenced. The analysis of the sequences of the 16S rRNA genes revealed that all of the bands were closely related to the genus *Nitrosomonas* spp. The sequencing results showed that the main populations present in the raw sewage were non-cultivated bacteria, while in the activated sludge the predominant populations belonged to the subclass β-*Proteobacteria*. Their numerical dominance suggests that the representatives of this subclass play an important role in the activated sludge process (removal of organic matter and nutrients as well as the formation of floc structure) (LIU et al. 2007).

Considering that the specific growth rates and activity of the ammonium-oxidizing bacteria (AOB) are very reduced at low temperatures, ISAKA et al. (2007) developed a technology for the immobilization of nitrifying bacteria on a support in the form of a polyethylene glycol (PEG) gel. By adopting this procedure, it was possible to reach high nitrification rates (0.71 kgN/(m^3 day)) in the treatment of leachate even at low temperatures (10 °C). The PEG supports were added to a 1.2 L air-lift reactor in which the filling fraction (ratio between the filling volume and total volume of the reactor) was 15% (180 mL). According to the authors, the immobilization of AOB is a key factor in the development of a nitrification process under low temperature conditions. The ammonium concentration of the effluent varied from 16 to 35 mg/L independently of the nitrogen load. Moreover, accumulation of nitrite was observed. In this study, in particular, the nitrite accumulation was not related to the low level of dissolved oxygen, which was greater than 7 mg/L, but may have been associated with the loss of activity of the nitrite-oxidizing bacteria (NOB) due to other factors, such as inhibition by free ammonia (ANTHONISEN et al. 1976). On applying the DGGE technique, four bands corresponding to ammonium-oxidizing bacteria were observed. The phylogenetic tree based on the sequences of the 16S rRNA gene from DGGE bands revealed that three of these bands were similar to those of *Nitrosomonas* and one band was similar to that of *Nitrosospira* spp. (95% similarity). Based on the DGGE results obtained, FISH was employed to identify the presence of these AOB. *Nitrosomonas* spp. and *Nitrosospira* spp. were detected. Nevertheless, direct cell counts showed that *Nitrosomonas* spp. was the dominant AOB group.

MONTRÀS et al. (2008) investigated the microbial community of the biofilm adhered to the spherical expanded polystyrene (Biostyr®) supports of a packed bed reactor containing two autotrophic species (*Nitrosomonas europaea* and *Nitrobacter winogradskyi*). The analysis were carried out after 4.8 years of continuous operation of reactor, which was achieving complete nitrification. It was observed that these two nitrifying species were, respectively, the only ammonia-oxidizing bacteria (AOB) and nitrite-oxidizing bacteria (NOB) present in the reactor. Quantitative real-time PCR (Q-PCR) was used to quantify the two abovementioned species along the vertical axis of the reactor (divided into eight fractions F_0 to F_7, from the bottom to the top), and the authors observed a spatial segregation of these two microbial lineages. The Q-PCR results showed a high degree of similarity at F_0 (89.7 ± 7.8% *N. europaea*) and F_1 (91.3 ± 7.1% *N. europaea*) positions. The relative quantities of *N. europaea* in F_2 (57.3 ± 8.3% *N. europaea*) and F_3 (58.7 ± 5.3% *N. europaea*) samples were also very similar. There was a gradual increase in *N. winogradskyi* with an increase in the bed height, as demonstrated by the relative quantities obtained at positions F_4 (61.0 ± 2.6% *N. winogradskyi*), F_5 (70.1 ± 1.3% *N. winogradskyi*), and F_6 (70.0 ± 1.0% *N. winogradskyi*).

In the last fraction of the packed bed (F_7), an increase in the *N. europaea*/*N. winogradskyi* ratio was observed, as the relative quantity of the latter species reduced by 53.9 ± 1.9% at this position in the reactor. The results have shown that spatial distribution of AOB and NOB occurred along the bed height, similarly to that observed in other nitrifying bioreactors with unidirectional flow, for instance, nitrifying trickling filters or nitrifying plug-flow bioreactors (HOLBEN et al. 1998 *apud* MONTRÀS et al. 2008).

MERTOGLU et al. (2008) monitored the changes in a population of nitrifying bacteria submitted to a cadmium load in a continuous flow nitrification system, fed with synthetic effluent. The authors used molecular microbiology tools, such as slot-blot hybridization (for the detection of nitrite-oxidizing bacteria), DGGE, and real-time PCR followed by melting curve analysis (for the detection of ammonium-oxidizing bacteria), cloning, and sequencing (for the identification of microbial species).

When the system was fed with doses equivalent to 1 mgCd/L, no decrease in the ammonium removal rate was observed, and even a stepwise increase to 10 mgCd/L did not cause inhibition of the microbial activity. However, after 15 days of operation at this concentration, a decrease in the ammonium removal efficiency from 99 to 10% was observed, which was accompanied by a reduction in the biomass concentration. On observing these results, the addition of cadmium was interrupted in order to allow the recovery of the microbial activity. Under these conditions, the inhibitory effect ceased. With a second application of cadmium at the same concentration (10 mg/L), it was observed that the biomass adapted to this metal and the inhibition reached 60%. In the next step, when the cadmium concentration was increased to 15 mg/L, the ammonium removal efficiency was 50% (MERTOGLU et al. 2008).

A change in the diversity of the ammonia-oxidizing bacteria was observed after the first application of 10 mg/L of Cd. The species resistant to this metal became dominant, and *Nitrosomonas* and *Nitrosococcus* spp. were replaced by *Nitrosospira* spp. This latter lineage was found to be resistant to high cadmium loads and gradually became dominant in the system. The results obtained indicated that the level of nitrification inhibition was not only related to the metal concentration and the quantity of microorganisms but was also dependent on the specific type of species present in the system (MERTOGLU et al. 2008).

DYTCZAK et al. (2008) observed that a laboratory-scale nitrifying sequencing batch reactor (SBR_1), operating under alternating anoxic/aerobic conditions, reached nitrification rates twice as high as those of another reactor (SBR_2), with the same dimensions and operational conditions, except that it was strictly aerobic. The microbial populations in the two reactors were investigated using the FISH technique, which revealed the predominance of rapid nitrifying microorganisms, such as *Nitrosomonas* and *Nitrobacter* (79.5% of the nitrifying population), in the anoxic/aerobic reactor, while in the strictly aerobic reactor, slow nitrifiers, such as *Nitrosospira* and *Nitrospira*, were predominant (78.2% of total nitrifiers). The proportion of ammonium-oxidizing bacteria (AOB) in relation to the total cells identified by DAPI was greater in SBR_1 (25.9%) compared with SBR_2 (19.9%), although the most significant difference was observed on comparing the AOB populations (particularly of *Nitrosomonas*) present in each reactor. In SBR_1, these microorganisms reached 81% of the total AOB, while in SBR_2 this percentage was only 19%. Similar behavior was observed for the NOB population in relation to the total bacteria, with a higher proportion in SBR_1.The largest discrepancy was related to the population of *Nitrobacter* in relation to the total NOB in the reactors, which was 76% in SBR_1 and 27% in SBR_2. In summary, both the AOB and the NOB were more abundant in SBR_1, which also had a greater AOB/NOB ratio.

The reactor SBR_2 (strictly aerobic) runs at the maximum rate and was negatively affected by ammonium and nitrite, while the nitrification rates obtained in the reactor submitted to alternating conditions (anoxic/aerobic) were proportional to the ammonium and nitrite concentrations. It was observed that the alternating conditions were more favorable for nitrification, since they allowed a more rapid selection of nitrifying microorganisms, characterized by higher oxidation, growth, and decay rates. These results are very useful for the operation of biological reactors aimed at nitrification, which are generally designed based on the growth rate of nitrifiers determined under strictly aerobic conditions (DYTCZAK et al. 2008).

MUNZ et al. (2008) compared a pilot-scale membrane bioreactor (MBR) with a conventional activated sludge (AS) system, in order to evaluate the global efficiency of the treatment, the presence and distribution of Gram-negative bacteria, and the kinetics of the nitrifying bacteria. Both systems were fed with industrial tannery wastewater and subjected to the same operating conditions. The MBR, in comparison with the AS reactor, presented a higher COD removal and a more stable nitrification process. These results were attributed to the differences in the composition and distribution of the microbial community.

The Gram-negative bacteria were detected by means of FISH. Phylogenetic probes specific for α-, β-, and *γ-Proteobacteria* of the main ammonia-oxidizing (AOB) and nitrite-oxidizing (NOB) bacteria belonging to the genera *Nitrobacter* and *Nitrospira* were used. The results showed that the main differences between the two reactor systems in terms of the biomass composition were the high abundance of α- and *β-Proteobacteria* in the MBR and the presence of aggregates of AOB only on the surface of the flocs present in this reactor. It should be noted that there was a predominance of *α-Proteobacteria* in the samples originating from the two reactor configurations and not of *β-Proteobacteria* as frequently reported in the literature. This finding may be related to the specific industrial wastewater fed to the system. The differences observed in the quantitative distribution of the three subdivisions of *Proteobacteria* (α, β, and γ) in the samples collected from the MBR and the AS reactor suggest that the biomass separation process, by filtration (MBR) or sedimentation (AS reactor), affected the selection of the specific microorganisms (MUNZ et al. 2008).

WEN et al. (2008) operated a fluidized bed reactor (FBR) without a separation between the aerobic and anoxic zones and achieved simultaneous nitrification-denitrification (SND) through control of the dissolved oxygen concentration. In this study, the efficiency and composition of the nitrifying bacteria were investigated. The volumetric nitrogen load varied between 0.12 and 0.20 kgN/(m^3 day). The ammonium removal efficiency was greater than 80% and the DO concentration, maintained within the range of 1–5 mg/L, did not appear to influence this performance. When the DO concentration was below 3 mg/L, the nitrogen removal efficiency was greater than 50%. With the use of the T-RFLP technique based on fragments of 16S rRNA, it was possible to characterize the diversity and distribution of the ammonium-oxidizing (AOB) and nitrite-oxidizing (NOB) bacteria in the FBR. The results indicated that the composition and quantity of AOB and NOB varied according to the position inside the reactor and time of operation. The dominant species of the AOB community were represented by *Nitrosomonas* spp. Also, *Nitrobacter* spp. were detected in the system. According to the authors, the SND mechanism in FBR was associated with the vertical stratification of the active populations. However, the possible presence of microenvironments within the biofilm was not discarded (WEN et al. 2008).

MOURA et al. (2007) employed the DGGE technique to estimate the bacterial diversity and monitor the microbial communities in two aerated lagoons (lagoon 1 and lagoon 2) of a plant treating urban and industrial wastewaters. Significant changes were detected on comparing the samples collected in the winter/spring and autumn/summer months. Thus, the microbial community showed seasonal variations. As previously described, OTAWA et al. (2006) also observed a pattern of seasonal variability in the bacterial community present in a sequencing batch reactor operated with intermittent aeration.

The variables which had the greatest impact on the bacterial communities were temperature, dissolved oxygen, and pH. The authors carried out a numerical analysis of the DGGE patterns based on two indicators, each one describing a different aspect of the diversity of the community of microorganisms. The diversity indices were obtained taking into consideration the relative intensity of each band of the DGGE gel. The reproducibility of the results was obtained by conducting replicates of the experiments (MOURA et al. 2007).

The Shannon index (*H*), used to calculate the diversity of microbial communities, varied from 0.717 to 1.042 and from 0.784 to 1.121 in lagoons 1 and 2, respectively. The highest *H* values were recorded in the months of August and March, indicating a greater number of species and higher relative abundance during these months. Another parameter evaluated was the equitability index (*E*), which can vary from values close to 0 (indicating the pronounced dominance of some microbial groups) to values close to 1 (indicating complete equality of abundance of all species). The analysis of the DGGE patterns provided values for the equitability index in the ranges 0.756–0.886 and 0.796–0.968 for lagoons 1 and 2, respectively. These values indicate a high degree of uniformity in the distribution of taxons between samples in terms of abundance in both lagoons (MOURA et al. 2007).

The phylogenetic affiliation of the predominant groups was determined through the sequencing of 16S rDNA fragments excised from the DGGE gel. The results for the phylogenetic analysis allowed classifying the microorganisms in the groups *Cytophaga-Flexibacter-Bacteroides* (CFB), *Firmicutes*, and β- and ε-*Proteobacteria* (MOURA et al. 2007).

Recently, it was discovered that the autotrophic oxidation of ammonia is not restricted only to the domain bacteria (Könneke et al. 2005; TREUSCH et al. 2005).

The ammonium-oxidizing archaea (AOA) called *Nitrosopumilus maritimus* was isolated from a marine aquarium (Könneke et al. 2005) and was the first representative of AOA. These microorganisms, as in the case of ammonium-oxidizing bacteria (AOB), grow under chemolithotrophic conditions oxidizing ammonium to nitrite. Characteristics such as high diversity, high nitrification rate, and low demand for dissolved oxygen mean that AOA can be important agents in the partial nitrification process.

In this context, ZHANG et al. (2009) conducted a study to characterize the AOA community both in an activated sludge bioreactor designed for nitrogen removal and several wastewater treatment plants. A pair of primers specific for the α subunit of ammonia monooxygenase (*amoA*) of archaea was used in the analysis. This study confirmed the occurrence of AOA in the systems evaluated. In total, 18 sequences of the functional gene *amoA* were recovered and compared with previously reported sequences. A comparison between the clone libraries obtained with the samples collected from one of the plants and from the laboratory scale bioreactor indicated that the community of AOA varied significantly only after 30 days of enrichment. The authors also observed that the community of AOA of the treatment plants studied was similar to those found in sediments and soils, but distinct from those present in other activated sludge systems. Finally, they also mentioned that for the detection of the different AOA communities present in environmental samples, a combination of different primers specific for the *amoA* gene is required (ZHANG et al. 2009).

Bassin et al. (2011) studied the dynamics of the microbial community structure of an aerobic granular sludge subjected to high salt concentrations. The bacterial population was evaluated by applying PCR-DGGE of 16S rRNA and *amoA* genes. The PCR products obtained from both genomic DNA and rRNA after reverse transcription were compared to determine the bacteria present as well as the metabolically active fraction of the bacteria. Fluorescence in situ hybridization (FISH) was used to validate the PCR-based results and to quantify the dominant bacterial populations. The authors observed that salt concentrations up to 33 gNaCl/L did not affect the ammonium removal efficiency. Conversely, a high accumulation of nitrite was observed above 22 gNaCl/L, which coincided with the disappearance of the only nitrite-oxidizing bacteria detected in the sludge (*Nitrospira* sp.). The increase in salt concentration had a strong impact on the phosphorus removal. Under steady-state operation conditions at 33 gNaCl/L, polyphosphate-accumulating organisms (PAOs) (*Candidatus Accumulibacter phosphatis*) were no longer detected by PCR-DGGE or by FISH. As a consequence, no P-release or P-uptake was observed at this salinity level. Glycogen-accumulating organisms (GAOs) became the dominant microorganisms at increasing salt concentrations, especially at 33 gNaCl/L. From the comparative analysis of the diversity (DNA-derived pattern) and the activity (cDNA-derived pattern) of the microbial population, it was observed that the microorganisms with the highest metabolic activity were those related to ammonium (*Nitrosomonas* spp.) and phosphate (*Candidatus Accumulibacter*) removal (Bassin et al. 2011).

6.4 Final Considerations

This chapter has demonstrated the large amount of information that can be provided by the molecular biology techniques. These approaches allow the study of the diversity of the microbial communities present in a variety of different environments, such as biological wastewater treatment systems.

A period in which an enormous and increasing quantity of information will become available has only just begun, and this will enable the evaluation of the taxonomy and the functional diversity of complex communities of microorganisms. This information is of particular interest since establishing a link between the experimental measurements and the microbiological analysis will elucidate the relations between the biodiversity and the functioning of natural or engineered ecosystems.

In the particular case of wastewater treatment systems, new techniques will be very useful for the diagnosis of the system and identification of problems related to its operation, such as bulking sludge and foaming. Currently, the treatment plants may operate with good or poor performance, and in most cases the reason for this is unknown.

Although a certain level of understanding has been gained regarding the structure of activated sludge and biofilm processes, we are still a long way from a complete understanding of the structure of microbial communities involved in these processes. The objective is to elucidate the origin of certain microorganisms, the factors and mechanisms which lead to changes in the bacterial populations (microbial population dynamics) and their relation with the performance of biological processes. In other words, it is necessary to gain an understanding of microbial populations and how they interact with the wastewater components. For this purpose, it is necessary to link a wide range of areas from the microbiology of a single cell to studies in laboratory scale. A better understanding of the microbiology and ecology of various groups of bacteria would provide insights essential for improving the performance and control of the processes, revealing the factors which lead to their proper functioning. This understanding represents the bridge which unites fundamental knowledge on microbial communities and the wastewater treatment processes.

Studies seeking to enhance this knowledge are already underway in several research centers around the world, and many are being carried out in laboratory systems, providing a rapidly growing understanding of the most important microorganisms involved in the wastewater treatment processes and the factors which alter their numbers and activity. However, there is a wide gap in the information available, particularly due to the fact that few studies have been carried out in full scale and the difficulty associated with extrapolating information gained in laboratory studies to full-scale plants.

With the aid of molecular biology techniques, it is hoped that it will be possible to manipulate and control the populations of microorganisms, to select only those which function under certain operational conditions, and thus to achieve high pollutant removal efficiencies and stable operation. Looking even further forward, real-time monitoring of biological treatment systems can be imagined, in which the diversity and activity of the microbial communities associated with certain operating conditions are obtained at any given time.

Therefore, it is necessary to encourage the process engineers to better appreciate the valuable information obtained at the molecular level when designing treatment plants, rather than simply considering the chemical and physical parameters. With this approach, the operation of these systems will be carried out in a less empirical way, contributing to the beginning of a new era of novel and advanced biological treatment processes.

References

ADAMCZYK, J., HESSELSOE, M., IVERSEN, N., HORN, M., LEHNER, A., NIELSEN, P. H., SCHLOTER, M., ROSLEV, P., WAGNER, M. The isotope array, a new tool that employs substrate-mediated labeling of rRNA for determination of microbial community structure and function. *Applied and Environmental Microbiology*, v. 69, p. 6875-6887, 2003.

ALBERTS, B., BRAY, D., LEWIS, J., RAFF, M., ROBERTS, K., WATSON, J. D. *DNA and RNA structure*, 1994. In: Molecular biology of the cell. Third edition. New York: Garland Science, p. 100-109.

AMANN, R. I., KRUMHOLZ, L., STAHL, D. A. Fluorescent-oligonucleotide probing of whole cells for determinative, phylogenetic, and environmental studies in microbiology. *The Journal of Bacteriology*, v. 172, p. 762-770, 1990.

AMANN, R. I., LUDWIG, W., SCHLEIFER, K. H. Phylogenetic identification and in situ detection of individual microbial cells without cultivation. *Microbiological Reviews*, v. 59, p. 143-169, 1995.

AMANN, R., LUDWIG, W., SCHULZE, R., SPRING, S., MOORE, E., SCHLEIFER, K. H. rRNA-targeted oligonucleotide probes for the identification of genuine and former pseudomonads. *Systematic and Applied Microbiology*, v. 19, p. 501-509, 1996.

ANDREOTE, F. D., AZEVEDO, J. L., ARAÚJO, W. L. Assessing the diversity of bacterial communities associated with plants. *Brazilian Journal of Microbiology*, v. 40, n. 3, p. 417-432, 2009.

ANTHONISEN, A. C., LOHER, R. C., PRAKASAM, T. B. S., SRINATH, E. Inhibition of nitrification by ammonia and nitrous acid. *Journal Water Pollution Control Federation*, v. 48, n. 5, p. 835-852, 1976.

AOI, Y., MIYOSHI, T., OKAMOTO, T., TSUNEDA, S., HIRATA, A., KITAYAMA, A., NEGAMUNE, T. Microbial Ecology of Nitrifying Bacteria in Wastewater Treatment Process Examined by Fluorescence *In Situ* Hybridization. *Journal of Bioscience and Bioengineering*, v. 90, n. 3, p. 234-240, 2000.

BASSIN, J.P., KLEEREBEZEM, R., MUYZER, G., ROSADO, A.S., VAN LOOSDRECHT, M.C.M., DEZOTTI, M. Effect of different salt adaptation strategies on the microbial diversity, activity, and settling of nitrifying sludge in sequencing batch reactors. *Applied Microbiology and Biotechnology*, v. 93, n. 3, p. 1281-1294, 2012.

BASSIN, J.P., PRONK, M., MUYZER, G., KLEEREBEZEM, R., DEZOTTI, M., VAN LOOSDRECHT, M.C.M. Effect of elevated salt concentrations on the aerobic granular sludge process: linking microbial activity with microbial community structure. *Applied and Environmental Microbiology*, v. 77, n. 22, p. 7942-7953, 2011.

BAYÉS, M., HEATH, S., GUT, I.G. Applications of Second Generation Sequencing Technologies in Complex Disorders. Current topics in behavioral neurosciences. In: CRYAN, J.F., REIF, A. Behavioral Neurogenetics. Springer-Verlag, Berlin Heidelberg, 2012.

BENTLEY, D.R., BALASUBRAMANIAN, S., SWERDLOW, H.P., SMITH, G.P., MILTON, J., BROWN, C.G., HALL, K.P., EVERS, D.J., BARNES, C.L., BIGNELL, H.R., et al. Accurate whole human genome sequencing using reversible terminator chemistry. *Nature*, v. 456, p. 53–59, 2008.

BOND, P. J., HUGENHOLTZ, P., KELLER, J., BLACKALL, L. L. Bacterial community structures of phosphate-removing and non-phosphate-removing activated sludges from sequencing batch reactors. *Applied and Environmental Microbiology*, v. 61, n. 5, p. 1910-1916, 1995.

BORNEMAN J., TRIPLETT, E. Molecular microbial diversity in soils from eastern Amazonia: evidence for unusual microorganisms and microbial population shifts associated with deforestation. *Applied and Environmental Microbiology*, v. 63, p. 2647-2653, 1997.

BRESLAUER, K. J., FRANK, R., BLOCKER, H., MARKY, L.A. Predicting DNA duplex stability from the base sequence. *Proceedings of the National Academy of Sciences - USA*, v. 83, p. 3746-3750, 1986.

BURRELL, P. C., KELLER, J., BLACKALL, L. L. Microbiology of a Nitrite-Oxidizing Bioreactor. *Applied and Environmental Microbiology*, v. 64, n. 5, p. 1878-1883, 1998.

CHANG, J.-J., LIANG, W., XIAO, E.-R., WU, Z.-B. Effect of intermittent aeration on the microbial community structure of activated sludge in a submerged membrane bioreactor. *Water and Environment Journal*, v. 25, p. 214-218, 2011.

CHEN, G. H., WONG, M. T., OKABE, S., WATANABE, Y. Dynamic response of nitrifying activated sludge batch culture to increased chloride concentration. *Water Research*, v. 37, p. 3125-3135, 2003.

CHEUNG, V. G., MORLEY, M., AGUILAR, F., MASSIMI, A., KUCHERLAPATI, R., CHILDS. Making and reading microarrays. *Nature Genetics*, v. 21, p. 15-19, 1999.

CHRISTENSEN, H., NORDENTOFT, S., OLSEN, J.E. Phylogenetic relationships of Salmonella based on rRNA sequences. *International Journal of Systematic Bacteriology*, v. 48, p. 605-610, 1998.

CIESIELSKI, S., BULKOWSKA, K., DABROWSKA, D., KACZMARCZYK, D., KOWAL, P., MOZEIKO, J. Ribosomal intergenic spacer analysis as a tool for monitoring methanogenic *Archaea* changes in an anaerobic digester. *Current Microbiology*, v. 67, n. 2, p. 240-248, 2013.

CORTÉS-LORENZO, C., MOLINA-MUÑOZ, M. L., GÓMEZ-VILLALBA, B., VILCHEZ, R., RAMOS, A., RODELAS, B., HONTORIA, E. GONZÁLEZ-López, J. Analysis of community composition of biofilms in a submerged filter system for the removal of ammonia and phenol from industrial wastewater. *The 11th Nitrogen Cycle Meeting*, p. 165-168, 2005.

COUTINHO, H. L. C., OLIVEIRA, V. M., MANFIO, G. P., ROSADO, A. S. Evaluating the microbial diversity of soil samples: methodological innovations. *Anais da Academia Brasileira de Ciências*, v. 71, p. 491-503, 1999.

CROCETTI, G. R., BANFIELD, J. F., KELLER, J., BOND, P. L., BLACKALL, L. L. Glycogen-accumulating organisms in laboratory-scale and full-scale wastewater treatment processes. *Microbiology*, v. 148, p. 3353-3364, 2002.

CROCETTI, G. R., HUGENHOLTZ, P., BOND, P. L., SCHULER, A., KELLER, J., JENKINS, D., BLACKALL, L. L. Identification of polyphosphate-accumulating organisms and design of 16S rRNA-directed probes for their detection and quantitation. *Applied and Environmental Microbiology*, v. 66, p. 1175-1182, 2000.

CROSS, N. C. P. Quantitative PCR techniques and applications. *British Journal of Haematology*, v. 89, p. 693-697, 1995.

DAIMS, H., BRUHL, A., AMANN, R. SCHLEIFER, K. H., WAGNER, M. The domain-specific probe EUB338 is insufficient for the detection of all bacteria: development and evaluation of a more comprehensive probe set. *Systematic and Applied Microbiology*, v. 22, p. 434-444, 1999.

DAIMS, H., MAIXNER, F., SCHMID, M. C. The nitrifying microbes: ammonia oxidizers, nitrite oxidizers, and anaerobic ammonium oxidizers. In: NIELSEN, P. H., DAIMS, H., LEMMER, H. FISH Handbook for Biological Wastewater treatment: Identification and Quantification of Microorganisms in Activated Sludge and Biofilms by FISH. London: IWA Publishing, 2009.

DAIMS, H., NIELSEN, J. L., NIELSEN, P. H., SCHLEIFER, K. H., WAGNER, M. In situ characterization of *Nitrospira*-like nitrite oxidizing active in wastewater treatment plants. *Applied and Environmental Microbiology*, v. 67, p. 5810-5818, 2001.

DALMA-WEISZHAUSZ, D. D., WARRINGTON, J., TANIMOTO, E. Y., MIYADA, C. G. The affymetrix GeneChip platform: an overview. *Methods in Enzymology*, v. 410, p. 3-28, 2006.
DAVENPORT, R. J., CURTIS, T. P. GOODFELLOW, M., STAINSBY, F. M., BINGLEY, M. Quantitative use of fluorescent in situ hybridization to examine relationships between mycolic acid-containing actinomycetes and foaming in activated sludge plants. *Applied and Environmental Microbiology*, v. 66, p. 1158-1166, 2000.
DEGRAVE, W., FERNANDES, O., CAMPBELL, D., BOZZA, M., LOPES, U. Use of molecular probes and PCR for detection and typing of *Leishmania* – a mini-review, *Memórias do Instituto Oswaldo Cruz*, v. 89, p. 463-469, 1994.
DENNIS, G., SHERMAN, B. T., HOSACK, D. A., YANG, J., GAO, W., LANE, H. C., LEMPICKI, R. A. DAVID: Database of annotation, visualization, and integrated discovery. *Genome. Biology*, v. 4, p. R60, 2003.
DÍAZ, E., AMILS, R., SANZ, J.L. Molecular ecology of anaerobic granular sludge grown at different conditions. *Water Science and Technology*, v. 48, n. 6., p. 57-64, 2003.
DIEFFENBACH, C. W., LOWE, T. M. J., DVEKSLER, G. S. General concepts for PCR *primer* design. *Genome Research*, v. 3, p. 30-37, 1993.
DIEHL, F., LI, M., HE, Y. KINZLER, K.W., VOGELSTEIN, B., DRESSMAN, D., BEAMing: single-molecule PCR on microparticles in water-in-oil emulsions. *Nature Methods*, v. 3, p. 551–559, 2006.
DIONISI, H. M., LAYTON, A. C., HARMS, G., GREGORY, I. R., KEVIN, G. R., SAYLER, G. S. Quantification of *Nitrosomonas oligotropha*-like ammonia-oxidizing bacteria and *Nitrospira* spp. from full-scale wastewater treatment plants by competitive PCR. *Applied and Environmental Microbiology*, v. 68, n. 1, p. 245-253, 2002.
DOLLED-FILHART, M.P., LEE, M. JR., OU-YANG, C.W., HARAKSINGH, R.R., LIN, J.C. Computational and Bioinformatics Frameworks for next-generation whole exome and genome sequencing. Scientific World Journal, Article ID 730217 p. 1-10, 2013.
DORIGO, U., VOLATIER, L., HUMBERT, J.F. Molecular approaches to the assessment of biodiversity in aquatic microbial communities. *Water Research*, v. 39, n. 11, p. 2207-2218, 2005.
DOWLING, T. E., MORITZ, C., PALMER, J. D., RIESEBERG, L. H. Nucleic acids III: Analysis of fragments and restriction sites, p. 249-320, 1996. In: HILLIS, D. M., MORITZ, C., MABLE, B. K. (Eds.). *Molecular systematic*, 2nd ed. Sunderland: Sinauer Associates Inc, 655p.
DYTCZAK, M. A., LONDRY, K. L., OLESZKIEWICZ, J. A. Activated sludge operational regime has a significant impact on the type of nitrifying community and its nitrification rates. *Water Research*, v. 42, p. 2320-2328, 2008.
EDEL, V. PCR in Mycology: an Overview. In: BRIDGE, P.D., ARORA, D.K., REDDY, C.A., ELANDER, R.P. (eds.) Applications of PCR in Mycology. CAB International, Wallingford, UK, p. 1–20, 1998.
EGERT, M., FRIEDRICH, M. W. Formation of pseudo-terminal restriction fragments, a PCR-related bias affecting terminal restriction fragment length polymorphism analysis of microbial community structure, *Applied and Environmental Microbiology*, v. 69, p. 2555-2562, 2003.
EGLI, K., BOSSHARD, F., WERLEN, C., LAIS, P., SIEGRIST, H., ZEHNDER, A. J. B., VAN DER MEER, J.R. Microbial Composition and Structure of a Rotating Biological Contactor Biofilm Treating Ammonium-Rich Wastewater without Organic Carbon. *Microbial Ecology*, v. 45, p. 419-432, 2003a.
EGLI, K., LANGER, C., SIEGRIST, H-R., ZEHNDER, A. J. B., WAGNER, M., ROELOF, J., VAN DER MEER, J. R. Community Analysis of Ammonia and Nitrite Oxidizers during Start-Up of Nitritation Reactors. *Applied and Environmental Microbiology*, v. 69, n. 6, p. 3213-3222, 2003b.
EKBLOM, R., WOLF, J.B.W. A field guide to whole-genome sequencing, assembly and annotation. *Evolutionary Applications*, v. 7, n. 9, p. 1026-1042, 2014.
ELLSWORTH, D. L., RITTENHOUSE, K. D., HONEYCUTT, R. L. Artifactual variation in randomly amplified polymorphic DNA banding patterns. *Biotechniques*, v. 14, p. 214-217, 1993.
ENGEBRETSON, J. J., MOYER, C. L. Fidelity of select restriction endonucleases in determining microbial diversity by terminal-restriction fragment length polymorphism. *Applied and Environmental Microbiology*, v. 69, n. 8, p. 4823-4829, 2003.

ERCOLINI D. PCR-DGGE fingerprinting: novel strategies for detection of microbes in food. *Journal of Microbiological Methods*, v. 56, p. 297-314, 2004.
FERREIRA, M. E., GRATTAPAGLIA, D. Introducción al uso de marcadores moleculares em el análisis genético. Brasilia: Embrapa-Cenargen, 1998, 200p.
FERRIS, M.J., MUYZER, G., WARD, D.M. Denaturing gradient gel electrophoresis profiles of 16S rRNA-defined populations inhabiting a hot spring microbial mat community. *Applied and Environmental Microbiology*, v. 62, p. 340–346, 1996.
FENG, W., SI-QING, X., YI, L., XUE-Song, C., JUN, Z. Community analysis of ammonia and nitrite oxidizers in start-up of aerobic granular sludge reactor. *Journal of Environmental Sciences*, v. 19, p. 996-1002, 2007.
FISHER, M. M., TRIPLETT, E. Automated approach for ribosomal intergenic spacer analysis of microbial diversity and its application to freshwater bacterial communities. *Applied and Environmental Microbiology*, v. 65, p. 4630-4636, 1999.
FOESEL, B. U., GIESEKE, A., SCHWERMER, C., STIEF, P., KOCH, L., CYTRYN, E., TORRÉ, J. R., VAN RIJN, J., MINZ, D., DRAKE, H. L., SCHRAMM, A. *Nitrosomonas* Nm143-like ammonia oxidizers and *Nitrospira marina*-like nitrite oxidizers dominate the nitrifier community in a marine aquaculture biofilm. *FEMS Microbiology Ecology*, v. 63, p. 192-204, 2008.
FRANKLIN, R., TAYLOR, D., MILLS, A. Characterization of microbial communities using randomly amplified polymorphic DNA (RAPD). *Journal of Microbiology Methods*, v. 35, p. 225-235, 1999.
FUNGARO, M. H. P., VIEIRA, M. L. C. Aplicações da PCR em ecologia molecular, 1998. In: MELO, I. S., AZEVEDO, J. L. (Ed.). *Ecologia microbiana*. Jaguariúna: Embrapa-CNPMA, p. 205-228.
GILLE, C., GILLE, A., BOOMS, P., ROBINSON, P. N. NURNBERG, P. Bipolar clamping improves the sensitivity of mutation detection by temperature gradient gel electrophoresis. *Electrophoresis*, v. 19, p. 1347-1350, 1998.
GIULIANO, L., DE DOMENICO, M., DE DOMENICO, E., HOFLE, M. G., YAKIMOV, M. M. Identification of culturable oligotrophic bacteria within naturally occurring bacterioplankton communities of the Ligurian sea by 16S rRNA sequencing and probing. *Microbial Ecology*, v. 37, p. 77-85, 1999.
GONZÁLEZ, J.M., MORAN, M. A. *Applied and Environmental Microbiology*, v. 63, p. 4237-4242, 1997.
GUSCHIN, D. Y., MOBARRY, B. K., PROUDINIKOV, D., STAHL, D. A., RITTMANN, B., IZARBEKOV, A. D. Oligonucleotide microchips as genosensors for determinative and environmental studies in microbiology. *Applied and Environmental Microbiology*, v. 63, p. 2397-2402, 1997.
HEAD, I. M., SAUNDERS, J. R., PICKUP, R. W. Microbial evolution, diversity, and ecology: a decade of ribosomal RNA analysis of uncultivated microorganisms. *Microbial Ecology*, v. 35, p. 1-21, 1998.
HERBERT, R. A. Methods for enumerating microorganisms and determining biomass in natural environments. *Methods in Microbiology*, v. 19, p. 1-40, 1990.
HESS, A., ZARDA, B., HAHN, D., HANER, A., STAX, D., HOHENER, P., ZEYER, J. In situ analysis of denitrifying toluene- and m-xylene-degrading bacteria in a diesel fuel-contaminated laboratory aquifer column. *Applied and Environmental Microbiology*, v. 63, p. 2136-2141, 1997.
HESSELMANN, R. P., WERLEN, C., HAN, D., VAN D.M., Jr., ZEHNDER, A. J. Enrichment, phylogenetic analysis and detection of a bacterium that performs enhanced biological phosphate removal in activated sludge. *Systematic and Applied Microbiology*, v. 22, p. 454-465, 1999.
HEIDM C.A., STEVENS, J., LIVAK, K.J., WILLIAMS, P.M. Real time quantitative PCR. *Genome Research*, v. 6, n. 10, p. 986-994, 1996.
Hillis, D. M.; Mable, B. K.; Moritz, C. Applications of molecular systematics: The state of the field and a look to the future. *Molecular Systematics*, p. 515-543, 1996. In: Molecular systematics. 2nd ed. D. M. Hillis, C. Moritz, and B. K. Mable (eds.). Sinauer Associates, Inc., Sunderland, MA.

HIRATA, M. H., HIRATA, R. D. C. Aplicação da PCR em laboratório clínico e medicina forense. In: Curso de Atualização, São Paulo. Laboratório de Biologia Molecular Aplicada ao Diagnóstico do Departamento de Análises Clínicas e Toxicológicas da Faculdade de Ciências Farmacêuticas da Universidade de São Paulo, p. 28-37, 1991.

HOLBEN, W. E., JANSSON, J. K., CHELM, B. K., TIEDJE, J. M. DNA probe method for the detection of specific microorganisms in the soil bacterial community. *Applied and Environmental Microbiology*, v. 54, p. 703-711, 1988.

HOLBEN, W. E., NOTO, K., SUMINO, T., SUWA, Y. Molecular analysis of bacterial communities in a three-compartment granular activated sludge system indicates community-level control by incompatible nitrification processes. *Applied and Environmental Microbiology*, v. 64, v. 7, p. 2528-2532, 1998.

HOSHINO, T., TERAHARA, T., TSUNEDA, S., HIRATA, A., INAMORI, Y. Molecular analysis of microbial population transition associated with the start of denitrification in a wastewater treatment process. *Journal of Applied Microbiology*, v. 99, p. 1165-1175, 2005.

HOVANEC, T. A., DELONG, E. F. Comparative analysis of nitrifying bacteria associated with freshwater and marine aquaria. *Applied and Environmental Microbiology*, v. 62, p. 2888-2896, 1996.

HUGENHOLTZ, P., PACE, N. R. Identifying microbial diversity in the natural environment: a molecular phylogenetic approach. *Trends in Biotechnology*, v. 14, p. 190-197, 1996.

HU, M., WANG, X., WEN, X., XIA, Y. Microbial community structures in different wastewater treatment plants as revealed by 454-pyrosequencing analysis. *Bioresource Technology*, v. 117, 72-79, 2012.

ISAKA, K., YOSHIE, S., SUMINO, T., INAMORI, Y., TSUNEDA, S. Nitrification of landfill leachate using immobilized nitrifying bacteria at low temperatures. *Biochemical Engineering Journal*, v. 37, p. 49-55, 2007.

JENKINS, D.; RICHARD, M. G.; DAIGGER, G. T *Manual on the causes and control of activated sludge bulking and foaming*. 2nd edition, Michigan, 1993.

JENSEN, M. A., WEBSTER, J. A., STRAUS, N. Rapid identification of bacteria on the basis of polymerase chain reaction-amplified ribosomal DNA spacer polymorphisms. *Applied and Environmental Microbiology*, v. 59, p. 945-952, 1993.

JONES, C. M., THIES, J. E. Soil microbial community analysis using two-dimensional polyacrylamide gel electrophoresis of the bacterial ribosomal internal transcribed, *Journal of Microbiological Methods*, v. 69, p. 256-267, 2007.

JURETSCHKO, S., LOY, A., LEHNER, A., WAGNER, M. The microbial community composition of a nitrifying-denitrifying activated sludge from an industrial sewage treatment plant analyzed by the full-cycle rRNA approach. *Systematic and Applied Microbiology*, v. 25, p. 84-99, 2002.

JURETSCHKO, S., TIMMERMANN, G., SCHMID, M., SCHLEIFER, K.H., POMMERENING-ROSER, A., KOOPS, H. P., WAGNER, M. Combined molecular and conventional analyses of nitrifying bacterium diversity in activated sludge: *Nitrosococcus mobilis* and *Nitrospira*-like bacteria as dominant populations. *Applied and Environmental Microbiology*, v. 64, n. 8, p. 3042-3051, 1998.

KARTAL, B., RATTRAY, J., VAN NIFTRIK, L. A., VAN DE VOSSENBERG, J., SCHMID, M. C., WEBB, R. I., SCHOUTEN, S., FUERST, J. A., DAMSTÉ, J. S., JETTEN, M. S., STROUS, M. *Candidatus* "Anammoxoglobus propionicus" a new propionate oxidizing species of anaerobic ammonium oxidizing bacteria. *Systematic and Applied Microbiology*, v. 30, p. 39-49, 2007.

KARTAL, B., VAN NIFTRIK, L., RATTRAY, J., VAN DE VOSSENBERG, J. L., SCHMID, M. C., SINNINGHE, D. J., JETTEN, M. S., STROUS, M. *Candidatus* 'Brocadia fulgida': an autofluorescent anaerobic ammonium oxidizing bacterium. *FEMS Microbiology Ecology*, v. 63, p. 46-55, 2008.

KAWAHARASAKI, M., KANAGAWA, T., TANAKA, H., NAKAMURA, K. Development and application of 16S rRNA-targeted oligonucleotide probe for detection of the phosphate-accumulating bacterium *Microlunatus phosphovorus* in an enhanced biological phosphorus removal process. *Water Science and Technology*, v. 37, p. 481-484, 1998.

KAWASAKI, E. S. *Amplification of RNA*. In: INNIS, M.A.; GELFAND, D.H.; SNINSKY, J.J.; WHITE, T.J. (Ed.) PCR protocols: a guide to methods and applications. San Diego: Academic Press, p. 21-27, 1990.

KENNEDY, A. C. Bacterial diversity in agroecosystems. *Agriculture, Ecosystems and Environment*, v. 74, p. 65-76, 1999.

KIM, D. J., KIM, T. K., CHOI, E. J., PARK, W. C., KIM, T. H., AHN, D. H., YUAN, Z., BLACKALL, L., KELLER, J. Fluorescence *in situ* hybridization analysis of nitrifiers in piggery wastewater treatment reactors. *Water Science and Technology*, v. 49, n. 5-6, p. 333-340, 2004.

KIRCHER, M., KELSO, J. High-throughput DNA sequencing—concepts and limitations. *Bioessays*, v. 32, n. 6, p. 524-536, 2010.

KIRCHMAN, D. L., YU, L., FUCHS, B. M., AMANN, R. Structure of bacterial communities in aquatic system as revealed by filter PCR. *Aquatic Microbial Ecology*, v. 26, p. 13-22, 2001.

KNIETSCH, A.; WASCHKOWITZ, T.; BOWIEN, S; HENNE, A.; DANIEL, R. Construction and screening of metagenomic libraries derived from enrichment cultures: generation of a gene bank for genes conferring alcohol oxidoreductase activity on *Escherichia coli*. *Applied and Environmental Microbiology*, v. 69, p. 1408-1416, 2003.

KONG, Y., NIELSEN, J. L., NIELSEN, P. H. Identity and ecophysiology of uncultured actinobacterial polyphosphate-accumulating organisms in full-scale enhanced biological phosphorous removal plants. *Applied and Environmental Microbiology*, v. 71, p. 4076-4085, 2005.

KONG, Y., XIA, Y., NIELSEN, J. L.; NIELSEN, P. H. Structure and function of the microbial community in a full-scale enhanced biological phosphorous removal plant. *Microbiology*, v. 153, p. 4061 - 4073, 2007.

KONG, Y. H., BEER, M., REES, G. N., SEVIOUR, R. J. Functional analysis of microbial communities in aerobic-anaerobic sequencing batch reactors fed with different phosphorous/carbon (P/C) ratios. *Microbiology-Sgm*, v. 148, p. 2299-2307, 2002.

Könneke, M., Bernhard, A.E., DE LA TORRE, J. R., WALKER, C. B., WATERBURY, J. B., STAHL, D.A. Isolation of an autotrophic ammonia-oxidizing marine archaeon. Nature, v. 437, p. 543-546, 2005.

KOZDRÓJ, J., VAN ELSAS, J. D. Response of the bacterial community to root exudates in soil polluted with heavy metals assessed by molecular and cultural approaches. *Soil Biology & Biochemistry*, v. 32, p. 1405-1417, 2000.

KUYPERS, M. M., SLIEKERS, A. O., LAVIK, G, SCHMID, M., JORGESEN, B. B., KUENEN, J. G. Anaerobic ammonium oxidation by anammox bacteria in the Black Sea. *Nature*, v. 422, p. 608-611, 2003.

LAJOIE, C. A., LAYTON, A. C., GREGORY, I. R., SAYLER, G. S., TAYLOR, D. E., MEYERS, A. J. Zoogleal clusters and sludge dewatering potential in an industrial activated-sludge wastewater treatment plant. *Water Environment Research*, v. 72, p. 56-64, 2000.

LAYTON, A. C., KARANTH, P. N., LAJOIE, C. A., MEYERS, A. J., GREGORY, I. R., STAPLETON, R. D., TAYLOR, D. E., SAYLER, G. S. Quantification of Hyphomicrobium populations in activated sludge from an industrial wastewater treatment system as determined by 16S rRNA analysis. *Applied and Environmental Microbiology*, v. 66, p. 1167-1174, 2000.

LEBLOND-BOURGET, N., PHILIPPE, H., MANGIN, I., DECARIS, B. 16S rRNA and 16S to 23S internal transcribed spacer sequence analyses reveal inter- and intraspecific *Bifidobacterium* phylogeny. *International Journal of Systematic Bacteriology*, v. 46, p. 102-111, 1996.

LEE, D.-H., ZO, Y.-G.., KIM, S.-J. Nonradioactive method to study genetic profiles of natural bacterial communities by PCR-single-strand-conformation polymorphism. *Applied and Environmental Microbiology*, v. 62, p. 3112-3120, 1996.

LEHNINGER, A.L. *Princípios de bioquímica*. São Paulo: Sarvier, 1985.

LI, J, JIN, Z., YU, B. Changes in the structure and diversity of bacterial communities during the process of adaptation to organic wastewater. Canadian Journal of Microbiology, v. 56, n.4, p. 352-355, 2010.

LIPSKI, A., FRIEDRICH, U., ALTENDORF, K. Application of rRNA-targeted Oligonucleotide Probes in Biotechnology. Applied Microbiology and Biotechnology, v. 56, n. 1-2, p. 40-57, 2001.

LIU, W. T., MARSH, T. L., CHENG, H., FORNEY, L. J. Characterization of microbial diversity by determining terminal restriction fragment length polymorphisms of genes encoding 16S rRNA. *Applied and Environmental Microbiology.*, v. 63, p. 4516-4522, 1997.

LIU, W. T., NIELSEN, A. T., WU, J. H., TSAI, C. S., MATSUO, Y., MOLIN, S. In situ identification of polyhydroxyalkanoate-accumulating traits for microbial populations in a biological phosphorous removal process. *Environmental Microbiology*, v. 3, p. 110-122, 2001.

LIMPIYAKORN, T., SHINOHARA, Y., KURISU, F., YAGI, O. Communities of ammonia-oxidizing bacteria in activated sludge of various sewage treatment plants in Tokyo. *FEMS Microbiology,* v. 54, p. 205-217, 2005.

LIU, X.-C., ZHANG, Y., YANG, M., WANG, Z.-Y., LV Q.-Z. Analysis of bacterial structures in two sewage treatment plants with different sludge properties and treatment performance by nested PCR-DGGE method. *Journal of Environmental Sciences*, v. 19, p. 60-66, 2007.

LOCKHART, D.J.; DONG, H.; BYRNE, M.C.; FOLLETTIE, M.T.; GALLO, M.V.; CHEE, M.S., 1996. Expression monitoring by hybridization to high-density oligonucleotide arrays. *Nature Biotechnology*, v. 14, p. 1675-1680.

LOGAN, J., EDWARDS, K., SAUNDERS, N. Real-time PCR: current technology and applications. Caister Academic Press, 2009.

LORENZ, T. C., 2012. Polymerase chain reaction: basic protocol plus troubleshooting and optimization strategies. Journal of Visualized Experiments, v. 63, e 3998, p. 1-15.

MACRAE, A. The use of 16S RDNA methods in soil microbial ecology, *Brazilian Journal of microbiology,* v. 31, p. 77-82, 2000.

MAIXNER, F., NOGUERA, D. R., ANNESER, B., STOECKER, K., WEGL, G., WAGNER, M., DAIMS, H. Nitrite concentration influences the population structure of *Nitrospira*-like bacteria. *Environmental Microbiology*, v. 8, p. 1487-1495, 2006.

MARSHALL, K. C. *Adsorption of microorganisms to soils and sediments.* In: BRITTON, G., MARSHALL, K. C., ed. Adsorption of microorganisms to surfaces. New York, N.Y. Wiley. 317-329, 1980.

MARTINS, A. M. P., PAGILLA, K., HEIJNEN, J. J., VAN LOOSDRECHT, M. C. M. Filamentous Bulking Sludge. A Critical Review. *Water Research*, v. 38. p. 793-817, 2004.

MELO, I.S.; SCHNEIDER, R.P.; INGLIS, P.W.; VALADARES-INGLIS, M.C. Biomarcas e Biossensores microbianos. In: Melo, I.S.; VALADARES-INGLIS, M.C.NASS, L.L.; VALOIS, A.C.C. (Ed.). *Recursos Genéticos e Melhoramento – Micrarganismos*, Jaguariúna: Embrapa Meio Ambiente, p. 164-169, 2002.

MERTOGLU, B., SEMERCI, N., GULER, N., CALLI, B., CECEN, F., SAATCI, A. Monitoring of population shifts in an enriched nitrifying system under gradually increased cadmium loading. Journal of Hazardous Materials, v. 160, p. 495–501, 2008.

MITCHELL, T. G., FREEDMAN, E. Z., MEYER, W., WHITE, T. J., TAYLOR, J. W. PCR detection of *Cryptococcus neoformans*. In: PERSING, D. H., SMITH, T. F., TENOVER, F. C., WHITE, T. J. (Ed.) Diagnostic molecular microbiology: principles and applications. Washington: American Society for Microbiology, p. 431-436, 1993.

MOBARRY, B. K., WAGNER, M., URBAIN, V., RITTMANNM B. E., STAHL, D. A. Phylogenetic probes for analyzing abundance and spatial organization of nitrifying bacteria. *Applied Environmental and microbiology*, v. 62, p. 2156-2162, 1996.

MONTRÀS, A., PYCKE, B., BOON, N., GODIA, F., MERGEAY, M., HENDRICKX, L. Distribution of *Nitrosomonas europaea* and *Nitrobacter winogradskyi* in an autotrophic nitrifying biofilm reactor as depicted by molecular analyses and mathematical modeling. *Water Research*, v. 42, pp. 1700-1714, 2008.

MOROZOVA, O., MARRA, M.A. Applications of next-generation sequencing technologies in functional genomics. *Genomics*, v. 92, p. 255-264, 2008.

MOTER, A., GOBEL, U.B. Fluorescence in situ hybridization (FISH) for direct visualization of microorganisms. *Journal of Microbiological Methods*, v. 41, p. 85-112, 2000.

MOURA, A. TACAO, M., HENRIQUES, I., DIAS, J., FERREIRA, P., CORREIA, A. Characterization of bacterial diversity in two aerated lagoons of a wastewater treatment plant using PCR-DGGE analysis. *Microbiological Research*, v. 164, p. 560-569, 2007.

MULLIS, K. B., FALOONA, F. A. Specific synthesis of DNA *in vitro* via a polymerase-catalyzed chain reaction. *Methods in Enzymology*, v. 55, p. 335-350, 1987.
MUNZ, G., GUALTIERO, M., SALVADORI, L., CLAUDIA, B., CLAUDIO, L. Process efficiency and microbial monitoring in MBR (membrane bioreactor) and CASP (conventional activated sludge process) treatment of tannery wastewater. *Bioresource Technology*, v. 99, p. 8559-8564, 2008.
MUYZER, G. DGGE/TGGE a method for identifying genes from natural ecosystems, *Current Opinion in Microbiology*, v. 2, p. 317-322, 1999.
MUYZER, G., DE WAAL, E. C., UITTERLINDEN, A. G. Profiling of complex microbial by denaturing gel electrophoresis analysis of polymerase chain reaction-amplified genes coding for 16S rRNA. *Applied and Environmental Microbiology*, v. 59, p. 695-700, 1993.
MUYZER, G., HOTTENTRAGER, S., TESKE, A., WAWER, C. In: AKKERMANS, A.D.L; VAN ELSAS, J.D; BRUJIN, F.J. Denaturing gradient gel electrophoresis of PCR-amplified 16S rDNA - A new molecular approach to analyze the genetic diversity of mixed microbial communities (2nd ed.), *Molecular Microbial Ecology Manual*, Kluwer, Dordrecht, The Netherlands, p. 3.4.4.1-3.4.4.22, 1996.
MUYZER, G., SMALLA, K. Application of denaturing gradient gel electrophoresis (DGGE) and temperature gradient gel electrophoresis (TGGE) in microbial ecology, Mini review. *Antonie van Leeuwenhoek*, Dordrecht, v. 73, p. 127-141, 1998.
MUYZER, G., BRINKHOFF, T., NUBEL, U., SANTEGOEDS, C., SCHAFER, H., WAWER, C. Denaturing gradient gel electrophoresis (DGGE) in microbial ecology. In AKKERMANS, A.D.L., VAN ELSAS, J.D., DE BRUIJN, F.J. Molecular Microbial Ecology Manual (eds.), p. 1–23. The Netherlands: Kluwer Academic, 1998.
MYERS, R. M., FISHER, S. G., LERMAN, L. S., MANIATIS, T. Nearly all single base substitutions in DNA fragments joined to a GC-clamp can be detected by denaturing gel electrophoresis. *Nucleic Acids Research*, v.13, p. 3131-3145, 1985.
NASIDZE, I., LI, J., QUINQUE, D., TANG, K., STONEKING, M., Global diversity in the human salivary microbiome. *Genome Research*, v. 19, n. 4, p. 636-643, 2009.
NEEF, A., AMANN, R., SCHLESNER, H., SCHLEIFER, K. H. Monitoring a widespread bacterial group: in situ detection of planctomycetes with 16S rRNA-targeted probes. *Microbiology*, v. 144, p. 3257-3266, 1998.
NEEF, A., ZAGLAUER, A., MEIER, H., AMANN, R., LEMMER, H., SCHLEIFER, K. H. Population analysis in a denitrifying sand filter: conventional and in situ identification of *Paracoccus* spp. in methanol-fed biofilms. *Applied and Environmental Microbiology*, v. 62, p. 4329-4339, 1996.
NG, K.K., SHI, X., ONG, S.L., NG, H.Y. Pyrosequencing reveals microbial community profile in anaerobic bio-entrapped membrane reactor for pharmaceutical wastewater treatment. *Bioresource Technology*, v. 200, p. 1076-1079, 2015.
NIELSEN, A. T., LIU, W. T., FILIPE, C., GRADY, L. Jr., MOLIN, S., STAHL, D. A. Identification of a novel group of bacteria in sludge from a deteriorated biological phosphorus removal reactor. *Applied and Environmental Microbiology*, . 65, p. 1251-1258, 1999.
NIELSEN, J. L. Protocol for Fluorescence in situ hybridization (FISH) with rRNA-targeted oligonucleotides. In: NIELSEN, P. H., DAIMS, H., LEMMER, H. FISH Handbook for Biological Wastewater treatment: Identification and Quantification of Microorganisms in Activated Sludge and Biofilms by FISH. London: IWA Publishing, 2009.
NIELSEN, J. L., HANSEN, A. A. Identification of denitrifying microorganisms in activated sludge by FISH. In: NIELSEN, P. H., DAIMS, H., LEMMER, H. FISH Handbook for Biological Wastewater treatment: Identification and Quantification of Microorganisms in Activated Sludge and Biofilms by FISH. London: IWA Publishing, 2009.
NIELSEN, P. H., DAIMS, H., LEMMER, H. Introduction. In: NIELSEN, P. H., DAIMS, H., LEMMER, H. FISH Handbook for Biological Wastewater treatment: Identification and Quantification of Microorganisms in Activated Sludge and Biofilms by FISH. London: IWA Publishing, 2009a.

NIELSEN, P. H., NGUYEN, H. T. T., McILROW, S. J., MIELCZAREK, A. T., SEVIOUR, R. Identification of polyphosphate-accumulating and glycogen-accumulating organisms by FISH. In: NIELSEN, P. H., DAIMS, H., LEMMER, H. FISH Handbook for Biological Wastewater treatment: Identification and Quantification of Microorganisms in Activated Sludge and Biofilms by FISH. London: IWA Publishing, 2009b.

NIEMI, R., HEISKANEN, I., WALLENIUS, K., LINDSTRÖM, K., Extraction and purification of DNA in rhizosphere soil samples for PCR-DGGE analysis of bacterial consortia. *Journal of Microbiology Methods*, v. 43, p. 155-165, 2001.

NÜBEL, U., GARCIA-PICHEL, F., KÜHL, M., MUYZER, G. Quantifying microbial diversity: morphotypes, 16S rRNA genes, and carotenoids of oxygenic phototrophs in microbial mats. *Applied and Environmental Microbiology*, n. 65, p. 422-430, 1999.

OELMÜLLER, U., FRÜGER, N., STEINBÜCHEL, A. FRIEDRICH, G. C. Isolation of prokaryotic RNA and detection of specific mRNA with biotinylated probes. *Journal of Microbiology Methods*, v. 11, p. 73-84, 1990.

OGRAM, A. Soil molecular microbial ecology at age 20: methodological challenges for the future. *Soil Biology and Biochemistry*, v. 32, p. 1399-1504, 2000.

OGRAM, A., SAYLER, G. S., BARKAY, T. J. The extraction and purification of microbial DNA from sediments. *Journal of Microbiological Methods*, v. 7, p. 57-67, 1987.

OKABE S, KAMAGATA Y. Wastewater treatment. In: Liu WT, Jansson JK, editors. Environmental Molecular Microbiology. Norfolk, UK: Caister Academic Press; 2010. pp. 191–210.

OKABE, S., SATOH, H., WATANABE, Y. In situ analysis of Nitrifying biofilms as determined by in situ hybridisation and the use of microelectrodes. *Applied and Environmental Microbiology*, v. 65, n. 7, p. 3182-3191, 1999.

OLSEN, G. J., WOESE, C. R., OVERBEEK, R. The winds of (evolutionary) change: breathing new life into microbiology. *Journal of Bacteriology*, v. 176, p. 1-6, 1994

ORITA, M., IWAHANA, H., KANAZAWAT, H., HAYASHI, K., SEKIYA, TAKAO. Detection of Polymorphisms of Human DNA by Gel Electrophoresis as SSCPs. *Proceedings of the National Academy of Sciences of the United States of America*, v. 86, p. 2766-2770, 1989.

OSTE, C. Product application focus: polymerase chain reaction. *BioTechniques*, v. 6, n. 2, p. 162-167, 1988.

OTAWA, K., ASANO, R., OHBA, Y., SASAKI, T., KAWAMURA, E., KOYAMA, F., NAKAMURA, S., NAKAI, Y. Molecular analysis of ammonia-oxidizing bacteria community in intermittent aeration sequencing batch reactors used for animal wastewater treatment. *Environmental Microbiology*, v. 8, n. 11, p. 1985-1996, 2006.

OUVERNEY, C. C., FUHRMAN, J. A. Increase in fluorescence intensity of 16S rRNA in situ hybridization in natural samples treated with chloramphenicol. *Applied and Environmental Microbiology*, v. 63, p. 2735-2740, 1997.

ØVREÅS, L., FORNEY, L., DAAE, F. L., TORSVIK, V. Distribution of Bacterioplankton in Meromictic Lake Sælenvannet, as Determined by Denaturing Gradient Gel Electrophoresis of PCR-Amplified Gene Fragments Coding for 16S rRNA. *Applied and Environmental Microbiology*, v. 63, p. 3367-3373, 1997.

PACE, N. R. A molecular view of microbial diversity and the biosphere. *Science*, v. 276, p. 734-740, 1997.

PACE, N. R. New perspective on the natural microbial world: molecular microbial ecology. *ASM News*, v. 62, p. 463-470, 1996.

PARK, H. D., REGAN, J. M., NOGUERA, D. R. Molecular analysis of ammonia-oxidizing bacterial populations in aerated-anoxic Orbal processes. *Water Science Technology*, v. 46, n. 1-2, p. 273-280, 2002.

PEIXOTO, R. S., ROSADO, A. S., TAKETANI, R. G. *Bioprospecção da diversidade microbiana cultivável e não cultivável*. In: Mello, I.S.; AZEVEDO, J.L. (Org.), Microbiologia Ambiental, 2. ed.. Jaguariúna: Embrapa Meio Ambiente, v. 1, p. 83-106, 2008.

PERSSON, F., WIK, T., SORENSSON, F., HERMANSSON, M. Distribution and activity of ammonia oxidizing bacteria in a large full-scale trickling filter. *Water Research*, v. 36, p. 1439-1448, 2002.

PICKUP, R. W. Development of molecular methods for the detection of specific bacteria in the environment. *Journal of General Microbiology*, v. 137, p. 1009-1019, 1991.
PROSSER, J. I. Autotrophic nitrification in bacteria. *Advances in Microbial Physiology*, v. 30, p. 125-181, 1989.
RABUS, R., WILKES, H., SCHRAMM, A., HARMS, G., BEHRENDS, A., AMANN, R., WIDDEL, F. Anaerobic utilization of alkylbenzenes and n-alkanes from crude oil in an enrichment culture of denitrifying bacteria affiliating with the beta-subclass of Proteobacteria. *Environmental Microbiology*, v. 1, p. 145-157, 1999.
RADAJEWSKI, S., MURRELL, J. C. Stable isotope probing for detection of methanotrophs after enrichment with $^{13}CH_4$, *Methods in Molecular Biology*, v. 179, p. 149-157, 2002.
REIS JUNIOR, F. B., MENDES, I.C., TEIXEIRA, K. R. S.; REIS, V. M. Uso de Ferramentas Moleculares em Estudos da Diversidade de Microrganismos do Solo. Planaltina, DF: Embrapa Cerrados, 33p, 2002.
REYSENBACH, A., GIVER, L., WICKMAN, G. & PACE, N. Differential amplification of rRNA genes by polymerase chain reaction. *Applied and Environmental Microbiology*, v. 58, p. 3417-3418, 1992.
RODÍCIO, M. R., MENDOZA, M. C. Identificación bacteriana mediante secuencación del ARNr 16S: fundamento, medología y aplicaciones en microbiología clínica. *Enferm. Infecc. Microbiol. Clin.*, v. 22, n. 4, p. 238-245, 2004.
RODRÍGUEZ-EZPELETA, N., HACKENBERG, M., ARANSAY, A.M. (Eds.) Bioinformatics for High Throughput Sequencing. Springer, 2012.
ROGERS, Y. H., VENTER, J. C. Genomics: massively parallel sequencing. *Nature*, v. 437, p. 326-327, 2005.
ROLLER, C., WAGNER, M., AMANN, R., LUDWIG, W., SCHLEIFER, K. H. In situ probing of Gram-positive bacteria with high DNA G+C content using 23S rRNA-targeted oligonucleotides. *Microbiology*, v. 140, p. 2849-2558, 1994.
ROONEY-VARGA, J. N., GENTHNER, B. R. S, DEVEREUX, R., WILLIS, S. G., FRIEDMAN, S. D., HINES, M. E.. Phylogenetic and physiological diversity of sulphate-reducing bacteria isolated from a salt marsh sediment. *Systematic and Applied Microbiology*, v. 21, p. 557-568, 1998.
ROSADO, A. S., DUARTE, G. F. *Utilização da eletroforese em gel com gradiente de desnaturantes (DGGE) e gel com gradientes de temperatura (TGGE) para estudar diversidade microbiana.* In: MELO, I.S.; VALADARES-INGLIS, M.C.; NASS, L.L.; VALOIS, A.C.C. Recursos genéticos e melhoramento – microrganismos. Jaguariúna: Embrapa Meio Ambiente, p. 97-128, 2002.
ROSADO, A. S., DUARTE, G. F.; SELDIN, L., VAN ELSAS, J. D. Molecular Microbial Ecology: a minireview. *Brazilian Journal of Microbiology*. v. 28, p. 135-147, 1997.
ROSADO, A. S., SELDIN, L., WOLTERS, A. C., VAN ELSAS, J.D. Quantitative 16S rDNA-targeted polymerase chain reaction and oligonucleotide hybridization for detection of *Paenibacillus azotofixans* in soil and the wheat rhizosphere. *FEMS Microbial Ecology*, v. 19, p. 153-164, 1996.
ROSSELLÓ-MORA, R. A., WAGNER, M., AMANN, R., SCHLEIFER, K. H. The abundance of *Zoogloea ramigera* in sewage treatment plants. *Applied and Environmental Microbiology*, v. 61, p. 702-707, 1995.
ROWAN, A. K., SNAPE, J. R., FEARNSIDE, D., BARER, M. R., CURTIS, T. P., HEAD, I. M. Composition and diversity of ammonia-oxidizing bacterial communities in wastewater treatment reactors of different design treating identical wastewater. *FEMS Microbiology Ecology*, v. 43, p. 195-206, 2003.
SAIKI, R. K., GELFAND, D. H., STOFFEL, S., SCHARF, S. J., HIGUCHI, R., HORN, G. T., MULLIS, K. B., ERLICH, H. A. Primer-directed enzymatic amplification of DNA with a thermostable DNA polymerase. *Science*, v. 239, n. 4839, p. 487-491, 1988.
Salomonová, S., LAMAČOVÁ, J., RULÍK, M., ROLČÍK, J., ČÁP, L., BEDNÁŘ, P., BARTÁK, P. Determination of phospholipid fatty acids in sediments. *Acta Universitatis Palackianae Olomucensis. Facultas Rerum Naturalium*, v. 42, p. 39-49, 2003.

SANZ, J. L., KÖCHLING, T. Molecular biology techniques used in wastewater treatment : An overview. *Process biochemistry*, v. 42, n. 2, p. 119-133, 2007.

SCHLOSS, P. D., HANDELSMAN, J. Biotechnological prospects from metagenomics. *Current Opinion in Biotechnology*, v. 14, p. 303-310, 2003.

SCHMID, M., SCHMITZ-ESSER, S., JETTEN, M., WAGNER, M. 16S-23S rDNA intergenic spacer and 23S rDNA of anaerobic ammonium-oxidizing bacteria: implications for phylogeny and in situ detection. *Environmental Biotechnology*, v. 3, p. 450-459, 2001.

SCHMID, M., TWACHTMANN, U., KLEIN, M., STROUS, M., JURETSCHKO, S., JETTEN, M., METZGER, J. W., SCHLEIFER, K. H., WAGNER, M. Molecular evidence for genus level diversity of bacteria capable of catalyzing anaerobic ammonium oxidation. *Systematic and Applied Microbiology*, v. 23, p. 93-106, 2000.

SCHMID, M., WALSH, K., WEBB, R., RIJPSTRA, W. I., PAS-SCHOONEN, K, VERBRUGGEN, M. J., HILL, T., MOFFETT, B., FUERST, J., SCHOUTEN, S., DAMSTEM, J. S., HARRIS, J., SHAW, P., JETTEN, M., STROUS, M. *Candidatus* "Scalindua brodae" sp. nov., *Candidatus* "Scalindua wagneri", sp. nov., two new species of anaerobic ammonium-oxidizing bacteria. *Systematic and Applied Microbiology*. v. 26, p. 529-538, 2003.

SCHRAMM, A., DE BEER, D., WAGNER, M., AMANN, R. I. Identification and activity in situ of *Nitrospira* and *Nitrospira spp.* as dominant populations in a nitrifying fluidized bed reactor, *Applied and Environmental Microbiology.*, v. 64, n. 9, p. 3480-3485, 1998.

SCHUTTE, U. M., ABDO, Z., BENT, S. J. Advances in the use of terminal restriction fragment length polymorphism T-RFLP analysis of 16S rRNA genes to characterize microbial communities. *Applied Microbiology Biotechnology*, v. 80, p. 365-380, 2008.

SEVIOUR, R. J., SODDELL, J. A., LIU, J.-R., McKENZIE, C. M., SAINT, C. The filamentous bacteria causing bulking and foaming in activated sludge plants. *Australian Biotechnology*, v. 10, n. 3, p. 24-29, 2000.

SHEFFIELD, V. C., COX, D. R., MYERS, R. M. Attachment of a 40bp G+C rich sequence (GC-Clamp) to genomic DNA fragments by polymerase chain reaction results in improved detection of single-base changes. *Proceedings of the National Academy of Sciences USA*, v. 86, p. 232-236, 1989.

SHENDURE, J., HANLEE, J. Next-generation DNA sequencing. *Nature Biotechnology*, v. 26, p. 1135–1145, 2008.

SIGLER, W. V., MINIACI, C., ZEYER, J. Electrophoresis time impacts the denaturing gel electrophoresis-based assessment of bacterial community structure. *Journal of Microbialogical Methods*, v. 57, p. 17-22, 2004.

SMALLA K., CRESWELL, L. C., MENDOÇA-HAGLER, L.C.; WOLTERS, A., VAN ELSAS, J.D. Rapid DNA extraction protocol from soil for polymerase chain reaction-mediated amplification. *Journal of Applied Bacteriology*, v. 74, p. 78-85, 1993.

SNAIDR, J., AMANN, R., HUBER, I., LUDWIG, W., SCHLEIFER, K.-H. Phylogenetic analysis and in situ identification of bacteria in activated sludge. *Appl Environmental Microbiology*, v. 63, p. 2884-2896, 1997.

SPECTOR, D. L., GOLDMAN, R. D., LEINWAND, L.A. *Cells: A Laboratory Manual*. Cold Spring Harbor Laboratory Press: Cold Spring Harbor, NY, 1998.

STACKEBRANDT, E., LIESACK, W., GOEBEL, B. M. Bacterial diversity in a soil sample from a subtropical Australian environment as determined by 16S rDNA analysis. *FASEB Journal*, v. 7, p. 232-236, 1993.

STAHL, D. A. Molecular approaches for the measurement of density, diversity and phylogeny. In: McINERNEY, M. J., WALTER, M. V., STETZENBACK, L. D. (Eds.) *Manual of environmental microbiology*, p. 102-114. ASM Press, Washington, D.C, 1997.

STEPHEN, J. R., MCCAIG, A. E., SMITH, Z., PROSSER, J. I.; EMBLEY, T. M. Molecular diversity of soil and marine 16S rRNA gene sequences related to β-subgroup ammonia-oxidizing bacteria. *Applied and Environmental Microbiology*, v. 62, n. 11, p. 4147-4154, 1996.

STRACHAN, T., READ, A. P. Genética Molecular Humana. Porto Alegre: Artmed, p. 119-122, 2002.

SUGIMOTO, N., NAKANO, S., YONEYAMA, M., HONDA, K. Improved thermodynamic parameters and helix initiation factor to predict stability of DNA duplexes. *Nucleic Acids Res.*, v. 24, p. 4501-4505, 1996.

SUSHCHIK, N. N., GLADYSHEV, M. I., KALACHOVA, G. S., KRAVCHUK, E. S., DUBOVSKAYA, O. P., IVANOVA, E. A. Particulate Fatty Acids in Two Small Siberian Reservoirs Dominated by Different Groups of Phytoplankton. *Freshwater Biology*, v. 48, n. 3, 394-403, 2003.

STRYER, L. *Explorando os genes*. In: Bioquímica (4ª ed.), New York: FREEMAN, W.H. & Company, 1996.

TACÃO, M., MOURA, A., ALVES, A., HENRIQUES, I., SAAVEDRA, M.J., CORREIA, A. Evaluation of 16S-rDNA and gyrB-DGGE for typing members of the genus *Aeromonas*. *FEMS Microbiology Letters*, v. 246, n.1, p. 11-18, 2005.

TAL, Y., WATTS, J. E. M., SCHREIER, S. B., SOWERS, K. R., SCHREIER, H. J. Characterization of the microbial community and nitrogen transformation processes associated with moving bed bioreactors in a closed recirculated mariculture system. *Aquaculture*, v. 215, p. 187-202, 2003.

TAN, B., NG, C., NSHIMYIMANA, J.P., LOH, L.L., GIN, K.Y.-H, THOMPSON, J.R., Next-generation sequencing (NGS) for assessment of microbial water quality: current progress, challenges, and future opportunities. Frontiers in Microbiology, v. 6., Article 1027, p. 1-20, 2015.

THOMSEN, T. R., NIELSEN, J. L., RAMSING, N. B., NIELSEN, P. H. Micromanipulation and further identification of FISH-labelled microcolonies of a dominant denitrifying bacterium in activated sludge. *Environmental Microbiology*, v. 6, p. 470-479, 2004.

TORSVIK, V., GOKSOYR, J., DAAE, F.L. High diversity in DNA of soil bacteria. *Applied And Environmental Microbiology*, v. 56, p. 782-787, 1990.

TREUSCH, A.H., LEININGER, S., KLETZIN, A., SCHUSTER, S.C., KLENK, H.P., SCHLEPER, C. Novel genes for nitrite reductase and Amo-related proteins indicate a role of uncultivated mesophilic Crenarchaeota in nitrogen cycling. *Environmental Microbiology*, v. 7, n. 12, p. 1985-1995, 2005.

TREVORS, J. T., VAN ELSAS, J. D. Nucleic Acids in the Environment: Methods and Applications. Heidelberg, Springer-Verlag, 1995.

TSUNEDA, S., MIKAMI, M., KIMOCHI, Y., HIRATA, A. Effect of salinity on nitrous oxide emission in the biological nitrogen removal process for industrial wastewater, *Journal of Hazardous Materials*, v. 119, p. 93-98, 2005.

VAN DER ZEE, H., HUIS, J. H. J. Rapid and alternative screening methods for microbiological analysis. *Journal of AOAC International*, v. 80, p. 934-940, 1997.

VAN ELSAS, J. D., DUARTE, G. F., ROSADO, A. S., SMALLA, K.. Microbial and molecular biological methods for monitoring microbial inoculants and their effects in the soil environment. *Journal of Microbiological Methods*, v. 32, p. 133-154, 1998.

VAN ELSAS, J.D., WELLINGTON, E. M. H., TREVORS, J. T. In: Modern Soil Microbiology. New York, Marcel Dekker, 1997.

VITZTHUM, F., GEIGER, G., BISSWANGER, H., BRUNNER, H., BERNHAGEN, J. A quantitative fluorescence-based microplate assay for the determination of double-stranded DNA using SYBR Green I and a standard ultraviolet transilluminator gel imaging system. *Analytical Biochemistry*, v. 276, p. 59-64, 1999.

VOELKERDING, K.V., DAMES, S.A., DURTSCHI, J.D. Next-generation sequencing: from basic research to diagnostics. *Clinical Chemistry*, v. 55, n. 4, p. 641-658, 2009.

WAGNER, M, LOY, A. Bacterial community composition and function in sewage treatment systems. Current Opinion Biotechnology, v. 13, p. 218-227, 2002.

WAGNER, M., AMANN, R., LEMMER, H., SCHLEIFER, K.-H. Probing activated sludge with oligonucleotides specific for proteobacteria: inadequacy of culture-dependent methods for describing microbial community structure. *Applied and Environmental Microbiology*, v. 59, p. 1520-1525, 1993.

WAGNER, M., AβMUS, B., HARTMANN, A., HUTZLER, P., AMANN, R. In situ analysis of microbial consortia in activated sludge using fluorescently labelled, rRNA-targeted oligonucleotide probes and confocal laser scanning microscopy. *Journal of Microscopy*, v. 176, p. 181-187, 1994.

WAGNER, M.; RATH, G.; AMANN, R.; KOOPS, H. P.; SCHLEIFER, K. H. In situ identification of ammonia-oxidizing bacteria. *Systematic and Applied Microbiology*, v. 18, p. 251-264, 1995.

WAGNER, M., RATH, G., KOOPS, H. P., FLOOD, J., AMANN, R. In situ analysis of nitrifying bacteria in sewage treatment plants. *Water Science and Technology*, v. 34, p. 237-244, 1996.

WENTZEL, M. C., LOEWENTHAL, R. E., EKAMA, G. A., MARAIS, G. R. Enhanced polyphosphate organism cultures in activated sludge systems. Part 1. Enhanced culture development. Water S. A. (Pretoria), v. 14, p. 81-92, 1988.

WETZEL, J., KINGS, C., POP, M. Assessing the benefits of using mate-pairs to resolve repeats in de novo short-read prokaryotic assemblies. BMC Bioinformatics, v. 12, n. 95, p. 1-14, 2011.

WOESE, C.R. Bacterial evolution. *Microbiology Review*, v. 51, p. 221-271, 1987.

WOESE, C. R., KANDLER, O.; WHEELIS, M. L. Towards a natural system for organisms: proposal for the domains Archaea, Bacteria and Eukarya. *Proceedings of National Academy of Sciences USA*, v. 87, p. 4576-4579, 1990.

WONG, M. T., TAN, F. M., NG, W. J., LIU, W. T. Identification and occurrence of tetrad-forming *Alphaproteobacteria* in anaerobic-aerobic activated sludge processes. *Microbiology-Sgm*, v. 150, p. 3741-3748, 2004.

YERGEAU, E., FILION, M., VUJANOVIC, V., ST-ARNAUD, M. A PCR-denaturing gradient gel electrophoresis approach to assess *Fusarium* diversity in asparagus. *Journal of Microbiological Methods*, v. 60, p. 143-154, 2005.

YERSHOV, G., BARSKY, V., BELGOVSKIY, A., KIRILLOV, E., KREINDLIN, E., IVANOV, I., PARINOV, S., GUSHIN, D., DROBISHEV, A., DUBILEY, S., MIRZABEKOV, A. *Proceedings of the National Academy of Sciences*, v. 93, p. 4913-4918, 1996.

YOO, H., AHN, K.H., LEE, H.J., LEE, K.H., KWAK, Y.J., SONG, K.G. Nitrogen Removal from synthetic wastewater by simultaneous nitrification and denitrification (SND) via Nitrite in an Intermittently-Aerated reactor. *Water Research*, v. 33, n. 1., p. 145–154, 1999.

YU, C.P., AHUJA, R., SAYLER, G, CHU, K.H. Quantitative molecular assay for fingerprinting microbial communities of wastewater and estrogen-degrading consortia. *Applied and Environmental Microbiology*, v. 71, p. 1433-1444, 2005.

ZAK, J. C., WILLIG, M. R., MOORHEAD, D. L., WILDMAN, H. G. Functional diversity of bacterial communities: a quantitative approach. *Soil Biology & Biochemistry*, v. 26, p. 1101-1108, 1994.

ZHANG, T., JIN, T., YAN, Q. Occurrence of ammonia-oxidizing Archaea in activated sludges of a laboratory scale reactor and two wastewater treatment plants. *Journal of Applied Microbiology*, v. 107, p. 907-977, 2009.

ZHAO, X., CHEN, Z., WANG, X., LI, J., SHEN, J., XU, H. Remediation of pharmaceuticals and personal care products using an aerobic granular sludge sequencing bioreactor and microbial community profiling using Solexa sequencing technology analysis. *Bioresource Technology*, v. 179, p. 104-112, 2015.

ZILLES, J. L., PECCIA, J., KIM, M. W., HUNG, C. H., NOGUERA, D. R. Involvement of *Rhodocyclus*-related organisms in phosphorous removal in full-scale wastewater treatment plants. *Applied and Environmental Microbiology*, v. 68, p. 2763-2769, 2002.

ZUCKERKANDL, E., PAULING, L. Molecules as documents of evolutionary history. *Journal of Theoretical Biology*, v. 8, p. 357-366, 1965.

FSC
www.fsc.org
MIX
Papier | Fördert gute Waldnutzung
FSC® C083411

Kolibri 360 GmbH
Ferdinand-Jühlke-Straße 7
99095 Erfurt, Deutschland
produktsicherheit@kolibri360.de